Ziegler   Numerical Differential Protection

Numerical Differential Protection

Principles and Applications

by Gerhard Ziegler

Second edition, 2012

Publicis Publishing

Bibliographic information published by the Deutsche Nationalbibliothek

The Deutsche Nationalbibliothek lists this publication in the Deutsche Nationalbibliografie; detailed bibliographic data are available in the Internet at http://dnb.d-nb.de.

www.publicis-books.de

ISBN 978-3-89578-351-7

Second edition, 2012

Editor: Siemens Aktiengesellschaft, Berlin and Munich
Publisher: Publicis Publishing, Erlangen
© 2012 by Publicis Erlangen, Zweigniederlassung der PWW GmbH

Printed in Germany

Foreword
to the First Edition

Differential protection provides absolute selectivity and fast operation, and is applied in numerous variations for the protection of electrical machines, transformers, busbars and feeders at all voltage levels.

Substantial progress has been made with numerical technology which has made this measuring principle even more attractive for the user, such as for example the integrated CT ratio adaptation and the large degree of CT saturation tolerance. The application of digital data exchange over interference free fiber optic cables has simplified the protection of cables and overhead lines in urban and industrial networks substantially, while also improving the security. Digital communication networks are finding increasing application for the transfer of protection data in overhead line networks. Thereby, the differential protection may also be applied on longer lines well exceeding 100 km as well as complex system configurations with multiple line terminals.

The book at hand initially conveys the basic principles of differential protection with analogy and digital technology. Special note will be taken of current transformers, data transfer and digital communication. Subsequently the various types of differential protection and the practical applications will be covered, using the Siemens SIPROTEC product range. In principle, the explanations however also apply to the devices of other manufacturers. Practical examples are calculated for illustration purposes.

This book is aimed at students and young engineers, who require an introduction to the topic of differential protection. However, users with practical experience, seeking an entry to digital technology of differential protection may also find this book to be a useful addition to their library. Furthermore, it may also be used as a reference for special application problems.

Nuremberg, March 2005   Gerhard Ziegler

Foreword
to the Second Edition

Differential protection is a fast and selective method of protection against short-circuits. It is applied in many variants for electrical machines, transformers, busbars, and electric lines.

Initially this book covers the theory and fundamentals of analog and digital differential protection. Current transformers are treated in detail including transient behaviour, impact on protection performance, and practical dimensioning. An extended chapter is dedicated to signal transmission for line protection in particular to modern digital communication and GPS timing.

The emphasis is then placed on the different variants of differential protection and their practical application illustrated by concrete examples. This is completed by recommendations for commissioning, testing and maintenance.

Finally the design and management of modern differential protection is explained by means of the latest Siemens SIPROTEC relay series.

A textbook and standard work in one, this book covers all topics which have to be paid attention to for planning, designing, configuring and applying differential protection systems. The book is aimed at students and engineers who wish to familiarise themselves with the subject of differential power protection, as well as the experienced user, entering the area of digital differential protection. Furthermore it serves as a reference guide for solving application problems.

For this second edition all contents have been revised, extended and updated to the latest state of protective relaying.

Nuremberg, August 2011   Gerhard Ziegler

Contents

1 Introduction

Differential protection was already applied towards the end of the 19th century, and was one of the first protection systems ever used.

Faults are detected by comparison of the currents flowing into and out of the protected plant item. As a result of the fast tripping with absolute selectivity it is suited as main protection of all important items of plant, i.e. generators, transformers, busbars as well as cables and overhead lines.

The protected zone is clearly defined by the positioning of the current transformers. The back-up protection function for external faults must therefore always be implemented with an additional time graded protection (over-current or distance protection).[1]

1.1 Protection principle

Differential protection calculates the sum of all currents flowing into and out of the protected object. Apart from magnetising currents and capacitive charging currents, this current sum must always be equal to zero (Kirchhoff's current law) if the protected object is un-faulted. Internal faults are therefore detected by the appearance of a differential current. For security against mal-operation due to CT transformation errors, the pick-up threshold of the protection is increased in proportion to the total current flow (stabilising or restraint current). Thereby, the protection sensitivity is automatically matched to the prevailing short circuit conditions.

Implementation of differential protection is simpler in the case of protected objects that are not geographically spread out (generators, transformers, busbars), where the current transformers are situated close together. In this case, the current transformers may be connected to the protection device directly via control cables.

In the case of HV cables and overhead lines, the measured currents must be transmitted over large distances to the corresponding opposite line end, for the comparison to take place. Utilising pilot wire connections (special protection cables), distances of approximately 25 km may be spanned. With modern relays, using digital communication via fiber optic cables, differential protection may also be implemented on long overhead lines of over 100 km.

---

[1] In most cases simple back-up protection is integrated in numerical devices, so that a separate device is not necessary for applications in distribution networks. A back-up protection for faults on the protected object must always be provided by a separate device in order to achieve hardware redundancy. This is particularly true for the transmission network.

High impedance differential protection is a particular variant of differential protection. It is adapted to the non-linear transformation characteristic of current transformers and achieves stability during CT saturation by means of a high series resistance at the differential relay.

Due to its simplicity, high impedance differential protection is relatively common in Anglo-Saxon influenced countries. It is suitable for protection of galvanically connected units such as busbars, generators, motors, compensating reactors and auto- transformers, but not for normal transformers with separate windings. A disadvantage lies in the fact that all current transformers must be identical.

1.2 Numerical differential protection

Towards the end of the 1980's, the numerical technology was introduced to protection applications. [1-1]

A number of general advantages are:

- Modern relays are multi-functional and, apart from the protection functions, capable of executing additional tasks such as operational measurement and disturbance recording.
- Due to integrated self-monitoring, event driven maintenance, instead of costly preventive routine maintenance may be applied.
- Devices may be operated locally and from remote with a PC via serial interface.
- All important measured values are indicated with the integrated measuring function. External measuring instruments during commissioning and testing are therefore not normally required.

Particular advantages are also obtained specifically for differential protection:

- Digital measuring techniques have substantially improved filters for the inrush stabilisation and intelligent measuring algorithms provide additional stabilisation during CT saturation.
- With conventional devices, additional matching transformers were required to adapt different CT ratios and transformer vector groups. Numerical relays implement this adaptation internally by computation.
- Phase segregated measurement can be implemented with moderate effort and therefore achieves the same pick-up sensitivity for all fault types as well as reliable pick-up in the event of multiple faults.
- Communication links are also covered by the continuous self-monitoring.
- Decentralised construction of the busbar protection with communication via fiber optic cables and PC-based configuration achieved a significant reduction in complexity.

2 Definitions

The following terms are used in this document.

If they correspond to the definitions of the International Electrotechnical Vocabulary Chapter 448: Power System Protection (IEC 60050-448), then the corresponding reference number is given:

Protection (in USA: Relaying)
The provisions for detecting faults or other abnormal conditions in a power system, for enabling fault clearance, for terminating abnormal conditions, and for initiating signals or indications. [448-11-01]

Protection relay (in USA: protective relay)
Measuring relay that, either solely, or in combination with other relays, is a constituent of a protection equipment. [448-11-02]

Protection equipment (in USA relay system)
An equipment incorporating one or more protection relays and, if necessary, logic elements intended to perform one or more specified protection functions. [448-11-03]

Protection system
An arrangement of one or more protection equipments, and other devices intended to perform one or more specified protection functions. [448-11-04]

NOTE 1: A protection system includes one or more protection equipments, instrument transformer(s), wiring, tripping circuit(s) and, where provided, communication system(s). Depending on the principle(s) of the protection system, it may include one end or all ends of the protected section and, possibly, automatic reclosing equipment.

NOTE 2: The circuit breaker(s) are excluded.

Protected section
That part of the system network, or circuit within a network, to which specified protection has been applied. [448-11-05]

Digital (Numerical) protection (relay)
Fully digital relays utilising microprocessor technology with analog to digital conversion of the measured values (current and voltage) and subsequent numerical processing by computer programs. Earlier the designation "computer relay" has been used.

In Europe, it has sometimes been distinguished between "digital" and "numerical" relays. The term "digital" has been used with the earlier relay generation using micro-processors instead of discrete static measuring and logic circuits. The term "numerical" has been reserved for the modern computer type relays. [A-15]

In the US, the term "digital" has always been used in this meaning of "numerical" protection.

Nowadays, both terms are used in parallel.

Unit protection
A protection whose operation and section selectivity are dependent on the comparison of electrical quantities at each end of the protected section.

NOTE: In the USA, the term "unit protection" designates the protection provided for an electrical generator.

Longitudinal differential protection (generally designated as differential protection)
Protection, the operation and selectivity of which depends on the comparison of the magnitude or the phase and magnitude of the currents at the ends of the protected section. [448-14-16]

Transverse differential protection
Protection applied to parallel connected circuits and in which operation depends on unbalanced distribution of currents between them. [448-14-17]

Biased (stabilised) differential protection (in USA: percentage differential relay)
Differential protection, with a pick-up threshold that increases proportional to the increasing through current (sum of the absolute values of the currents from all the line ends).

Operating (tripping) current (of a differential relay)
Current difference that tends to initiate operation (in general tripping).

Restraint (stabilising) current (of a differential relay)
Current proportional to through current that tends to inhibit differential relay operation.

Variable slope characteristic
Operating characteristic of a differential relay with an increasing slope (percentage) dependent on increasing restraint current.

High impedance differential protection
Current differential protection using a current differential relay whose impedance is high compared with the impedance of the secondary circuit of a saturated CT. [448-14-22]

Low impedance differential protection (generally designated differential protection)
Current differential protection using a current differential relay whose impedance is not high compared with the impedance of the secondary of a saturated CT. [448-14-23]

Phase comparison protection
Protection whose operation and selectivity depends on the comparison of the phase of the currents at each end of the protected section. [448-14-18]

Discriminative zone
The selective part of a multi-zone busbar protection, generally supervising current flow into and out of a single section of busbar. [448-14-24]

Check zone
The non-selective part of a multi-zone busbar protection, generally supervising current flow at the complete station. [448-14-25]

NOTE: Tripping of the busbar protection is conditional on the operation of both the check and a discriminating zone.

Restricted earth fault protection (in USA: ground differential protection)
Protection, in which the residual current from a set of three-phase current transformers is balanced against the residual output from a similar set of current transformers located on the earthing connection, if any, of a neutral point. [448-14-29]

NOTE: This term is also used when the neutral of the protected plant is unearthed i.e. neither a second set of three-phase current transformers nor a CT in the neutral connection is needed to restrict the protected section.

Partial Differential Protection
Protection circuit which is often used in regions with Anglo-Saxon history. It is applied to busbars with bus section coupler and parallel in-feed. Current relays are connected to measure the differential current between the in-feed and the section coupler. One time grading step can be saved in the grading of the overcurrent relays. (see section 3.5)

Pilot wire protection
Protection associated with telecommunication using metallic wires. [448-15-04]

Short circuit loop (fault loop)
The circuit path in the energy system to and from the fault location as seen from the source.

Short circuit (fault) impedance
The impedance at the point of fault between the faulted phase conductor and earth (ground) or between the faulted phase conductors themselves. [448-14-11]

Source impedance
For a particular fault location, the impedance in the equivalent circuit of the fault current path between the point where the voltage is applied to the measuring relay and the EMF in the equivalent circuit producing the fault current in the same path. [448-14-13]

Fault resistance
The resistance at the point of fault between the faulted phase conductor and earth (ground) or between the faulted phase conductors themselves.

Phasor
In this book, the phasor representation is used for electrical signals:

$$\underline{A} = A \cdot e^{j\varphi} = A \cdot (\cos\varphi + j\sin\varphi) = B + jC$$

$$A = \sqrt{B^2 + C^2}$$

Thereby A represents the *RMS* value of current, voltage or power and φ is the phase angle referenced to the time $t = 0$.

The representation is extended for impedances which are not time dependent.

Vector
This designation is often used instead of phasor (in this case, A may also represent the peak value of the electrical AC signal)

α- and β-plane
The operating characteristic of the differential protection may be visualised in the complex plane (polar characteristic) using the current ratios $\underline{\alpha} = \underline{I}_A / \underline{I}_B$ and $\underline{\beta} = \underline{I}_B / \underline{I}_A$. In this context $\underline{I}_A$ is the current at the local terminal and $\underline{I}_B$ the current at the remote terminal. This is primarily applied to feeder protection.

Polar Characteristic
Representation of the operating characteristic of the differential protection in terms of the ratio of the compared currents. (Refer to *α*- and *β*-plane)

Current sign convention for differential protection
In this book, currents that are flowing into the protected object will be designated as positive. Therefore the vectorial current sum for internal faults is $\underline{I}_1 + \underline{I}_2 + \ldots$ $\ldots + \underline{I}_n$.

According to this convention the vectorial current sum corresponds to the differential current.

Polarity of current transformers
If no indication to the contrary exists, the following polarity rules for transformers and instrument transformers apply in this book (Refer to section 5.5, Figure 5.10).

Primary winding, secondary winding and possible further windings are wound in the same direction.

Voltages on the windings have the same polarity, i.e. they are in phase.

The currents have opposite polarity, i.e. they flow through the windings in opposite directions ($i_1 \cdot w_1 + i_2 \cdot w_2 + \ldots + i_n \cdot w_n = 0$).

Conductor/phase designation
In this book the standardised IEC designation L1, L2, L3 is generally used. In equations or diagrams, where clarity demands, the designations a, b, c are alternatively used.

Earth current (in USA: ground current)
Current (I_E) flowing from a neutral point to earth and current flowing through earth.

Zero sequence current
According to the computation with symmetrical components, the zero sequence current is one third of the sum of the phase currents, i.e. one third of the neutral current $I_0 = 1/3 \cdot I_N$. In three-phase HV systems (no neutral conductor), the neutral current corresponds to the earth current, therefore also $I_0 = 1/3 \cdot I_E$.

Residual current
Current (I_N) equal to the vector sum of the phase currents. [448-14-11]

Through fault current

A current due to a power system fault external to that part of the section protected by the given protection and which flows through the protected section [448-14-13]

3 Mode of Operation

The principle of differential protection is initially described in this section. Subsequently different protection principles and measuring techniques are covered in the further chapters. Current transformer response and signal transmission are then covered in detail as the reliability of the complete protection system is reliant thereon. Based on this, the application specific protection systems for generators, motors, transformers, feeders and busbars are described. In each case, application related questions are discussed.

3.1 Introduction

The differential protection is 100% selective and therefore only responds to faults within its protected zone. The boundary of the protected zone is uniquely defined by the location of the current transformers. Time grading with other protection systems is therefore not required, allowing for tripping without additional delay. Differential protection is therefore suited as fast main protection for all important plant items.

Due to the simple current comparison, the principle of differential protection is very straight forward. Stability in the event of external faults however demands an adequate dimension and matching of the current transformers. To ensure acceptable cost for the current transformers, the differential protection must however tolerate a fair degree of current transformer saturation. Determining the degree of saturation and providing adequate stabilisation against the false differential currents arising, are therefore important additional tasks of this measuring principle.

Generators, motors and transformers are often protected by differential protection, as the high sensitivity and fast operation is ideally suited to minimise damage. On feeders the differential protection is mainly used to protect cables, particularly on short distances where distance protection cannot be readily applied. On applications over longer distances of up to 25 km, the interference on the pilot wires by the earth fault current must be considered. Additional screening and isolation may be necessary. Numerical differential protection with serial data transfer via fiber optic communication is not affected in this manner, and distances of more than 100 km may be bridged. With the introduction of digital data networks by the utilities, serial data connections between all important substations have become available. This provides a further application opportunity for differential protection. The interfaces, protocols and procedures for the information exchange between protection device and data transfer system must be exactly matched and must correspond with the appropriate standards (open communication). When transferring data, using multiplexing, together with other services, the time response of the data channel must also be considered, in particular in the case of communication path switching. Furthermore, the availability of the communication system must generally be checked. The differential protection with digital communication improves the protection quality in the transmission system, as the measurement is strictly per phase allowing phase selective tripping with subsequent AR. This also applies to difficult fault constellations, such as for example multiple faults on double circuit lines which cannot be cleared phase selectively by the distance protection. The differential protection is ideally suited for the increasing number of three terminal lines.

The prime objective of busbar differential protection is fast, zone selective clearance of busbar faults to prevent large system outages and to ensure system stability. Mal-operation must be avoided at all cost as these could result in extensive supply interruption.

Extremely fast operating times of less than one cycle along with a high degree of stability against current transformer saturation have been state of the art for quite some time. Security against mal-operation due to hardware failure is achieved by AND combination of several independent tripping criteria.

In large substations the busbar configuration is often complex with numerous busbar sections as well as several bus-sectionalisers and bus couplers. This demands numerous measuring circuits and a complex isolator replica for the co-ordination of the bus section specific current differential protection. The switching of analog measuring circuits, that is necessary in conventional systems, is not required in the numerical busbar protection, where this is done by logic allocation of signals via an isolator replica within the software. A de-centralised configuration with bay units and communication via fiber optic reduces the previously very comprehensive substation cabling to a minimum.

The statements made so far refer to the "normal" current differential protection based on low impedance measuring technique.[1]

Apart from this the high impedance differential protection also exists. In this case, the measuring relay in the differential path has high impedance in comparison with the secondary impedance of a saturated current transformer. Stability is thereby automatically achieved in the event of through-fault currents with current transformer saturation. The current of the non-saturated CT in this case will not flow via the measuring relay but rather through the saturated CT which acts as a low impedance shunt.

This technique is frequently used outside continental Europe to protect galvanically connected circuits, primarily generators, motors and busbars as well as restricted earth fault protection on transformers. This method requires special current transformer cores (Class PX according to IEC 60044-1, formerly Class X to BS 3938) having the same ratio.

---

[1] In numerical relays the differential circuit only exists within the software. The differential protection however corresponds to the low impedance measuring technique.

During internal faults, when all CT's are feeding current to the high impedance relay, large voltage peaks in the CT secondary circuits arise. These must be restricted by means of a varistor. This measuring technique is not suited for the protection of multiple busbars, as the current transformer secondary circuits must be switched over. Some users apply this technique to dual busbar systems using the isolator auxiliary contacts for direct switching over of CT secondary circuits. The high impedance differential protection is however primarily used when isolator replicas are not required e.g. for one-and-a-half circuit breaker applications.

3.2 Basic principles

The basic principles, which have been known for decades, are still applicable and independent of the specific device technology. [3-2]

The differential protection compares the measured values with regard to magnitude and phase. This is possible by direct comparison of instantaneous values or by vector (phasor) comparison. In each case the measurement is based on Kirchhoff's laws which state that the geometric (vector) sum of the currents entering or leaving a node must add up to 0 at any point in time. The convention used in this context states that the currents flowing into the protected zone are positive, while the currents leaving the protected zone are negative.

3.2.1 Current differential protection

This is the simplest and most frequently applied form of differential protection. The measuring principle is shown in Figure 3.1. The current transformers at the extremities of the differential protection zone are connected in series on the secondary side so that the currents circulate through the current transformers during an external fault (Figure 3.1a) and no current flows through the differential measuring branch where the differential relay is situated. In the event of an internal fault (Figure 3.1b) the fault currents flow towards the fault location so that the secondary currents add up and flow via the differential branch. The differential relay picks up and initiates tripping.

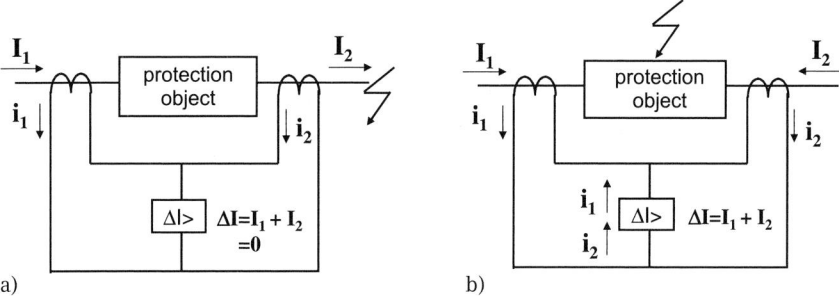

a)   b)

Figure 3.1 Measuring principle: External fault or load (a), Internal fault (b)

This simple circuit principle (non-biased current differential protection) may be used on all non-distributed protection objects where the current transformers are located in close physical proximity to each other.

The simplest arrangement results with generators or motors (Figure 3.2a), in particular when the current transformers have the same ratio.

The transformer protection requires interposing current transformers for the vector group and ratio correction of the currents used for the comparison (Figure 3.2b).

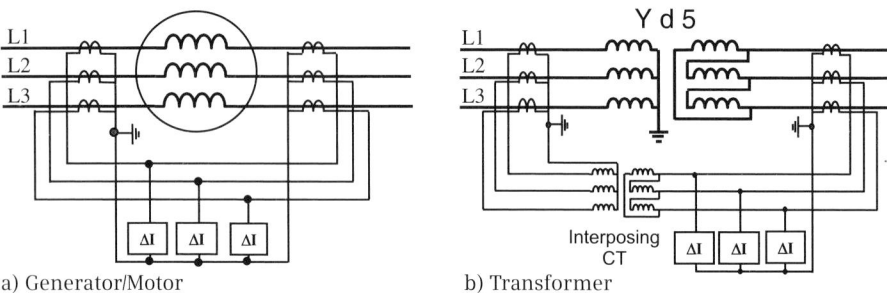

a) Generator/Motor   b) Transformer

Figure 3.2 Differential protection, three phase basic principle

For busbar protection, the currents from a number of feeders must be summated (Figure 3.3). In the case of load and external faults, the vector sum of the feeder currents is equal zero, so that no differential current flows in the relay. During internal faults, the currents however add up to a large differential current.

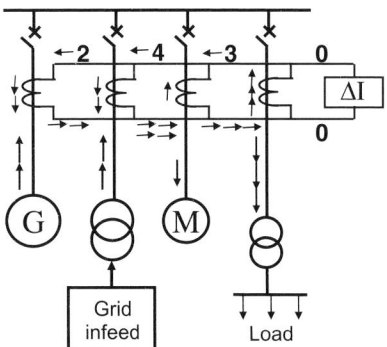

Figure 3.3 Busbar protection
(load or through fault condition)

For feeder differential protection, the current transformers at the two terminals of the protected object are far apart. In this case, the connection circuit according to Figure 3.4 is used (three core pilot differential protection). Three pilot wire cores are required for the connection between the two stations, which typically are provided as a "twisted triplet" via a communication cable. Current differential relays

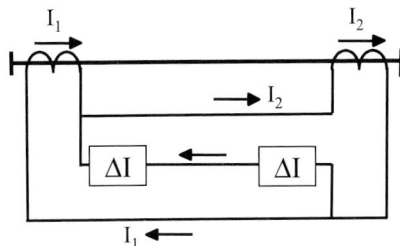

Figure 3.4
Line differential protection

are connected at both terminals in the differential core which, in the event of an internal fault, trip the circuit breakers in their respective stations. No further trip command communication between the stations is therefore required. In practice, the current transformer secondary currents (1 or 5 A) are converted to 100 mA by interposing current transformers to reduce the burden of the pilot wire cores. The current differential protection may be used over distances of approximately 10 km due to this reduced current transformer burden. Over short distances of 1 to 2 km, control cables (2 kV isolation) may be used.

When the pilot wire cables are in close proximity to power cables or overhead lines, adequate screening against fault currents via earth is required. On longer distances, high voltages of several kV may be induced in the pilot wires. This affects the isolation of the pilot wires against earth, and requires special pilot wire cables with higher insulation (e.g. 8 kV) and may necessitate barrier transformers to prevent the high voltages from reaching the protection relays (refer to section 6.1.1).

To further reduce the number of pilot wire cores required, the interposing current transformers are also summation transformers, whereby the phase currents are combined to a single (summated) composite current.

Current comparison with digital measured value transmission

The principle of current differential protection was so far described, based on the classic mode of 50/60 Hz analog measured value transmission via pilot wire communication. With numerical protection, the application of serial data transfer is increasingly used.

Thereby the measured values are digitally coded and transmitted via a dedicated fiber optic core or via a digital data communication system. Despite the numerical measured value transmission and processing, the basic principle remains the same.

The digital feeder differential protection 7SD52 and the decentralised numerical busbar protection 7SS52 are examples for this.

The comparison protection circuits described above are also designated as "longitudinal differential protection". For the sake of completeness the transverse differential protection that was used previously, must also be mentioned. It compared the current at the terminals of two or more circuits connected in parallel. This type of protection is hardly ever used anymore with lines in particular because the circuits must be connected in parallel for this type of protection and may not be operated independently. Only with generators that have parallel (split) windings in each phase brought out to separate terminals, the transverse differential protection is still used against turn faults.

3.2.2 Biased (stabilized) differential protection

So far, for the sake of simplicity, a fixed pick-up threshold was assumed for the relay measuring the current in the differential circuit. In practice however, a false differential current resulting from transformation errors of the current transformers must be considered.

In the linear range of the current transformers, this error is proportional to the through current. In the event of large fault currents, CT saturation may be the result, causing a rapid increase of this false differential current.

Additionally, transformer tap changers will cause a false current due to the modification of the transformation ratio.

Figure 3.5 shows the differential current measured by the relay, related to the through current (I_{through}) during load or external faults.

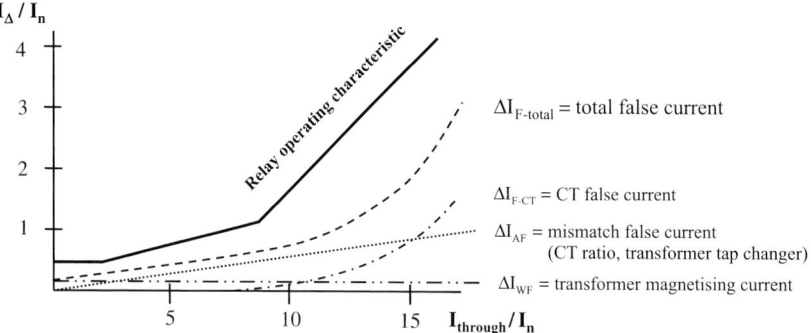

Figure 3.5 False differential current during load and external faults with adapted relay characteristic

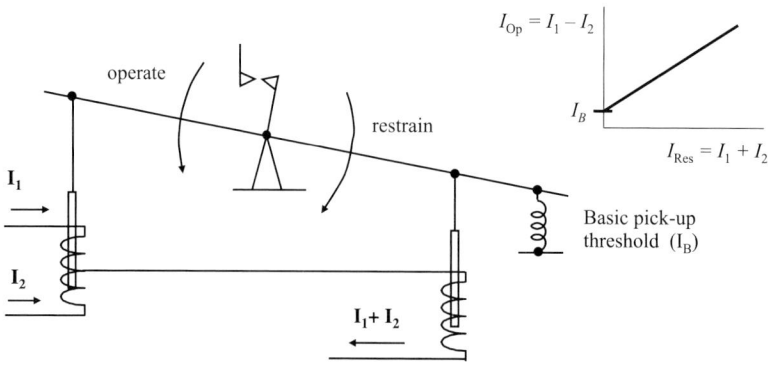

Figure 3.6 Biased differential relay according to McCroll

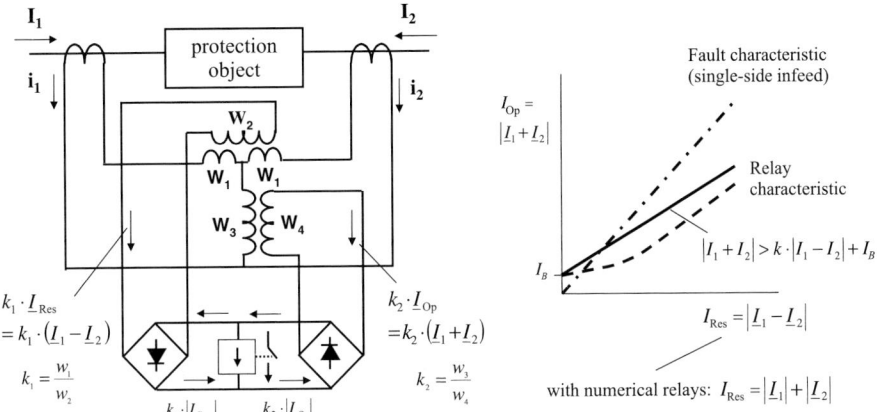

Figure 3.7 Differential protection with bridge rectifying circuit in the measuring path

It is apparent that the pick-up threshold should be increased when the through current increases. This results in high sensitivity during load and small fault currents, while at the same time providing improved stability against mal-operation with large currents when CT saturation is expected.

In the early days of protection this was achieved by increasing the pick-up threshold proportionally to the through current. This method was already suggested in 1920 as a biased differential relay [3-1, 3-2]. The principle of operation is shown in Figure 3.6: Biased differential relay according to McCroll.

Electromechanical and static relays implemented this method using a rectifier bridge comparator Figure 3.7. The measuring path was implemented with a polarised moving coil relay having high sensitivity and later with an electronic trigger circuit.

Bias (stabilisation) was provided by the signal $\underline{I}_{Bias} = k_1 \cdot (\underline{I}_1 - \underline{I}_2)$ which corresponds to the "sum" of the CT currents in the event of a through current. In this regard, the chosen sign convention for the currents must be observed; it designates the currents as positive when they flow into the protected object. Operation is effected by the "difference" of the CT currents $\underline{I}_{Op} = k_2 \cdot (\underline{I}_1 + \underline{I}_2)$.

The following states result:

	$\underline{I}_{Bias} = k_1 \cdot (\underline{I}_1 - \underline{I}_2)$	$\underline{I}_{Op} = k_2 \cdot (\underline{I}_1 + \underline{I}_2)$
External fault	$\underline{I}_{Bias} = 2 \cdot k_1 \cdot \underline{I}_F$	$\underline{I}_{Op} = 0$
Internal fault with single end in-feed	$\underline{I}_{Bias} = k_1 \cdot \underline{I}_F$	$\underline{I}_{Op} = k_2 \cdot \underline{I}_F$
Internal fault with in-feed from both ends	$\underline{I}_{Bias} = 0$	$\underline{I}_{Op} = 2 \cdot k_2 \cdot \underline{I}_F$

The pick-up criterion is:

$$\underline{I}_{Op} > \underline{I}_{Bias} \quad \text{i.e.} \quad k_2 \cdot |\underline{I}_1 + \underline{I}_2| > k_1 \cdot |\underline{I}_1 - \underline{I}_2|$$

By means of a restraint spring on the pick-up relay, a minimum pick-up threshold B can also be applied.

The principle equation for the biased differential protection is thus obtained:

$$|\underline{I}_1 + \underline{I}_2| > k \cdot |\underline{I}_1 - \underline{I}_2| + B \quad \text{whereby} \quad k = k_1/k_2 \tag{3-1}$$

Later, the measuring circuit was further refined, and supplemented with an additional diode resistor combination. Thereby, the restraint with small currents sets in slowly and only starts increasing strongly above a threshold value (variable restraint) as illustrated by the dotted characteristic in Figure 3.7. Numeric protection then implemented a characteristic made up of several sections. This allows better adaptation to the area of false current measurement that must be excluded.

In newer protection devices, the threshold B is no longer added to the restraining side, but is provided as a separate setting value: $I_{Op} > B$. As a result the biased characteristic $I_{Op} > k \cdot I_{Res}$ is no longer displaced by the initial value B, but instead passes through the origin of the coordinates. Consequently an increased sensitivity with small currents is achieved (refer to Figure 3.14 below).

The described measuring principle may also be applied to protection objects having more than two terminals (three winding transformer or busbar protection). Thereby the sum of the current magnitudes (arithmetic sum) is used for the restraint[1] while the magnitude of the geometric (vectorial) sum of the currents is used for operation:

$$I_{Res} = |\underline{I}_1| + |\underline{I}_2| + |\underline{I}_3| + \dots + |\underline{I}_n| \tag{3-2}$$

$$I_{Op} = |\underline{I}_1 + \underline{I}_2 + \underline{I}_3 + \dots + \underline{I}_n| \tag{3-3}$$

The conditions stated above apply as pick-up criterion:

$$I_{Op} > k \cdot I_{Res} \quad \text{and} \quad I_{Op} > B \tag{3-4}$$

The bias factor k (%bias /100), which defines the slope of the bias characteristic, can be set in the range from $k = 0.3$ to 0.8, depending on the application and the dimensions of the current transformers. The threshold B may be set to 10% I_N for a generator, while 130% of the maximum feeder current is typical for busbar protection.

This is referred to at length in the section regarding the individual protection systems.

The corresponding circuit based on analog signal processing is shown in Figure 3.8. The magnitude computation is achieved with the rectification.

In the event of an external fault, the operating current I_{Op} must be zero, i.e. the current vectors must add up to zero. The restraint current corresponds to the sum of the current magnitudes.

---

[1] With numerical relays, this formula also applies to two ended differential protection.

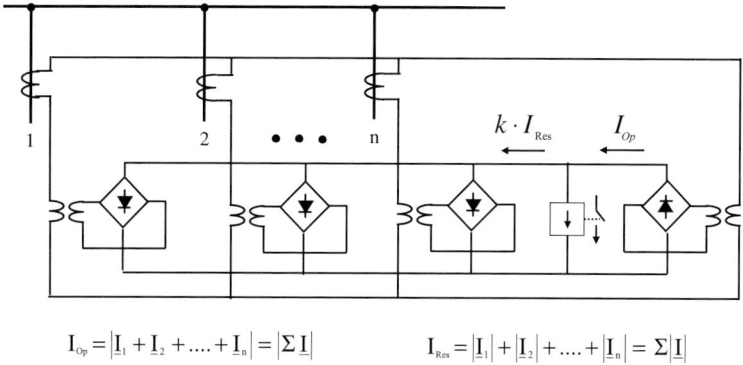

$$I_{Op} = \left| \underline{I}_1 + \underline{I}_2 + ... + \underline{I}_n \right| = \left| \Sigma\, \underline{I} \right| \qquad\qquad I_{Res} = \left| \underline{I}_1 \right| + \left| \underline{I}_2 \right| + ... + \left| \underline{I}_n \right| = \Sigma\left| \underline{I} \right|$$

Figure 3.8 Multi-terminal differential protection – schematic

In case of an internal fault, the operating current is the result of the summated current vectors. In its simplest form, when the in-feeds and consequently also the associated fault currents are all approximately in phase, the vector and the magnitude sums are equal i.e. $I_{Op} = I_{Res}$.

Under normal circumstances (low resistance fault and phase-equivalent in-feeds) the following may be noted:

External fault	$I_{Res} = 2 \cdot I_{\text{F-thru}}$	$I_{Op} = 0$	$I_{\text{F-thru}}$ is the fault current flowing through the protected object
Internal fault	$I_{Res} = I_{\text{F-int}}$	$I_{Op} = I_{\text{F-int}}$	$I_{\text{F-int}}$ is the sum of the fault currents at the fault location

In the event of internal faults with relatively large fault resistance it must however be considered that part of the load current may still be flowing through the protected object during the fault. The through flowing load current is superimposed onto the fault currents flowing into the protected object. The ratio I_{Op}/I_{Res} correspondingly reduces.

Example 3-1:

Short circuit with fault resistance (Figure 3.9)

$I_{Op} = 2300 - 300 = 2000$

$I_{Res} = 2300 + 300 = 2600$

$I_{Op}/I_{Res} = 0.77$

In expansive transmission systems or in the event that power swings or even out of step conditions arise, the fault currents flowing into the fault location may however have substantial phase angle differences. In this case the vectorial sum of the currents is smaller than the sum of the current magnitudes and therefore $I_{Op} < I_{Res}$. For a two ended in-feed, the conditions according to Figure 3.10 will result.

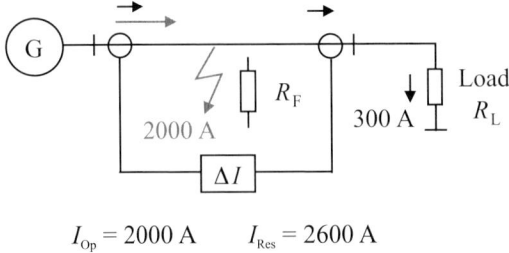

$I_{\text{Op}} = 2000$ A $I_{\text{Res}} = 2600$ A

Figure 3.9 Internal fault with fault resistance, current distribution

If for the sake of simplicity, the two currents are assumed to have equal magnitude, then the following applies:

$$I_{\text{Res}} = 2 \cdot |I_{\text{F}}| \quad \text{and} \quad I_{\text{Op}} = 2 \cdot |I_{\text{F}}| \cdot \cos\frac{\delta}{2}.$$

When $\delta = 30°$, the lower ratio $I_{\text{Op}}/I_{\text{Res}} = 0.87$ results.

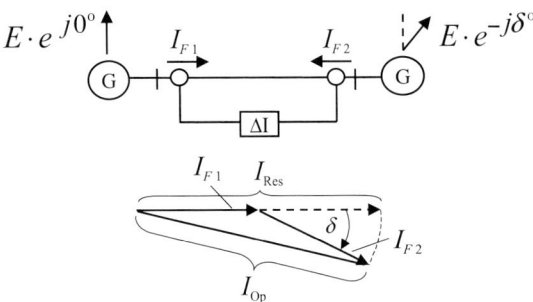

Figure 3.10 Internal fault with phase shift between in-feeds

The observed effects may of course also be compounded. The bias (stabilisation) factor k should therefore not be set above 0.8. On the contrary the current transformers should be chosen such that a setting above 0.7 is not necessary.

Note 1: In the protection literature and relay manuals, very often the through flowing current is taken as reference and counted positive. In this case, the operating current corresponds to $I_{\text{Op}} = |I_1 - I_2|$ (differential current) and the restraint current to $I_{\text{Res}} = |I_1 + I_2|$ with traditional relays or $I_{\text{Res}} = |I_1| + |I_2|$ with numerical relays. This rule, which fits for protection of two-terminal objects, however is impractical in the case of multiple end protection objects like busbar protection. Therefore the sign rule, that currents flowing into the protection object are counted positive, is uniformly applied in this book (also corresponds to the Siemens relaying conventions).

Note 2: In this book, the restraint quantity corresponds to the sum of the current magnitudes $I_{\text{Res}} = |\underline{I}_1| + |\underline{I}_2|$. The $I_{\text{Op}}/I_{\text{Res}}$ locus for internal faults is in this case a straight line with 45° inclination (100% slope) in the operating/restraint diagram (see Figure 3.7). This also applies to all Siemens relays.

Some relay manufacturers only use one half of the current sum as restraint quantity: $I_{\text{Res}} = (|\underline{I}_1| + |\underline{I}_2|)/2$, even with multi-terminal protection, i. e. $I_{\text{Res}} = (|\underline{I}_1| + |\underline{I}_2| + |\underline{I}_3| + \ldots + |\underline{I}_n|)/2)$. In this case, the internal fault locus has a 200% slope! This has to be kept in mind for comparing relays of different make and setting the bias factor (slope percentage).

3.2.3 Differential Protection with two pilot wire cores

The pilot wire differential protection (twisted pair pilot wires) was developed for the application with communication cables having twisted pilot wire pairs (telephone cables). It is primarily used outside continental Europe, where the twisted pairs are often leased from telephone companies.

Two variants are in essence possible:

- opposed voltage principle (tripping pilot scheme)
- circulating current principle (blocking pilot scheme)

Both variants were developed and applied in practice [A-13, A-22]. The relays supplied by Siemens, described in detail in section 9.2, operate with the opposed voltage principle.

Voltage comparison (opposed voltage principle)

With this technique, the current at each line terminal is routed through a shunt resistance (R_{Q}) thereby producing voltages U_1 and U_2 each proportional to the corresponding current Figure 3.11. These two voltages are then compared via the pilot wire pair. The connection is chosen such that in the event of load current or external fault current flowing through the line, the voltages are in opposition and no current flows via the pilot wire pair. During internal faults, the two voltages however are in phase and drive a current through the pilot wire loop. This current, which is only a few mA referred to nominal current of the current transformers, results in tripping via the sensitive current relay (ΔI).

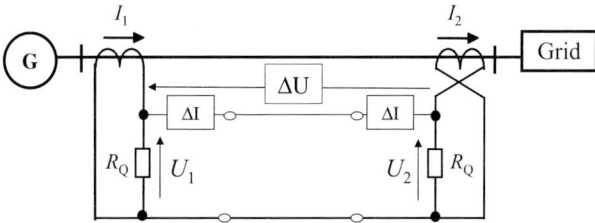

Figure 3.11 Line differential protection, voltage comparison principle

The voltage on the pilot cores is only a few volts when nominal current flows via the CTs but rises in case of large fault currents. The maximum transverse voltage on the pilot cores may however not be greater than 60% of the rated insulation voltage of the telephone cable (500 V), in other words 300 V. To restrict the voltage in the event of heavy internal fault currents, a varistor is provided. During external faults, the voltage limiting threshold should not be reached. The small burden imposed by these devices permits distances of up to approximately 25 km to be bridged. With regard to insulation and screening of the pilot wire cores, the statements made in section 3.2.1 with respect to three core pilot wire apply here also.

In addition, the pilot wire cores must be properly twisted so that the transverse voltage induced by the fault current in the earth, which influences the measurement, is reduced to a minimum. This is referred to at length in section 6.1.1.

The measuring circuit applied in practice operates with the biased differential protection principle. The operating current $I_{Op} = I_1 + I_2$ [1] is thereby proportional to the pilot wire current. The biasing current $I_{Res} = I_1 - I_2$ is obtained from the current in the shunt branch together with a supplementary component from the pilot wire current (Figure 3.12).

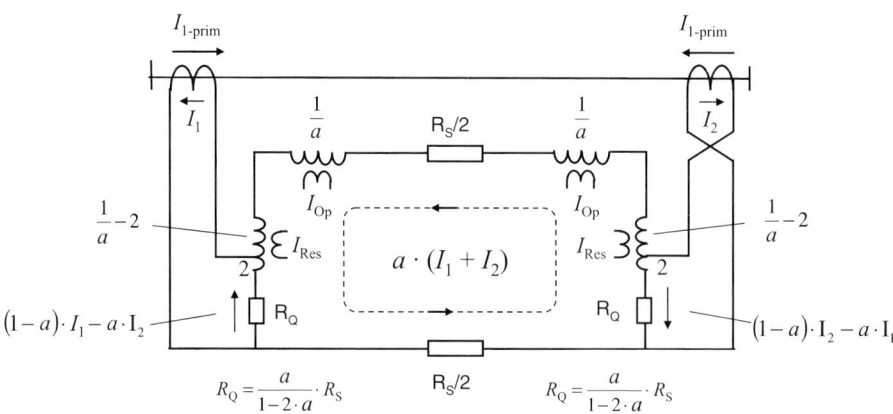

Figure 3.12 Two core pilot wire differential protection – opposed voltage principle

The ratio of shunt resistance R_Q to loop resistance R_S determines the current distribution factor k (ratio of pilot current to total current $I_1 + I_2$).

R_S is determined by the loop resistance of the pilot wires. R_Q is set on the relay to obtain the desired k value:

$$R_Q = \frac{k}{1 - 2 \cdot k} \cdot R_S \qquad (3\text{-}5)$$

---

[1] The sign convention must be noted : Currents flowing into the line are designated as positive

To obtain I_{Op} and I_{Res}, the pilot wire and shunt currents are summated in the relay with the indicated weighting factors. In analog devices, this is done by internal interposing transformers with corresponding tappings. (Figure 3.12)

$$\underline{I}_{Op} = \frac{1}{k} \cdot [k \cdot (\underline{I}_1 + \underline{I}_2)] = \underline{I}_1 + \underline{I}_2 \tag{3-6}$$

$$\begin{aligned}\underline{I}_{Res} &= 2 \cdot [(1-k) \cdot \underline{I}_1 - k \cdot \underline{I}_2] - \left(\frac{1}{k} - 2\right) \cdot \\ &\quad \cdot [k \cdot (\underline{I}_1 + \underline{I}_2)] = \underline{I}_1 - \underline{I}_2\end{aligned} \tag{3-7}$$

The analog relays provided by Siemens always implement a fixed value $k = 1/8$, in other words a setting $R_Q = 1/6 \cdot R_S$ is applied for purposes of alignment (refer to section 9.2, Figure 9.2).

With numerical protection the alignment (R_Q setting) is not necessary. The current distribution is always calculated, for each application from the value of pilot wire loop resistance R_S, which must be set at the relay, and the fixed value of R_Q. This is described in detail in section 9.2.

Circulating current principle

The circuit is similar to that applied for voltage comparison. Operation and restraint are however swapped. The auxiliary transformer in the pilot wire loop now supplies the current $\underline{I}_{Res} = \underline{I}_1 - \underline{I}_2$ and the auxiliary transformer in shunt branch supplies $\underline{I}_{Op} = \underline{I}_1 + \underline{I}_2$. The CTs are connected with phase opposition to the pilot wire cores as is usually done with differential protection. Consequently the cross-over of the CT connection on the right hand side, as shown in Figure 3.12 is not applied. Accordingly, the secondary voltages of the CTs are in phase when current flows through the feeder. A circulating current is thus driven through the pilot wire loop. During an internal fault, the two voltages are in opposition, resulting in a reduction of the pilot wire current. With the same in-feed at both sides, the pilot wire current is theoretically zero.

Comparison of the measuring principles

With the opposed voltage principle, the pilot wires do not carry current during normal operation. In case of an internal fault, current must however flow through the pilot wire loop to cause tripping. This operating mode therefore uses the release principle (tripping pilot scheme). If the pilot wires are interrupted, no tripping is possible. In the event of a pilot wire short circuit, tripping would occur during external faults. A separate overcurrent condition must therefore be applied to prevent incorrect tripping during load when pilot wire short circuits occur.

With the circulating current principle, restraint current flows through the pilot wire loop and prevents relay pick-up during load and external faults. During an internal fault, the restraining pilot currend is reduced and allows tripping. This operating mode is therefore based on the blocking principle (blocking pilot scheme). An interruption of the pilot wires results here in tripping with large through flowing currents. An additional overcurrent criterion should therefore also be applied here. A short circuit of the pilot wires results in over-restraint and blocking.

3.2.4 Operating characteristics

Different representations are possible.

The following diagram of operating versus restraint current is the most useful and generally applied form. Other versions based on the terminal currents are also discussed below but are only used for special applications.

I_{Op}/I_{Res}-diagram (Scalar diagram)

The response of the differential protection can be explained by means of the current diagram in which the operating current (differential current[1]) $I_{Op} = |\sum I|$ is shown on the vertical axis and the restraint current (summated currents) $I_{Res} = \sum |I|$ is shown on the horizontal axis (Figure 3.13).

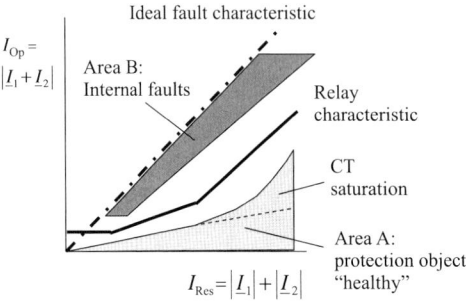

Ideal fault characteristic

$I_{Op} = |\underline{I}_1 + \underline{I}_2|$

Area B:
Internal faults

Relay characteristic

CT saturation

Area A:
protection object
"healthy"

$I_{Res} = |\underline{I}_1| + |\underline{I}_2|$

Figure 3.13
I_{Op}/I_{Res}-diagram of the differential protection

Initially two states can be distinguished:

Healthy protected object

Ideally no differential current is present; therefore during load and through fault currents (external faults) only the summated currents should be present. The ideal healthy state is therefore represented by the horizontal axis of the diagram. Current transformer inaccuracies and mismatches, caused for example by tap changers, result in false differential currents that are proportional to the current flowing through the protected object.

Above a threshold, depending on the CT dimensions, current transformer saturation may take place and the false differential current will rise rapidly.

The area A consequently is the range defining a "healthy" protected object.

Short circuit in the protected object:

In the ideal fault condition, the currents flow into the protected object and are in phase. Operating and restraint current are therefore equal in magnitude as shown above.

---

[1] The name differential current is derived from the fact that the vectorial current sum has a non-zero result during faults.

This is represented by a 45°-line in the diagram that is designated as the ideal fault characteristic. As a result of the discussed phase angle difference of the in-feeds and the load currents flowing through the protected object during internal faults with fault resistance, the ratio of I_{Op}/I_{Res} may be smaller than 1. In practice, internal faults therefore appear in a range below the 45°-line.

Relay characteristic

The ranges for healthy and faulted states of the protected object have been defined.

Between these, there is a zone in which the relay characteristic may be placed to form the boundary of discrimination between operation and restraint (non-operation).

Conventional devices usually had a fixed characteristic that started off flat and had a steeper slope from a designated threshold (variable slop). Numerical relay characteristics may usually be applied in piecewise liner shape, e.g. three zones as indicated in Figure 3.14:

Zone a: $I_{Op} > I_B$

Zone b: $I_{Op} > k_1 \cdot I_{Res}$

Zone c: $I_{Op} > k_2 \cdot (I_{Res} - I_{R0})$

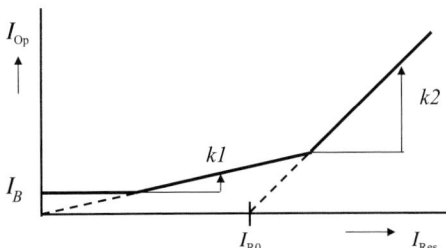

Figure 3.14
Numerical relay characteristic

The setting parameters for this purpose are: I_B, k_1, k_2, I_{R0}

Initially the diagram may be set up and tested with steady state RMS currents.

Dynamic response must however be analysed with the instantaneous values (each sample in numerical relays) or with correspondingly filtered values (according to the appropriate data window of the numerical protection). In principle, the measuring algorithm must be replicated. Appropriate simulation software is available for this purpose.

In any event, the points defined by I_{Res} and I_{Op} must be situated above the relay characteristic during internal faults and below during external faults. The progression in time (locus) can for example also indicate the presence of CT saturation. (See section 4.2.4).

I_1/I_2-Diagram

In this case the pick-up threshold is shown in relation to the currents at the terminals of the protected object. This representation is only suitable for protected objects having two terminals and is usually only used for the line differential protection.

Figure 3.15 shows an example of a characteristic with the I_{Op}/I_{Res}-diagram along with the corresponding representation in the I_1/I_2-diagram for comparison.

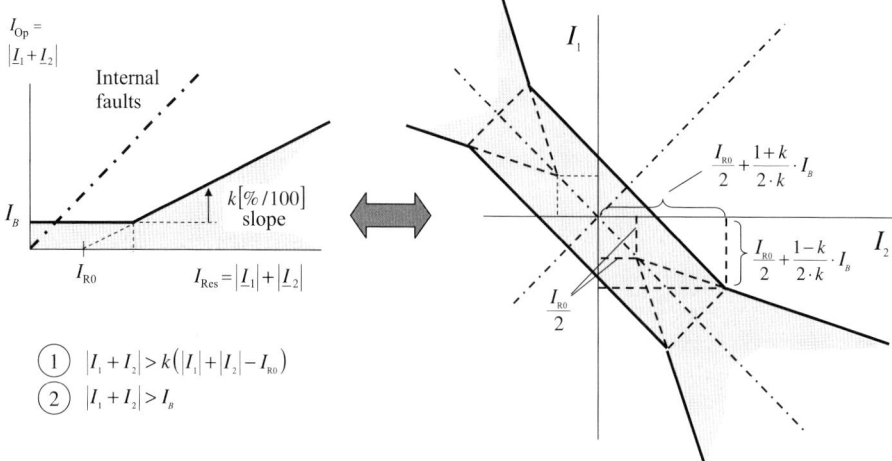

Figure 3.15 I_{Op}/I_{Res} diagram and I_1/I_2- diagram for comparison

The stable condition with current flowing through the protected object, i.e. when $\underline{I}_1 = -\underline{I}_2$, is given by the axis of symmetry between the 2nd and 4th quadrant. In this context the applied sign convention, whereby the currents flowing into the protected object are positive, must be observed. A deviation from the axis of symmetry is tolerated in the designated area, for example in the case of current transformer saturation during external faults.

Internal faults appear in the 1st and 3rd quadrant. In this case $\underline{I}_1$ and $\underline{I}_2$ will have the same sign, i.e. both currents flow into the protected object. A fault that is only fed from one terminal will either be on the horizontal or vertical axis depending on where the fault current is coming from.

This diagram is usually applied with the RMS current values. Dynamic behaviour may however also be shown with the instantaneous measured values, as was described in connection with the I_{Op}/I_{Res}-diagram.

Polar characteristic (Vector diagram)

This type of diagram with vector quantities is used to consider the influence of the current signal phase angles. [3-3, 3-4] It can only be applied to protected objects with two terminals. By combination of terminals the typical response of multi-terminal protection systems can however also be examined. The polar diagram is most commonly applied with line differential protection.

The ratio of the currents at the extremities of the protected object $\underline{\alpha} = \underline{I}_1/\underline{I}_2$ and $\underline{\beta} = \underline{I}_2/\underline{I}_1$ is represented in the complex plane (according to [A-22] designated as α- and β-plane). $\underline{I}_1$ is the current at the local end and $\underline{I}_2$ the current at the remote end.

Figure 3-16 is an example of the β-plane representation.

From the pick-up criterion (equation 3-4) $I_{Op} > k \cdot I_{Res}$ the response of conventional relays is obtained, whereby $\underline{I}_{Op} = \underline{I}_1 + \underline{I}_2$ and $\underline{I}_{Res} = \underline{I}_1 - \underline{I}_2$. By neglecting the minimum pick-up threshold B, we can write

$$\frac{\left|1 + \dfrac{\underline{I}_2}{\underline{I}_1}\right|}{\left|1 - \dfrac{\underline{I}_2}{\underline{I}_1}\right|} > k \quad \text{or} \quad \frac{|1 + \underline{\beta}|}{|1 - \underline{\beta}|} > k \quad \text{with} \quad \underline{\beta} = \frac{\underline{I}_2}{\underline{I}_1} \tag{3-8}$$

This inequality gives rise to a circular characteristic as shown in Figure 3.16. The inside of the circle corresponds to the stable condition with current flowing through the protected object. Ideally $\underline{I}_2 = -\underline{I}_1$, so that $\underline{\beta} = -1$. Tripping results outside the circle. The level of permissible error current, produced by current transformer errors or internal consumers of real or reactive power, can be seen.

The response during current transformer saturation can also be visualised in the polar diagram when we consider that the fundamental component of the saturated secondary current has a smaller amplitude and is phase-shifted with a leading sense in relation to the primary current.

If the local CT saturates, the β-value therefore drifts in a lagging sense upwards and assumes a larger amplitude. On the other hand, during saturation of the CT at

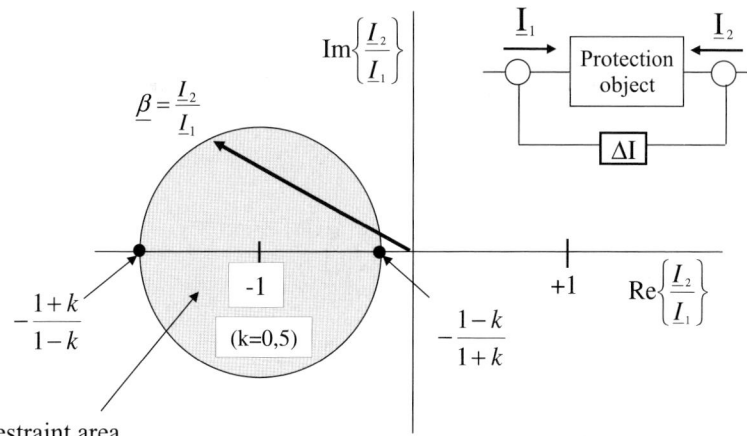

restraint area

Figure 3.16 Polar characteristic of the differential protection

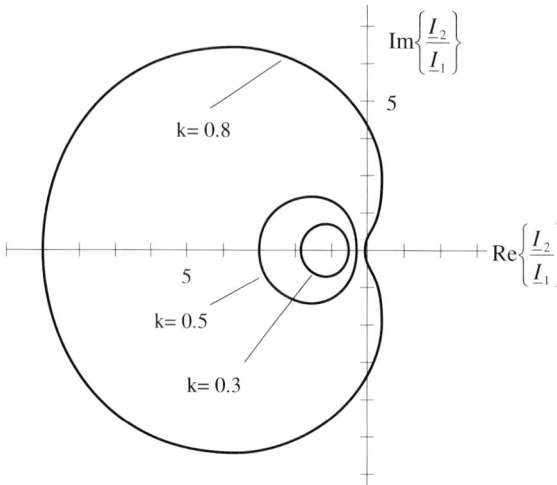

Figure 3.17
Polar diagram of the differential protection with different stabilising factors k
$I_B/I_1 = 0.3$ in all cases,
i.e. the relay pick-up value corresponds to about 30% of the fault current

the remote terminal, the β-value departs from the ideal point -1 downwards and assumes a new point with a smaller amplitude.

The influence of pilot wire capacitance on feeder differential protection can also be clearly represented. This is shown in section 9.2.

Internal faults appear on the right hand side of the diagram in the 1st and 4th quadrant. During ideal fault conditions $\beta = +1$. When the generator in-feeds have differing phase angles the vector β rotates away from the real axis.

In Siemens numerical relays the restraint current for two terminal differential protection systems is calculated with the sum of current magnitudes ($I_{Res} = |I_1| + |I_2|$, instead of $I_{Res} = I_1 - I_2$). In practice the basic pick-up threshold (I_B) has also to be considered. Consequently equation 3-8 changes to:

$$\left|1 + \frac{I_2}{I_1}\right| > k \cdot \left(1 + \left|\frac{I_2}{I_1}\right|\right) + \frac{I_B}{|I_1|} \tag{3-9}$$

In this case the pick-up characteristic is a conchoid which assumes the shape of a cardioid (heart shape) with large k settings as shown in Figure 3.17. With k settings smaller than $k = 0.5$, the characteristic approaches a circle.

3.3 Measuring circuit for three phase systems

In three phase systems the numerical relays will generally carry out a per-phase measurement, so that three independent comparison systems exist. This provides the following advantages:

– clearly identifiable protection response,
– the same pick-up threshold for all fault types
  (for the special case of Y-Δ-transformers, refer to section 5.8)

– same burden on all current transformers

– redundancy in case of multiple phase faults.

With conventional technology the complexity and cost is however almost three times as high, both in terms of the implemented devices and the amount of required wiring. To simplify this, summation transformer connections were therefore introduced. Typical application areas were line and busbar differential relaying. In two cases they are still applied with numerical protection: for pilot wire differential protection (7SD600) and for the simple busbar protection system 7SS600.

Line differential protection

In this case three pilot wire connections (each consisting of one pilot wire pair or triplet) would be required for phase segregated comparison of the currents between the two line ends. Due to cost constraints and the limited availability of pilot wires, this is not possible in most cases. Line differential protection with pilot wires has therefore been traditionally applied with a summation transformer connection. Only under exceptional circumstances with expensive protected objects (e.g. 400 kV cables) would a per phase protection have been applied.

The situation has undergone a dramatic change with the advent of digital information exchange via fiber optic cables or microwave radio, and even over pilot wires of short length. The three phase currents may now be transferred in digitally coded form in a single channel with serial communication. Modern numerical differential protection relays with digital communication therefore all utilise phase segregated measurement. The processor hardware is common to all three phases. The absence of hardware redundancy is compensated by the continuous self-monitoring implemented in the numerical devices, so that there is no reduction of the availability.

Busbar protection

High impedance differential protection may by principle only be implemented in a phase segregated manner. It has a simple construction, may however only be applied to single busbars (e.g. 1-1/2 circuit breaker applications). In substations with multiple busbars and numerous feeders, the "normal" (low impedance) current differential protection is applied. With analog measurement this results in a complex arrangement with a large number of devices and considerable amount of wiring, with the associated costs. Phase segregated measurement was therefore only justified in extra high voltage substations. In HV and MV substations, the summation transformer application was accordingly preferred.

A whole new set of circumstances arise with numerical busbar protection. Cyclic sampling of the measured values and independent processing in the software also allows for the implementation of a cost effective three phase measurement with this technology. In particular, with the decentralised busbar protection, a per phase measurement is therefore preferred (refer to section 10.1.2). Only the basic version 7SS600 is usually configured with summation current transformers.

Two features of the summation CT connection are worth noting:

- In systems with non-effective neutral earthing (isolated or resonant grounded system neutral), the different weighting of the phase currents with the summation CT connection results in a phase preference when double earth faults (e.g. cross country faults) occur. This is similar to the desired response of the distance relays used in these networks, which shall only trip one preferred earth footing point. [9-13] In this way, a uniform response to double earth faults is achieved. (The network can be operated for some time with the remaining single phase earth fault until cleared by controlled network switching.)

- Due to the higher weighting placed on earth currents, an increased sensitivity for earth faults is achieved. This may be of advantage in systems with earth fault current restriction.

3.3.1 Measurement per phase

In this case, a separate comparison of the measured signals is carried out for each phase.

It is however recommended to connect the neutral point at the protection relay with the neutral of the current transformers so that a common return circuit (neutral conductor) is achieved. Apart from the saving of two control cable cores, the burden on the current transformer is halved in the event of three phase faults.

The same applies to the configuration of the measuring circuits.

With conventional transformer differential protection, the CT secondary wiring produced a connection between the phases with the Y-Δ interposing current transformers. This is generally no longer the case with numerical protection as the Y-Δ-conversion is implemented with software inside the device.

With pilot wire line differential protection, separate return circuits per phase are automatically present due to the separation of the pilot wire cores into twisted pairs.

3.3.2 Composite current version

The phase currents are combined to a single phase equivalent AC current (composite current).

The mix of the currents must be chosen so that sufficient composite current will result for each type of fault (1-, 2- and 3-pole). The difference in the composite current magnitude for the various fault types, and under consideration of the affected phases, should not vary too much.

A number of theoretical analyses have been undertaken to investigate the influence of the in-feed, earthing and load distribution on the mixing ratio [3-2].

In practice, the following composition of the composite (mixed) current has proven to be successful:

$$\underline{I}_M = 5 \cdot \underline{I}_{L1} + 3 \cdot \underline{I}_{L2} + 4 \cdot \underline{I}_{L3} \tag{3-10}$$

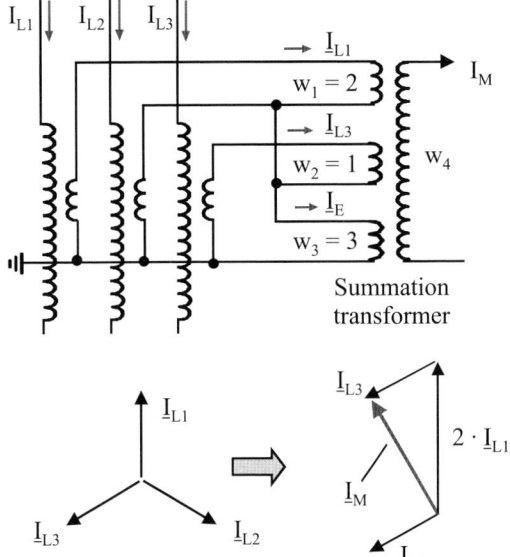

Figure 3.18
Summation CT for line
differential protection

When considering that $\underline{I}_{L1} + \underline{I}_{L2} + \underline{I}_{L3} = \underline{I}_E$, then (3-10) is also

$$\underline{I}_M = 2 \cdot \underline{I}_{L1} + 1 \cdot \underline{I}_{L3} + 3 \cdot \underline{I}_E \qquad (3\text{-}11)$$

The addition of the currents with this equation is done magnetically in a summation CT with three input windings. The weighting of the phase currents and the earth current thereby corresponds to the winding ratio: (Figure 3.18)

$$w_1 \cdot \underline{I}_{L1} + w_2 \cdot \underline{I}_{L3} + w_3 \cdot \underline{I}_E = w_4 \cdot \underline{I}_{M\text{-sec}} \qquad (3\text{-}12)$$

or:

$$\underline{I}_{M\text{-sec}} = \frac{w_1}{w_4} \cdot \underline{I}_{L1} + \frac{w_2}{w_4} \cdot \underline{I}_{L3} + \frac{w_3}{w_4} \cdot \underline{I}_E \qquad (3\text{-}13)$$

with $w_1 = 2 \cdot w$, $w_2 = 1 \cdot w$ and $w_3 = 3 \cdot w$ the following results:

$$\underline{I}_{M\text{-sec}} = \frac{w}{w_4}(2 \cdot \underline{I}_{L1} + 1 \cdot \underline{I}_{L3} + 3 \cdot \underline{I}_E) \qquad (3\text{-}14)$$

When symmetrical three phase current is applied in Figure 3.18 then the composite current according to (3-11), referred to the primary side of the summation CT, can be calculated with the following equation:

$$I_M = |2 \cdot \underline{I}_{L1} + 1 \cdot \underline{I}_{L3}| = \sqrt{3} \cdot \underline{I}$$

The associated secondary composite current according to (3-14) is then:

$$I_{M\text{-sec}} = \frac{w}{w_4} \cdot \sqrt{3} \cdot I \qquad (3\text{-}15)$$

The transformation ratio of the summation CT can be derived as the CT secondary nominal current and the nominal input current of the protection device are known:

$$\frac{w}{w_4} = \frac{I_{\text{M-sec-n.}}}{\sqrt{3} \cdot I_n} \tag{3-16}$$

For the two-core pilot wire differential protection (7SD600) the nominal (composite) current for example is 20 mA while the three phase CT nominal current is 1 or 5 A.

The corresponding nominal (composite) current of the busbar protection is 100 mA.

After substituting (3-16) into (3-14) the composite current as a p.u. value referred to the CT nominal current is obtained:

$$\underline{I}_{\text{M-p.u.}} = \frac{\underline{I}_{\text{M-sec}}}{I_{\text{M-sec-n.}}} = \frac{1}{\sqrt{3} \cdot I_n} \cdot (2 \cdot \underline{I}_{L1} + 1 \cdot \underline{I}_{L3} + 3 \cdot \underline{I}_E) \tag{3-17}$$

The following composite currents can then be calculated for the various types of fault:

Table 3.1 Composite currents and pick up sensitivity for various fault types using standard connection mode L1-L3-E

Fault type	$\underline{I}_{L1}$	$\underline{I}_{L3}$	$\underline{I}_E$	$|\underline{I}_M|$	Relative pick-up threshold
L1-L2-L3	1	$1 \cdot e^{-j240°}$	–	1	1.00
L1-L2	1	–	–	$2/\sqrt{3} = 1.15$	0.87
L2-L3	–	−1	–	$1/\sqrt{3} = 0.58$	1.73
L3-L1	−1	1	–	$1/\sqrt{3} = 0.58$	1.73
L1-E	1	–	1	$5/\sqrt{3} = 2.89$	0.35
L2-E	–	–	1	$3/\sqrt{3} = 1.73$	0.58
L3-E	–	1	1	$4/\sqrt{3} = 2.31$	0.43

From Table 3.1 it can be seen that a large summation current arises during earth faults. Correspondingly, the pick-up threshold for this type of fault is more sensitive. On the other hand, the protection is relatively insensitive to the phase-phase faults L2-L3 and L3-L1.

The earth fault current in solidly earthed networks has the same order of magnitude as the phase-phase fault current. The over-proportional weighting of earth currents therefore leads to very large currents in the secondary circuit of the summation CT and severely loads the measuring circuits (e.g. high voltages on the pilot wires). Furthermore, a large burden appears at the main current transformers as the impedance of the summation CT secondary circuit is transformed to the primary side with the squared turns ratio. A summation CT connection with reduced earth current sensitivity should be considered particularly in those cases where the earth fault current may be greater than the three phase fault current. For this purpose, the summation CT may be connected only to the three phases. Equation 3-17 is then modified as follows:

$$\underline{I}_{\text{M-p.u.}} = \frac{\underline{I}_{\text{M-sec}}}{I_{\text{M-sec-n.}}} = \frac{1}{\sqrt{3} \cdot I_{\text{n}}} \cdot (2 \cdot \underline{I}_{\text{L1}} + 1 \cdot \underline{I}_{\text{L2}} + 3 \cdot \underline{I}_{\text{L3}}) \qquad (3\text{-}18)$$

The composite currents and pickup thresholds are indicated in Table 3.2 for this case:

Table 3.2 Composite currents and pick-up threshold sensitivity for various fault types using the connection L1-L2-L3

Fault type	$\underline{I}_{\text{L1}}$	$\underline{I}_{\text{L2}}$	$\underline{I}_{\text{L3}}$	$|\underline{I}_{\text{M}}|$	Relative pick-up threshold
L1-L2-L3	1	$1 \cdot e^{-j120°}$	$1 \cdot e^{-j240°}$	1	1.00
L1-L2	1	−1	–	$1/\sqrt{3} = 0.58$	1.73
L2-L3	1	−1	−1	$2/\sqrt{3} = 1.15$	0.87
L3-L1	−1	–	1	$1/\sqrt{3} = 0.58$	1.73
L1-E	1	–	1	$2/\sqrt{3} = 1.15$	0.87
L2-E	–	–	1	$1/\sqrt{3} = 0.58$	1.73
L3-E	–	1	1	$3/\sqrt{3} = 1.732$	0.58

The earth fault currents may be very small in systems using earth current limiting devices (neutral earthing via an impedance). The normal earth current sensitivity according to Table 3.1 may in this event be not sufficient. The earth current may be raised with an interposing CT in this instance. The relative weighting of the earth current is then increased proportionally.

Phase preference for double earth faults in systems with resonant grounded neutral

In resonant grounded systems only one of the two earth fault locations must be switched off to interrupt the short circuit loop. The remaining earth fault, in which only the compensated earth fault current flows may remain for several hours without the need to interrupt supply to the consumer. The earth fault is then localised with directional earth current relays and isolated with coordinated system switching. The traditional switched distance relays used in these systems were designed for this mode of operation and contained a special circuit that switched the measuring system to the earth fault loop of the preferred phase. Modern numerical relays contain corresponding logic algorithms which only release one phase-earth measuring loop corresponding to the set double earth-fault preference.

As a result of the difference in the relative weighting of the phase currents in the summation CT the differential protection also has a certain degree of phase preference for double earth faults, i.e. preference of the phase with the highest degree of current weighting. Due to the relatively high weighting of the earth current, both fault locations are however generally tripped.

Example 3-2: Pick-up for double earth fault (High impedance neutral earthing)

Given: A radial line with a single in-feed is protected by a differential protection using summation CT connection (Figure 3.19). A cross country fault occurs. The first earth fault is in phase L1 on the protected circuit, while the second earth fault is in phase L3 on a further feeder beyond the next substation. The fault current equals 5 times the nominal current.

The protection is set as follows: Pick-up threshold $I_B = 1.5 \cdot I_n$ and biasing (stabilising) factor $k = 0.5$.

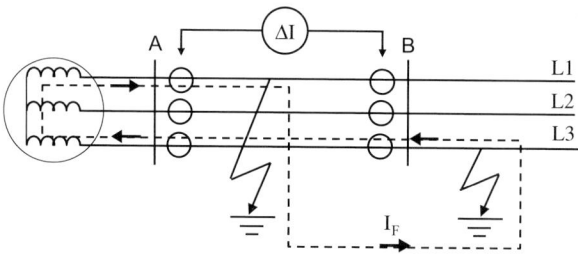

Figure 3.19 Fault condition for example 3-2

Task: Does the protection trip?

What is the response of the protection if the internal earth fault is on phase L2 instead of L1 (external fault remains on L3)?

Solution: According to equation (3-10): $\underline{I}_M = 5 \cdot \underline{I}_{L1} + 3 \cdot \underline{I}_{L2} + 4 \cdot \underline{I}_{L3}$ the composite currents at the two line terminals may initially be calculated:

$$I_{M\text{-}A} = 5 \cdot I_F - 4 \cdot I_F = +1 \cdot I_F \text{ and } I_{M\text{-}B} = +4 \cdot I_F$$

accordingly:

$$I_{Op} = |I_{M\text{-}A} + I_{M\text{-}B}| = 5 \cdot I_F \text{ and } I_{Res} = |I_{M\text{-}A}| + |I_{M\text{-}B}| = 5 \cdot I_F.$$

$I_{Op}/I_{Res} = 5/5 = 1.0$, i.e. greater than the set stabilising factor ($k = 0.5$).

The protection safely trips.

If the internal earth fault is on phase L2, the following conditions exist:

$$I_{M\text{-}A} = -1 \cdot I_F \text{ and } I_{M\text{-}B} = +4 \cdot I_F$$

therefore:

$$I_A = |I_{M\text{-}A} + I_{M\text{-}B}| = 3 \cdot I_F \text{ and } I_S = |I_{M\text{-}A}| + |I_{M\text{-}B}| = 5 \cdot I_F.$$

In this case $I_{Op}/I_{Res} = 3/5 = 0.6$, i.e. smaller than before.

The feeder differential protection that is set to $k = 0.5$ also securely trips in this case.

A busbar protection with a setting of $k = 0.65$ would however no longer trip.

The resultant phase preference of the summation CT, which is effective with large k settings, can be observed in this example.

A differential protection on the next feeder would trip and isolate the second foot point, as a single-phase earth fault with single-end in-feed and $I_{Op} = I_{Res}$ is detected.

Internal fault with superimposed load current

A degree of load current may continue to flow even when an internal fault is present:

- via the healthy phases in case of a non-symmetrical fault
- via the fault location itself, if it is not a solid short circuit
- during single-pole faults with high impedance neutral earthing.

With summation CT connection this results in a reduction of the pick-up sensitivity of the differential protection due to the additional stabilisation caused by the load current flowing through the protected object.

This is illustrated by means of the following example:

Example 3-3: Pick-up in the event of an earth fault (low impedance neutral earthing)

Given: In-feed to a load via a cable according to Figure 3.20.

The main protection of the cable is provided by a differential protection with wire bound connection and summation CT connection.

The earth current is limited to 1000 A by a neutral reactor.

The load current amounts to 500 A three phase symmetrical (ohmic load).

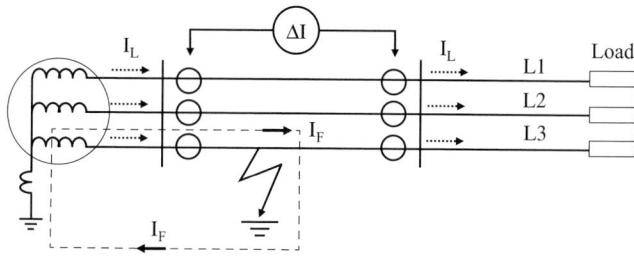

Figure 3.20 Fault condition for example 3-2

Task: Does the differential protection trip in the event of a single phase cable earth short circuit if the biasing factor is set to $k = 0.5$?

Solution: As the load and fault currents do not have the same phase angle, the summation currents must be determined in vectorial form.

The load currents in the three phases are as follows:

$$\underline{I}_{L\text{-}L1} = |\underline{I}_L| \cdot e^{j0°} = +500$$

$$\underline{I}_{L\text{-}L2} = |\underline{I}_L| \cdot e^{-j120°} = (-1/2 - j\sqrt{3}/2) \cdot 500$$

$$\underline{I}_{L\text{-}L3} = |\underline{I}_L| \cdot e^{-j240°} = (-1/2 + j\sqrt{3}/2) \cdot 500$$

The reactive earth fault current lags the phase current in L3 by 90°:

$$\underline{I}_{F\text{-}L3} = \underline{I}_E = |\underline{I}_E| \cdot e^{-j330°} = (\sqrt{3}/2 + j1/2) \cdot 1000 = 866 + j500 \text{ A}$$

The load component of the summation current may be summarised as follows:

$$\underline{I}_{M\text{-}L\text{-}3ph} = \frac{3}{2} \cdot |\underline{I}_L| + j\frac{\sqrt{3}}{2} \cdot |\underline{I}_L| = 750 + j433$$

Correspondingly the summation currents are:

$$\underline{I}_{M\text{-}A} = \underline{I}_{M\text{-}L\text{-}3ph} + 4 \cdot \underline{I}_{F\text{-}L3} = 4214 + j2433$$

$$\underline{I}_{M\text{-}B} = -\underline{I}_{M\text{-}L\text{-}3ph} = -750 + -j433$$

$$I_{Op} = |\underline{I}_{M\text{-}A} + \underline{I}_{M\text{-}B}| = \sqrt{3464^2 + 2000^2} = 4000 \text{ A}$$

$$I_{Res} = |\underline{I}_{M\text{-}A}| + |\underline{I}_{M\text{-}B}| = \sqrt{4214^2 + 2433^2} + \sqrt{750^2 + 433^2} = 5732 \text{ A}$$

$$I_A = |\underline{I}_{M\text{-}A} + \underline{I}_{M\text{-}B}| = \sqrt{3464^2 + 2000^2} = 4000 \text{ A}$$

$$I_{Op}/I_{Res} = 4000 \text{ A}/5732 \text{ A} = 0.7$$

The protection trips, as the ratio I_{Op}/I_{Res} is greater than the set bias factor $k = 0.5$.

3.4 High Impedance (HI) Differential Protection

High impedance differential protection is a wide-spread method to protect busbars and windings of generators, motors or reactors [3-5]. Main advantages are the simple setup and the absolute security against CT saturation. Special CT cores (Class PX to IEC 60044-1) are however required, and heavy voltage limiters (varistors) are needed in most cases to protect the secondary circuits against overvoltages.

To explain the principle, an object with two ends is initially assumed Figure 3.21.

The CT secondary circuits are connected so that through-fault currents circulate, as is also the case with normal (low impedance) current differential protection. The differential relay is connected as a shunt, and is made up of a sensitive current relay with a high resistance connected in series. The current relay and series resistance basically correspond to a voltage relay.

In the shunt path, in which the differential relay is located, the voltage is small (theoretically zero) during load and through fault current conditions, if the inter-

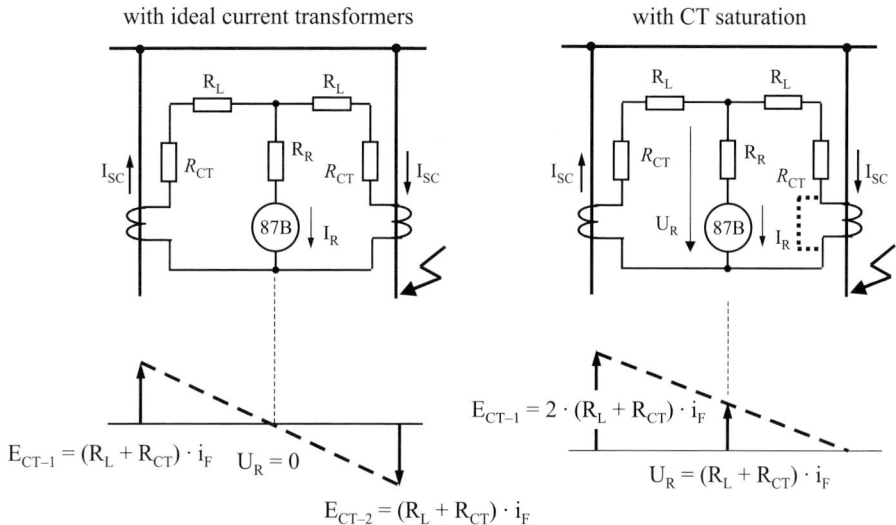

Figure 3.21 High impedance differential protection Response during external fault with and without CT saturation

nal resistances of the CT secondary windings (R_{CT}) and the CT connection cable burdens (R_L) on both sides of the resulting bridge circuit are equal. The current transformers must all have the same transformation ratio.

Stability during external faults

In the event of CT saturation, the balance point is however dislocated. Assuming the worst case, one CT is completely saturated, while the other CT is transforming current without any saturation. The saturated CT can then simply be substituted by its secondary internal resistance (R_{CT}) (Figure 3.21). CTs with low secondary leakage reactance must be used and the magnetising reactance in the saturated state should become negligibly small. Furthermore, if it is assumed that the series resistance is high in comparison to the resistance of the CT cables and the CT internal resistances, then the voltage distribution shown on the right hand side of Figure 3.21 applies.

The voltage across the shunt path then is:

$$U_{\Delta\text{-F-thr}} = \frac{I_{\text{F-thr}}}{r_{CT}} \cdot (R_L + R_{CT}) \tag{3-19}$$

with: $I_{\text{F-thr}}$ = Through flowing primary fault current
      r_{CT} = CT transformation ratio

The current relay with series resistance corresponds to a high resistance voltage relay as stated above. By setting the voltage pick-up $U_R = I_R \cdot R_R$ above $U_{\Delta\text{-F-thr}}$ stability is obtained even in the event of most extreme CT saturation. The highest through fault current that can arise should be used for $I_{\text{F-thr}}$ in this calculation.

Example 3-4: Setting of the high impedance relay

Given: $I_{\text{F-thr}} = 10$ kA

          CT: $r_{\text{CT}} = 400/1$ A; $R_{\text{CT}} = 3\ \Omega$

          CT connection cable: $R_{\text{L}} = 2\ \Omega$

Sought: Relay setting U_{R}

Result:
$$U_{\Delta\text{-F-thr}} = \frac{10,000}{400} \cdot (2 + 3) = 125\ \text{V}$$

A 20% security margin is included in the setting:

$$U_{\text{R}} = 1.2 \cdot U_{\text{D-F-thr}} = 1.2 \cdot 125 = 150\ \text{V}$$

Pick-up sensitivity during internal faults

The pick-up sensitivity of the protection may be checked by means of the determined pick-up threshold of the differential relay.

The minimum primary fault current of an internal fault that is fed from one end only must provide at least the magnetising current for the CTs connected in parallel and the pick-up current of the shunt connected relay. The leakage current of a voltage limiting varistor may also have to be considered:

$$I_{\text{min.}} = r_{\text{CT}} \cdot (n \cdot I_{\text{mR}} + I_{\text{R}} + I_{\text{V}}) \tag{3-20}$$

with: n =   number of feeders connected in parallel (only 2 in the above example, for busbar protection correspondingly more in accordance with the number of bays)

       I_{mR} =   secondary CT magnetising current at the relay pick-up voltage. (see Figure 3.22)

       I_{R} =   relay pick-up threshold

       I_{V} =   leakage current of the varistor (at relay pick-up voltage), if it is applied

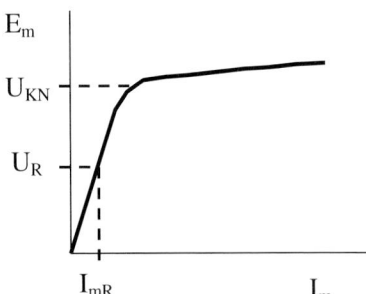

Figure 3.22
CT magnetising characteristic

Example 3-5: Sensitivity of high impedance protection

Given:	CT and relay settings as above
	Magnetising current at the pick-up voltage of 150 V: I_{mR} = 30 mA
	Relay pick-up current: I_R= 20 mA
	No varistor provided
Sought:	The minimum pick-up current
Result:	

$$I_{\text{Min.}} = \frac{400}{1} \cdot (2 \cdot 0.030 + 0.020) = 32 \text{ A}$$

The required CT knee-point voltage

The CTs go into saturation during internal faults, as they feed onto the high imped-ance of the relay. A minimum dimension of the CTs is required to achieve tripping by the differential relay despite this.

Based on theory [3-5] and practical experiments a CT knee point voltage of at least twice the relay pick-up threshold was found to be sufficient for secure operation (Figure 3.22):

$$U_{KN} \geq 2 \cdot U_R$$

Example 3-6: Required knee-point voltage

Given:	Relay setting U_R = 150 V according to example 3-4
Sought:	Required knee-point voltage of the CTs.
Result:	Corresponding to the setting above, the following is obtained:
	$U_{KN} \geq 2 \cdot 150 \text{ V} = 300 \text{ V}$

Application notes

In practice, HI-relays either make use of a current relay (pick-up threshold 20 mA for example) with an external resistor connected in series, or consist of a voltage relay (calibrated in V) with integrated series resistance. A fundamental component filter (50 or 60 Hz resonant circuit) is usually applied to suppress DC components and higher order harmonics.

The CTs must have an identical transformation ratio. If possible, CTs with identical construction should be used.

The following CT characteristics are of relevance:

- The secondary leakage reactance should be negligibly small. A uniform distribu-tion of the windings on the CT core is of importance in this regard.
- The secondary internal resistance must be kept as small as possible so that the voltage across it is minimised in the event of CT saturation. By increasing the diameter of the conductor used in the secondary winding the internal resistance can be reduced.

- Small magnetising currents result in a sensitive pick-up threshold, particularly if many CTs are connected in parallel. To achieve this, a large core cross section is however required, which on the other hand increases the length of the conductor used for the secondary winding, whereby the resistance of the secondary winding is also increased.

The dimensions of the CT are restricted by space limitation (particularly in gas encapsulated switchgear) as well as cost.

In the Anglo Saxon influenced regions where the high impedance protection is often applied for motors, generators, reactors, busbars and autotransformers, the current transformer class PX is defined in IEC 60044-1 for use with high impedance differential protection. This Class PX of the IEC standard corresponds to the earlier Class X of British Standard 3938: "Current Transformers".
The definition is as follows:

- Knee point voltage of the magnetising curve U_{KN}

- Magnetising curve (magnetising current at relay pick-up threshold)

- and resistance of the secondary winding R_{CT}.

Example 3-7: Specification of CTs for HI protection

For the example shown above, the CTs would have to be specified as follows:

Class PX according to IEC 60044-1:

$U_{KN} = 300$ V; $I_m \leq 25$ mA at $U_m = 150$ V; $R_{CT} \leq 3\ \Omega$.

The dimensions would approximately correspond to a CT class 5P10, 30 VA.

Applying class P CTs with high impedance differential protection

Often class PX current transformers according to IEC60044-1 (or Class X to BS 3839) are not available and Class 5P or 10P CTs have to be used instead. This can be the case particularly when retro-fitting or extending existing substations.

Here an equivalent knee point voltage may be calculated with the following equation (see section 5.1):

$$U_{KN}^* = 0.8 \cdot E_{al} = 0.8 \cdot \text{ALF} \cdot I_{2n} \cdot (R_{CT} + R_{Bn}) \qquad (3\text{-}21)$$

Whereby:
$$\begin{aligned}
U_{KN}^* &= \text{Equivalent knee point voltage} \\
\text{ALF} &= \text{Accuracy limit factor of the CT} \\
R_{CT} &= \text{Resistance of the CT secondary winding} \\
R_{Bn} &= \text{Rated burden resistance } R_{BN} = P_{BN}/I_{2N}^2 \\
P_{Bn} &= \text{CT rated burden} \\
I_{2n} &= \text{Secondary CT nominal current}
\end{aligned}$$

The magnetising current at the relay pick-up threshold may be estimated with the error angle δ_N if a magnetising curve is not available:

The magnetising current with nominal conditions and nominal current and rated burden is given by $I_{\text{m-n}} = I_{\text{2n}} \cdot \text{tg}\,\delta_{\text{n}}$. The corresponding internal "EMF" is calculated by $E_{\text{CT-n}} = (R_{\text{CT}} + R_{\text{Bn}}) \cdot I_{\text{2n}}$.

If it is assumed that the CT curve below the knee point is linear, then the magnetising current at the relay pick-up voltage may be calculated with the ratio of the relay pick-up voltage to the rated EMF as follows:

$$I_{\text{mR}} = I_{\text{m-n}} \cdot \frac{U_{\text{R}}}{E_{\text{CT-n}}} = I_{\text{2n}} \cdot \text{tg}\,\delta_{\text{n}} \cdot \frac{U_{\text{R}}}{E_{\text{CT-n}}} \tag{3-22}$$

Example 3-8: Use of P class CTs for high impedance protection

Given:	CT: 5P20, 30 VA, $R_{\text{CT}} = 6$ W,
	Error angle in accordance with IEC 60044-1: $\delta_{\text{n}} \leq 1°$ el.
	Relay setting: $U_{\text{R}} = 150$ V
Sought:	Equivalent knee point voltage and magnetising current at relay pick-up threshold.
Result:	$U_{\text{KN}}^{*} \approx 0.8 \cdot 20 \cdot (30 + 6) \cdot 1 = 576$ V
	$E_{\text{CT-n}} = (30 + 6) \cdot 1 = 36$ V
	$I_{\text{mR}} = 1 \cdot \text{tg}\,(1°) \cdot \dfrac{150}{36} = 0.073$ A

Response with heavy internal faults

During internal faults all the CTs feed current onto the high impedance relay in the common shunt path. As a result of this the voltage on all the CTs increases sharply until CT saturation ensues – large voltage spikes arise which endanger the insulation of the secondary circuits. A voltage limitation with varistors is usually required. The size of these varistors may be considerable with busbar protection, as many CTs in parallel feed onto the relay which may have a setting of more than 100 V which means that a large amount of energy must be dissipated. As a rule, the varistor consists of one or more disks with a diameter of typically 6 inches or more, weighing several kilograms.

The varistor-characteristic is defined by the following equation:

$$U_{\text{V}} = K \cdot I_{\text{V}}^{\text{B}}$$

K and B are constants depending on the varistor type.

The varistors used with the Siemens relay 7VH60 have the following parameters:

Varistor type	K	B	Applicable for relay setting U_{R}
600A/S1/S256	450	0.25	up to 125 V
600A/S1/S1/1088	900	0.25	125-240 V

The following equation is generally applied to calculate the maximum voltage that arises during internal faults: [3-5]

$$U_{peak} = 2 \cdot \sqrt{2 \cdot U_{KN} \cdot (U_F - U_{KN})} \qquad (3\text{-}23)$$

Whereby: U_{KN} = Knee point voltage of the CT
U_F = Maximum voltage in the shunt path which would arise if no CT saturation took place

The equation above only applies if the value of U_{KN} is smaller than $U_F/2$.

If U_{KN} is much smaller than U_F, which is normally the case, then the equation may be simplified as follows:

$$U_{peak} = 2 \cdot \sqrt{2 \cdot U_{KN} \cdot U_F} \qquad (3\text{-}24)$$

Example 3-9: Determine whether a varistor is required

Given: Fault current during internal faults: 10 kA

CT: 1000/1, U_{KN} = 400 V

Relay setting: U_R = 150 V, relay current at pick-up threshold: 20 mA

Relay internal resistance: R_R = 150 V/20 mA = 7500 Ω.

Solution: The voltage during an internal fault without CT saturation would then be:

$$U_F = 10{,}000 \cdot \frac{1}{1000} \cdot 7500 = 75{,}000 \text{ V}.$$

With equation (3-24)) the following results:

$$U_{peak} = 2 \cdot \sqrt{2 \cdot 400 \cdot 75{,}000} = 15.5 \text{ kV}$$

With a normal insulation voltage of 2 kV a varistor would therefore be required.

Busbar protection

The basic principle of high impedance differential protection may also be applied as busbar protection. [3-6]

In this case, the CTs of all bays are connected in parallel separately for each phase. (Figure 3.23)

In extensive outdoor substations the connection of the CTs should be radial at a central point so that the CT connection resistances are as symmetrical as possible. Balancing resistors may be necessary.

The calculation should be done in a similar way as shown above.

To determine the pick-up threshold voltage of the relay, the worst case external fault with CT saturation must be assumed, i.e. the case which induces the largest voltage across the shunt connected relay.

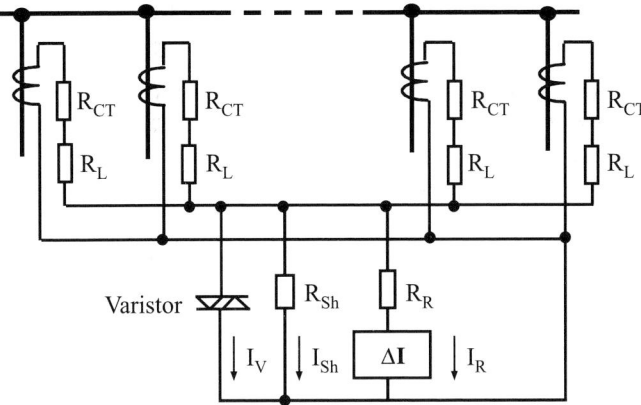

Figure 3.23 High impedance busbar protection

For pick-up the magnetising current of all the CTs must of course be provided so that $n \cdot I_{mR}$ must be applied in the equation above as already indicated.

Usually, the protection operates a master trip relay when it picks up. This master trip relay latches and also short circuits the shunt path thereby short circuiting the relay series resistance and varistor to protect these against thermal overload in the event of a breaker failure condition.

The pick-up threshold of the differential current relay may be increased by adding a shunt resistor connected in parallel. This is applied to reduce the sensitivity of the protection and to for example increase the pick-up threshold to above the maximum load current.

To monitor the secondary circuits, a current relay with more sensitive pick-up threshold is connected in parallel.

The high impedance protection is generally applied to single busbars (e.g. with one-and-a-half circuit breaker plants) where no isolator replica is required. Only in rare cases it is also applied on double busbars with bus couplers, in which case the CT secondary currents are switched directly by isolator auxiliary contacts. Due to the relatively high voltage in the secondary circuit, this may pose a problem with regard to dependability and a higher degree of insulation voltage may be necessary (with NGC in England for example 3.5 kV).

A second system used as check zone is commonly applied in this event.

For more complex busbars, the high impedance protection is not suitable for the reasons mentioned above.

Summarised evaluation of the high impedance differential protection

The high impedance differential protection is a typical Anglo Saxon technology. The protection equipment is not complex and is secure in the presence of CT saturation.

The disadvantages are:

- The CTs must all have the same ratio and construction (normally class PX).
- No further protection devices may be connected in series to the same CT core as the CTs are driven into severe saturation during internal faults.
- Additional voltage limiting devices (varistors) are required.
- An isolator replica with interposing relays is not suitable, but the CT secondary circuits must be directly switched with the isolator auxiliary contacts.
- Finally, a numerical version with isolator replica in the software is by definition not possible.

In Germany high impedance differential protection has practically no application.

The stabilising effect of a resistance connected in series with the differential relay in the shunt path is however well known and has often been applied in critical situations with electromechanical protection. A rare application example for a "moderately" high impedance protection is the fast three core line differential relay RN22 manufactured by Siemens which was until recently applied. [3-7]

3.5 Partial Differential Protection

This type of differential protection is often used outside continental Europe in distribution substations with double busbars and parallel in-feed (Figure 3.24).

On each transformer in-feed an overcurrent protection with time delay is connected in the differential mode (51Δ) to a CT in the in-feed and a further CT in the bus section coupler. It serves as busbar protection and back-up protection for faults in the outgoing feeders, and trips the associated in-feed circuit breaker and section coupler breaker. For example for the fault shown on feeder F3 in Figure 3.24, only

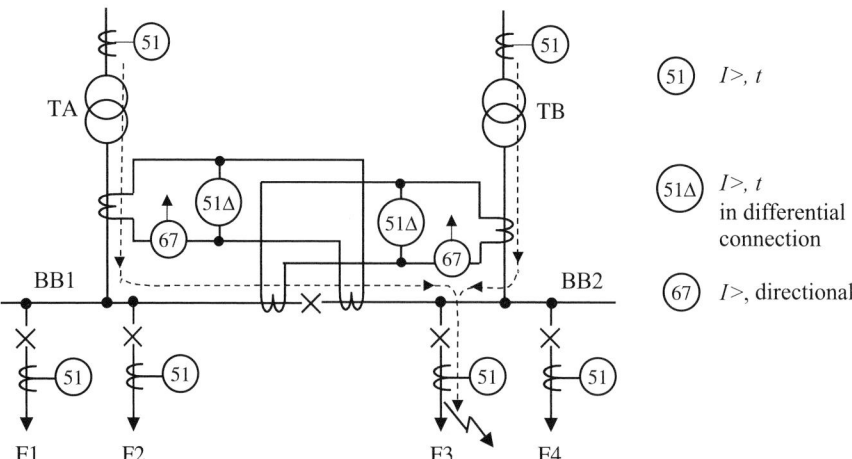

Figure 3.24 Partial Differential Protection

the differential relay 51Δ connected to the in-feed transformer TB will operate. The relay 51Δ connected to transformer TA does not pick up as the in-feed current is compensated by the current flowing via the section coupler, so that no current flows via the differential path. In this manner, an overcurrent relay in the bus section coupler is superfluous, and consequently, one time grading step can be saved. Faults on the busbars are therefore cleared faster.

The relays 51Δ must be applied with one time grade stage longer delay than the overcurrent relays 51 in the feeders F1 to F4. The additional directional relays 67 (voltage connection not shown) must have their direction towards the transformer. They clear faults in the transformer with the appropriate breaker connected to the busbar without delay. In this way, tripping of the parallel transformer is avoided and uninterrupted supply to the consumer ensured. The back-up overcurrent relays 51 on the high voltage side of the transformers must be applied with one time grading stage longer delay than the partial differential protection 51Δ.

In practice, inverse time delay relays are used for this mode of operation. As the partial differential protection is not stabilised, the current transformers must be matched. The pick-up threshold should be set above the combined rating of the two transformers.

4 Measuring Technique

Comparison protection was already applied around the turn of the century. In 1904 C.H. Merz and B. Price obtained the British patent number 3896 for a line differential protection. [4-1] The proposed protection system is shown in Figure 4.1.

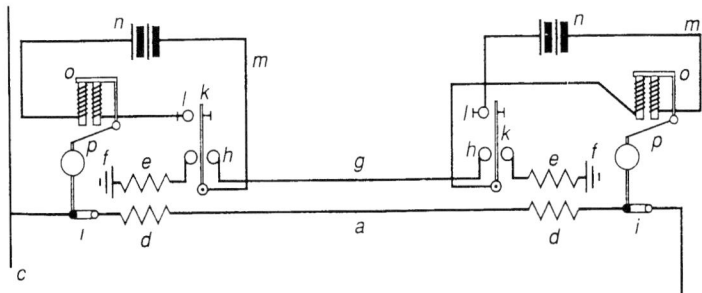

a: feeder, b: generator, c: substation, d: primary winding of CT, e: secondary winding of CT, f: earth or return conductor, g: pilot wire, h: relay windings, i: circuit breakers, k,l: movable and fixed relay contacts, m: circuit, n: battery, o: electromagnetic device with armature p.

Figure 4.1 Comparison protection according to Merz und Price

Based on these simple beginnings, the measuring systems were developed further using electro-mechanical induction and moving coil relays and later static and numerical technology.

4.1 Classical analog systems

In this system a current comparison with analog signals is applied. All electro-mechanical relays and the static analog generation of relays fall into this category.

Electro-mechanical measuring system

The current sum and difference is generated by electro-magnetic transformers in the device.

The comparison is then done using the induction principle with a Ferraris disk/cup or a rectifier bridge with moving coil relay.

The induction relays' mode of operation is based on the resultant force between a fixed coil through which current flows and a conductor which can move and into

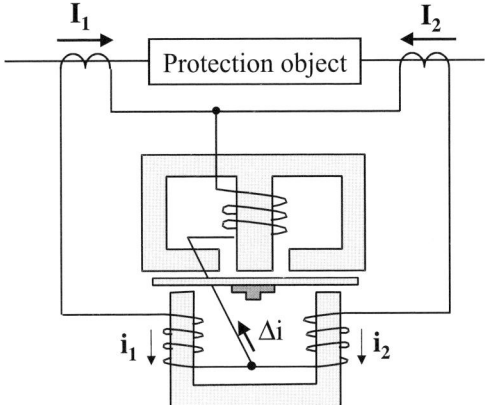

Figure 4.2
Differential protection
with induction relay

which the fixed coil induces current. The moving conductor is either a disk or a cylinder. The differential protection generates opposing alternating flux with the restraint and operating (differential) current via the corresponding excitation coils. The operating current acts as an accelerating torque on the disk and is opposed by the torque produced by the restraint current. The Ferraris disk is restrained towards its initial position with a restraining spring. When the relay operates, the contact that rotates with the disk, moving contact, touches the fixed opposing contact. (Figure 4.2)

With the advent of the moving coil relay, which has a pick-up burden in the microwatt range, and which therefore allows very sensitive pick-up thresholds whilst causing a very low CT burden, a significant step in progress was achieved.

It consists of a permanent magnet with a moving excitation coil in its magnetic field. The moving coil is held in the initial position by means of a restraint spring. When the pick-up threshold current is exceeded, the coil rotates in the operating

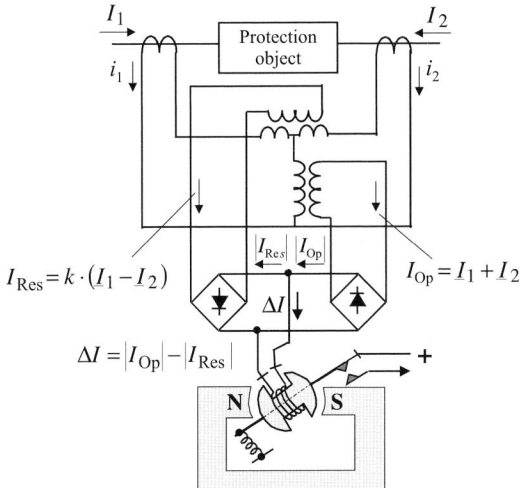

Figure 4.3
Differential protection
with bridge rectifier circuit
and moving coil relay

direction and activates a contact. The moving coil relay is a polarised DC current relay and only operates when the current flows in the tripping direction through the core of the moving coil relay.

The comparison of differential and summated current is in this case implemented with a rectifier bridge circuit (Figure 4.3).

The inertia of the mechanical system results in a smoothening and filtering of the current, which however for small fault currents also results in an increase of the response time. The fastest tripping times of 30–40 ms were therefore only achieved for fault currents that are greater than 5 times the nominal current.

The induction principle was mainly used overseas (USA). In Germany the DC current bridge rectifier circuit with moving coil relay was commonly applied after the Second World War from the 1950's onward.

Analog static measuring technique

This technology replaced the moving coil relay with a static threshold comparator (trigger) in the basic circuit design. Initially transistors were used and later on operational amplifiers. More complex and exact measuring circuits could be implemented in this manner [4-2]. The inertia-free measurement also permitted a reduction of the response time to below one cycle.

In Figure 4.4 the typical measuring circuit as used for differential protection is shown. In this case, the operational amplifiers are used as amplifiers (V1 and V2) and threshold stage (V3). The restraint and operating current are initially routed via a shunt R_S and then rectified. The rectified shunt voltages are then amplified via the operational amplifiers V1 and V2 and compared with the operational amplifier V3. The amplification of the stage V1 corresponds to the bias factor k that is set in the protection. During the unfaulted state the following is true: $I_{Op} < I_{Res}$ and input of V3 positive. Consequently, the voltage at the output of V3 is negative and

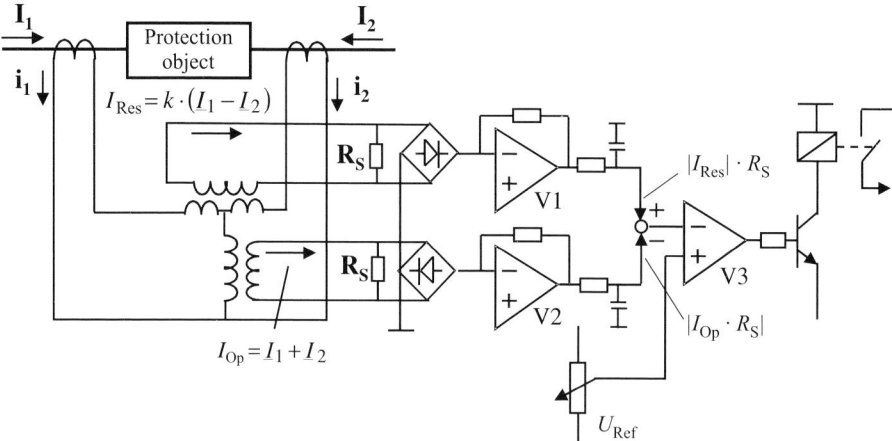

Figure 4.4 Differential protection with analogstatic measuring circuit (simplified)

the output transistor is blocked. During an internal short circuit, the voltage at input V3 will change to $U < U_{Ref}$ and the trigger will operate so that the output of V3 jumps to positive voltage. The output transistor will therefore conduct and operate the tripping relay. At the same time, the re-set level is reduced, by means of an additional circuit (not shown), to below the pick-up threshold (Hysteresis), to avoid "chattering" when the measured values are close to the pick-up threshold.

4.2 Numerical measuring technique

With modern numerical relays, the matching and processing of the measured values is done numerically. Thereby significant advantages in comparison with analog measuring techniques are gained. Numerical filters and intelligent protection algorithms facilitate high measuring accuracy and flexible setting characteristics. The application of adaptive measuring techniques results in short operating times during unambiguous internal faults, high stability against over-function in the event of internal faults with CT saturation as well as during system switching (e.g. in-rush).

4.2.1 Acquisition of measured values

The measuring inputs of the numerical relays are also designed for 1 or 5 A CT (usually selectable) secondary nominal current, as was the case with conventional relays. Input transformers provide galvanic separation. Thereafter the currents are converted to proportional shunt voltages (Figure 4.5).

Prior to sampling, the analog input signal must be band-width limited to half the sampling rate in order to ensure a single resultant for the discrete sample sequence f(n) in relation to the original time function *f(t)* (Shannon's sampling

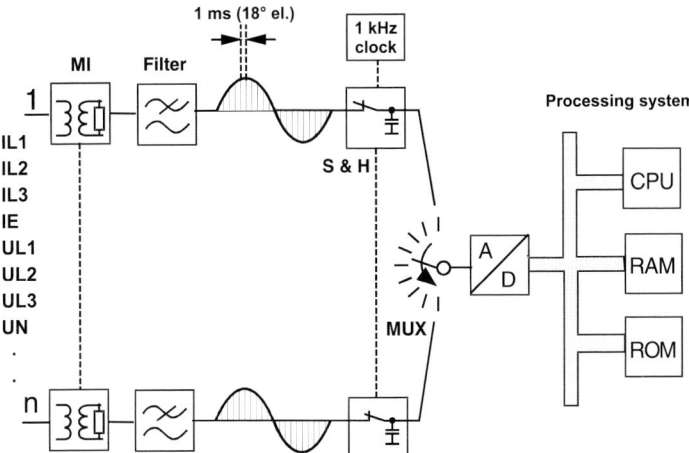

Figure 4.5 Numerical relays: measured value acquisition and processing

theorem [4-5]). In theory, filters with very sharp cut-off would be required for this purpose. In practice however, simple passive RC filters are used. These are reliable and sufficiently damp the high frequencies to facilitate sampling in the Kilohertz range. A sampling rate of 1 kHz, in other words a sampling interval of 1 ms (= 18° el. related to the 50 Hz fundamental wave) is in any event sufficient for differential protection.[1]

The sampled instantaneous values are then converted from analog to digital and stored in a memory (buffer) of the processing device. The measured values (currents, and voltages if these are used), are then available in the form of a sequence of discrete instantaneous values (set of numbers) for further processing.

In this context it is important that the measured values are always derived at the same point in time, in other words that they are sampled synchronously, as this is the only way to allow for a direct comparison (special asynchronous sampling techniques are not considered here). In concentrated individual devices (e.g. transformer differential protection) or distributed arrangements in a substation (e.g. busbar protection) a synchronisation with micro-second accuracy is easily achieved via the internal data bus or the serial data interfaces. For feeder differential protection, where the terminal devices are far apart, additional measures are required. This is dealt with in section 4.2.3.

The large dynamic range of the measuring inputs is significant. On one hand, acceptable accuracy (better than 5%) must be guaranteed for small pick-up currents, while on the other hand, a correct image of large currents up to the maximum short circuit current must be ensured. The following example shows this:

Example 4-1: A/D-Conversion

Given: A differential protection shall have a current range (dynamic range) of $\pm 50 \cdot I_n$. The accuracy must still be 5% of the set value at $0.1 \cdot I_n$ (smallest setting value).

Task: What must the resolution (how many bits) of the A/D converter be?

Solution: 5% accuracy requires a resolution of 100/5 = 20 steps. In this case it applies to the smallest setting value, in other words the 20 steps correspond to the stated $0.1 \cdot I_n$.

              $50 \cdot I_n$ results in $(50/0.1) \cdot 20 = 10{,}000$ steps based on the RMS value. The maximum instantaneous value is then $2 \cdot \sqrt{2} \cdot 50 = 141$ A, which corresponds with $2 \cdot \sqrt{2} \cdot 10{,}000 = 28{,}284$ steps.

              The next higher power of 2 is $2^{15} = 32{,}768$.

              A 16 bit (15 bit plus sign bit) A/D converter must therefore be applied.

---

[1] Often 600 Hz (corresponds to 30° el.) is used, as the phase rotation in the three phase system (e.g. the transformer vector group adaptation by transformer differential protection) can simply be implemented.

4.2.2 Differential protection with instantaneous value comparison

As Kirchhoff's current law applies at all times, the digitised current instantaneous values may be compared at each sampling instant. Naturally, the sampling must be synchronised as described above. For security reasons, tripping will not be made dependent on a single sampling instant, but rather tripping will only be released if the measuring criterion is fulfilled for n subsequent sampling instances.

If a fixed pick-up threshold is defined, then the measurement will pick up only if all n instantaneous values are above the pick-up threshold (Figure 4.6).

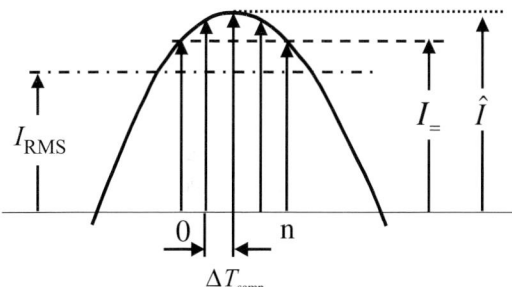

Figure 4.6
Calculation of the RMS value from n sampled values

If sinusoidal measured values are assumed, the corresponding maximum or RMS value can be calculated:

$$\hat{I} = \frac{I_=}{\cos\left(\frac{n-1}{2} \cdot \Delta\varphi\right)}$$

(4-1)

with

$$\Delta\varphi = \omega \cdot \Delta T_{\text{samp}} \cdot \frac{360°}{2\pi} = 2\pi \cdot f_{\text{N}} \cdot \Delta T_{\text{samp}} \cdot \frac{360°}{2\pi} = f_{\text{N}} \cdot \Delta T_{\text{samp}} \cdot 360° = \frac{f_{\text{N}}}{f_{\text{samp}}} \cdot 360°$$

In this context, f_{samp} is the sampling frequency and ΔT_{samp} is the corresponding sampling interval, while f_{N} is the system frequency.

The resultant RMS value then is $I_{\text{RMS}} = \dfrac{\hat{I}}{\sqrt{2}}$

Example 4-2: Calculation of the RMS value from instantaneous values (sampled values)

Given: The busbar protection 7SS5 (50 Hz) uses a sampling rate of 1 kHz and releases tripping with 3 successive sampled values.

            The internal pick-up threshold for the instantaneous values is $I_=$.

            The pick-up value is given as RMS value based on sinusoidal currents.

Task: How does the pick-up RMS value relate to the internal instantaneous value threshold?

Solution:
$$\Delta\varphi = \omega \cdot \Delta T_{samp} \cdot \frac{360°}{2\pi} = f_N \cdot \Delta T_{samp} \cdot 360° = 50 \cdot 0.001 \cdot 360 = 18°$$

With (4-1) we get:
$$\hat{I} = \frac{I_=}{\cos\left(\dfrac{3-1}{2} \cdot 18°\right)} = \frac{I_=}{0.95}$$

The RMS value therefore must be: $I_{RMS} = \dfrac{I_=/0,95}{\sqrt{2}} = 0.74 \cdot I_=$

Evaluation of a small number of sampled values (a short data window) permits very short tripping response times with differential protection. It however demands special measures to prevent an incorrect operation in the presence of CT saturation. A comparison of instantaneous values may only be done at a point in time where the CT transformation is correct. The range where saturation is present must be eliminated or be bridged with an additional stabilisation. Modern differential protection devices include a saturation detector for this purpose.

A typical application for instantaneous value comparison is the busbar protection where extremely fast operating times are required. (Refer to section 10.1.2)

Combination of momentary value and fundamental component measurement

The momentary value based algorithm is also used for high set overcurrent protection functions to ensure fast tripping in the case of heavy CT saturatuion.

An example for this is the very high set, non-biased tripping function $\Delta I>>>$ in the transformer differential protection. It supplements the normal high set function $\Delta I>>$ which responds to the fundamental component. Both functions are usually set above the maximum through flowing fault current. The pick-up of momentary value based measurement must be set to twice the AC peak current value considering a possible full offset of the through flowing current.

The two measuring methods are compared in Figure 4.7 for a sinusoidal short circuit current.

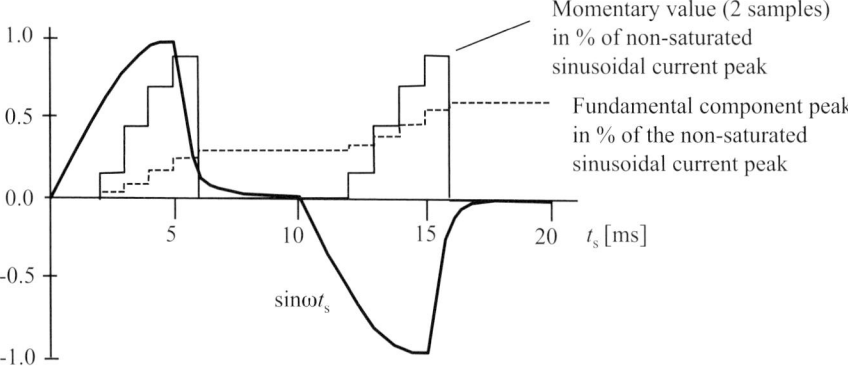

Figure 4.7 Momentary value measurement (2 samples, 1 ms sampling interval) versus fundamental component measurement

In this example, the fast measurement $\Delta I\ggg$ evaluates two consecutive samples which must both exceed the set pick-up limit.

The measuring time (necessary saturation free time of the CT) therefore depends on the length of the sampling interval and varies in the range of one sampling interval $(0 \ldots 1 \cdot \Delta T_{samp})$ dependent on the relative phase position of the sampling instants on the sine wave.

$$\sin\{\omega \cdot [t_S - 1 \cdot \Delta T_{samp.} - (0\ldots 1 \cdot \Delta T_{samp.})]\} > I_{set}/I_{SC\text{-Peak.}} \qquad (4\text{-}2)$$

This algorithm is simple and needs only little processing time. It can be repeated after each sample. Short operating times of a few milliseconds are achieved with high short circuit currents.

The fundamental component measurement $\Delta I\gg$ uses the Fourier integral. (See next paragraph) The filtered output quantity is accurate and nearly independent of transient oscillations but increases slowly after fault inception due to the long integration interval. The full value of the fundamental component is only reached after the integration of a complete cycle (two current half waves). This may cause one half cycle tripping delay with low current overshoot.

The processing time of the Fourier analysis is more time consuming and is therefore repeated only after about every quarter cycle (at least with older relay generations). This adds some additional time delay.

One cycle to one-and-a-half cycles operating times have therefore been typical. The pick-up level can however be set close to the maximum though flowing fault current. Inrush currents have low impact as the fundamental component is only about 50% in this case. DC component and harmonics are filtered out. The combination of fast momentary value and precise fundamental component measurement has therefore decisive advantages.

4.2.3 Differential protection with phasors

In this case, the sampled values are not compared directly. Current phasors (vectors), that are calculated with a number of sampled values (according to the length of the selected data window), are compared. [4-6]

The basic calculation technique used in this context is the discrete Fourier-Transform (DFT). It derives a function in the frequency domain $F(j\omega)$ from the time domain function $f(t)$, whereby the frequency components (included harmonics) appear explicitly [4-5]. Based on this, the fundamental component of the measured signal can be extracted while effectively suppressing the interference signals (higher order harmonics). The sampled values are correlated (convoluted) with the sine and cosine shaped filter coefficients, in other words multiplied and summated over one cycle[1] (Figure 4.8).

---

[1] The Fourier-computation of the fundamental component with integration of a full cycle suppresses the DC component and all high order harmonics. With shorter integration periods, the filtering becomes less effective. For example, using a data window of only half a cycle, the DC component and uneven harmonics are not eliminated. [4-3].

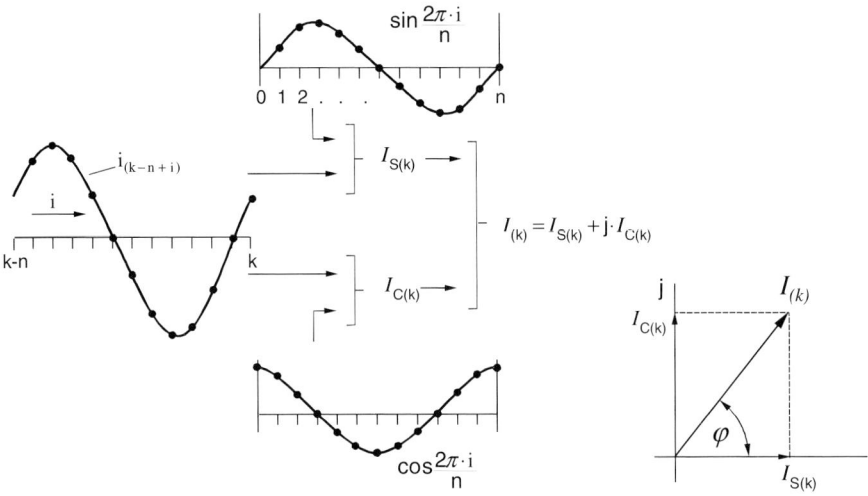

Figure 4.8 Discrete Fourier –Transform (principle)

The equations for determining the fundamental component of a measured value are as follows:

$$I_{S(k)} = \frac{2}{N} \cdot \left[\sum_{n=1}^{N-1} i_{k-N+n} \cdot \sin\left(2\pi \cdot \frac{n}{N}\right) \right] \tag{4-3}$$

$$I_{C(k)} = \frac{2}{N} \cdot \left[\frac{i_{k-N}}{2} + \frac{i_k}{2} + \sum_{n=1}^{N-1} i_{k-N+n} \cdot \cos\left(2\pi \cdot \frac{n}{N}\right) \right] \tag{4-4}$$

$$I_{(k)} = \sqrt{I_{S(k)}^2 + I_{C(k)}^2} \tag{4-5}$$

$$\Phi_{(k)} = \arctan\left(\frac{I_{C(k)}}{I_{S(k)}}\right) \tag{4-6}$$

With these equations it must be noted that a data window having the length $n \cdot \Delta T$ will have a total of $n + 1$ sampled values (Figure 4.9).

The discrete Fourier-Transform returns the fundamental component of the measured value in the form of a phasor (vector) with magnitude and angle.

The initial result of the orthogonal sine and cosine filter may also be considered as the real and imaginary component of the complex phasor $\underline{I}_{(k)}$ at the instant $k \cdot \Delta T$.

$$\underline{I}_{(k)} = I_{S(k)} + j \cdot I_{C(k)} \tag{4-7}$$

The ratio of $I_{S(k)}/I_{C(k)}$ changes depending on the position of the data window (the range of the integration) relative to the analog measured value sine wave. This is shown in Figure 4.10 based on the case where the current phasor is re-calculated at

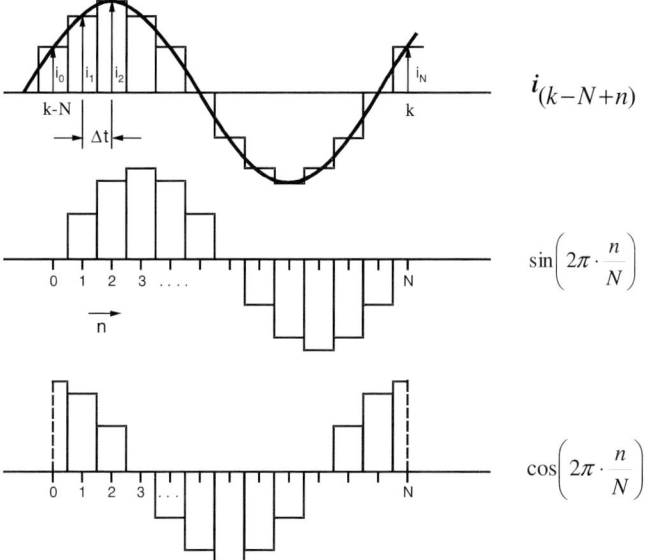

$i_{(k-N+n)}$

$\sin\left(2\pi \cdot \dfrac{n}{N}\right)$

$\cos\left(2\pi \cdot \dfrac{n}{N}\right)$

Figure 4.9 Discrete Fourier-Transform (DFT)

intervals of 1.67 ms (30° el. with $f_N = 50$ Hz). In the first window, only the real component I_S is present. In the following windows, the real component decreases and the imaginary component $j \cdot I_C$ increases, until only an imaginary component remains in the final window.

The phasors are only available at the end of the data window. The complete short circuit values are therefore only available at this time for the algorithm used by the differential protection (current comparison). As a rule, the phasors are re-computed at fixed time intervals (for example every quarter cycle) and used by the protection algorithm. At fault inception or following a change in the fault condition an initializing transient of the measurement that corresponds to the inertia of a conventional measuring relay (moving coil or induction relay), arises. To achieve extremely short tripping times of below one cycle, additional protection criteria must be considered (refer to section 9.3, Line Differential Protection 7SD61/84).

In Figure 4.11, the filter characteristic of the Fourier algorithm using a full cycle data window is shown. The zeroes at the natural harmonics and the low leakage of intermediate frequencies illustrate the excellent filtering that is obtained.

If the data window is shortened, the quality of the filtering reduces so that a greater degree of interference by noise and an increase of the measuring error in the presence of distorted measured values must be considered. Short data windows are therefore only applicable for estimations, as for example used in the distance protection for fast clearance of close-in faults. With differential protection, an instantaneous value measurement (refer to busbar protection 7SS5, section 10.1) or alternatively a charge comparison (refer to Line Differential Protection 7SD61/84, section 9.3) are recommended for short data windows.

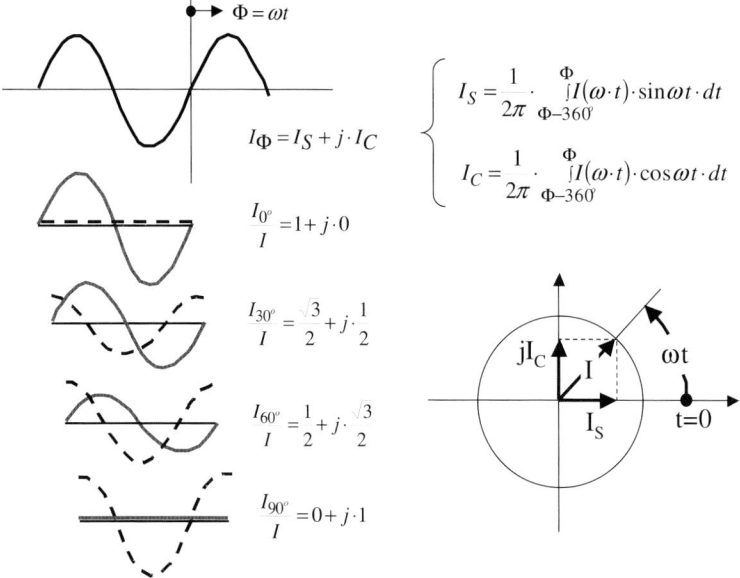

$I_\Phi = I_S + j \cdot I_C$

$$\left\{ \begin{array}{l} I_S = \dfrac{1}{2\pi} \cdot \displaystyle\int_{\Phi-360^\circ}^{\Phi} I(\omega \cdot t) \cdot \sin\omega t \cdot dt \\[2ex] I_C = \dfrac{1}{2\pi} \cdot \displaystyle\int_{\Phi-360^\circ}^{\Phi} I(\omega \cdot t) \cdot \cos\omega t \cdot dt \end{array} \right.$$

$\dfrac{I_{0^\circ}}{I} = 1 + j \cdot 0$

$\dfrac{I_{30^\circ}}{I} = \dfrac{\sqrt{3}}{2} + j \cdot \dfrac{1}{2}$

$\dfrac{I_{60^\circ}}{I} = \dfrac{1}{2} + j \cdot \dfrac{\sqrt{3}}{2}$

$\dfrac{I_{90^\circ}}{I} = 0 + j \cdot 1$

Figure 4.10 Orthogonal components of the current phasor, depending on the position of the measuring window

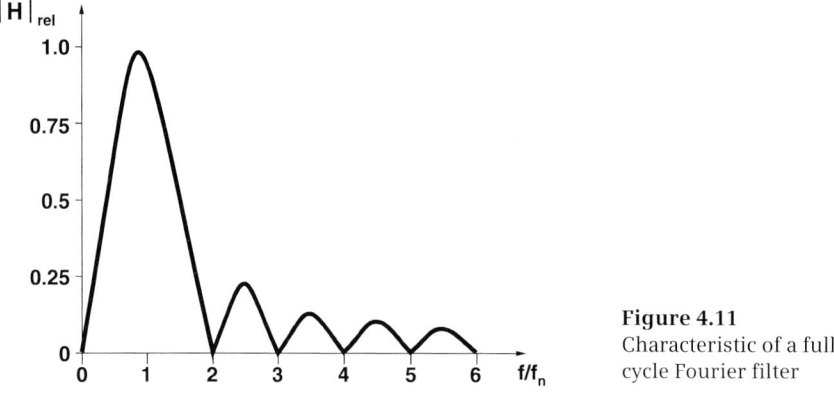

Figure 4.11
Characteristic of a full cycle Fourier filter

Comparison of the phasors

If the sampling of the various measured values (i_{L1}, i_{L2}, etc.) is done simultaneously, in other words, if the data windows are coincidental, then synchronous measured value phasors are obtained. These may be directly compared and processed using complex number calculation rules. This applies to protection devices that derive the measured values at a central location (for example with transformer differential protection).

In the case of line differential protection, the measured value capturing is carried out in devices that are separated by a large distance. Synchronisation of the sampling in all devices is possible with modern technology using GPS; this however requires an additional complexity and expense and is therefore only provided as an optional extra to the standard devices. Furthermore it must be considered that if the protection is synchronised from an external source the protection is dependent on the availability of the GPS system.

Most manufacturers therefore apply a technique whereby the time difference of the asynchronously sampled signals is determined via the communication channel. This time difference is then used to correct the angle deviation of the phasor by means of computation.

Synchronisation of the measured value phasors via the communication channel:

The sampling at the two line ends is done asynchronously under control of the internal clock of the micro-processor in the devices.

The time difference between the sampling of the two devices is determined by means of time stamps applied at each device and the communication delay (Figure 4.12).

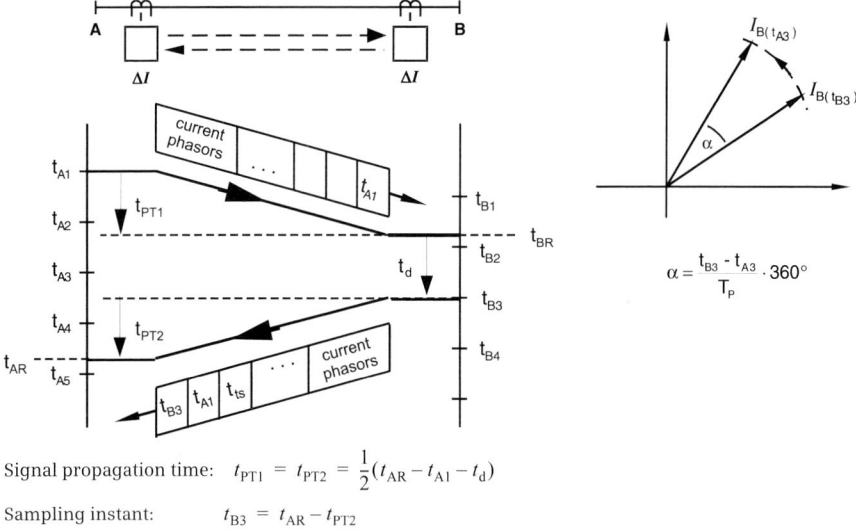

Signal propagation time: $\quad t_{PT1} = t_{PT2} = \frac{1}{2}(t_{AR} - t_{A1} - t_d)$

Sampling instant: $\qquad t_{B3} = t_{AR} - t_{PT2}$

Figure 4.12 Measurement and correction of the sampling time difference

The applied principle is known as "Ping Pong" method and used in the Internet for time synchronisation. [4-7] It is illustrated by means of the following example (simplified):

Device A transmits the current phasor data (real and imaginary component) applicable at the instant t_{A1} with a time stamp t_{A1}. The device at B receives the telegram at the instant t_{BR} based on its local time measurement. After internal processing, the device at B transmits a telegram with a time shift t_d back to A. It contains the current phasor data of the relay B that are valid at the instant t_{B3}, the received value

t_{A1} and the time shift t_d. Device A receives this answer telegram at the time t_{AR}, based on its own clock.

If it is assumed that the channel propagation time in the transmit and receive path are the same, then it can be calculated with the following simple equation:

$$t_{PT1} = t_{PT2} = \frac{1}{2} \cdot (t_{AR} - t_{A1} - t_d) \qquad (4\text{-}8)$$

The equation only contains times that are measured at A and a time difference that was measured at B. The sampling instant t_{B3} that applies for the current phasor at B can therefore be related to the clock at A:

$$t_{B3} = t_{AR} - t_{PT2} \qquad (4\text{-}9)$$

The current phasor $I_B(t_{B3})$ that is received at A from sending end B must be rotated ahead to $I_B(t_{A3})$ by an angle that corresponds with the time difference $\Delta t = t_{B3} - t_{A3}$, so that a synchronous comparison with the current phasor $I_{A(t_{A3})}$ is possible (Figure 4.12).

Synchronisation of the measured value phasors by GPS

The synchronisation of sampled currents by the described ping pong method is only feasible with symmetrical data transmission times. Unequal transmit and receive delays result in an angle error of the current phasors and consequently in false differential currents. (Refer to section 6.3)

Small propagation time differences up to about one millisecond may be acceptable with more insensitive relay setting.

Higher channel delay asymmetry however requires external synchronisation.

Modern line differential relays are designed for the optional use of GPS synchronisation through corresponding optical or static inputs.

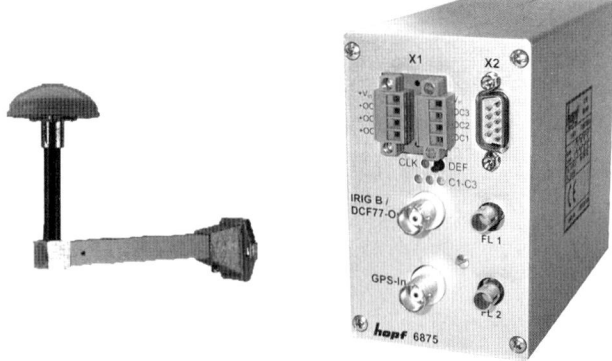

Figure 4.13 GPS antenna and receiver (www.hopf.com)

The Global Positioning System (GPS) is based on 24 satellites orbiting the earth at a height of about 20 kms. At least four satellites are in view from any location on earth.

The satellites send signals at frequencies between 1 and 2 GHz from which special GPS receivers can derive precise timing information. (Figure 4.13) The outdoor antenna of the receiver must be placed carefully to allow line-of-sight to the satellites. A coaxial cable is normally used to transmit the GPS signals from the antenna to the receiver which may be located in the telecommunications room of the substation.

The receiver output signals are further transferred to the relay room interference proof via optic fiber connection. Near the protection panels a transceiver converts the optic signals into electronic TTL signals (24 Volt) which are then distributed to the synchronising ports of the relays via bus cable.

The GPS receiver, specially configured for line differential protection, provides two output signals (Figure 4.14):

a. *IRIG B coded time telegram*
   (digital coded or as 1 kHz amplitude modulated signal) [4-8]

b. *Highly accurate second impulse (1 PPS).*
   The rising edge of this pulse has an accuracy within one micro-second.

The IRIG-B coded time telegram is used for synchronising the relay clocks to normal time and enables accurate (to the ms) time tagging of event and fault reports.

The highly accurate second impulse (1 PPS) is specially provided for the highly precise (to below one el. degree) phasor synchronisation of line differential protection or phasor measurement units (PMUs).

This allows the universal use of line differential protection also via data networks with unsymmetrical data propagation times. (Refer to section 9.3)

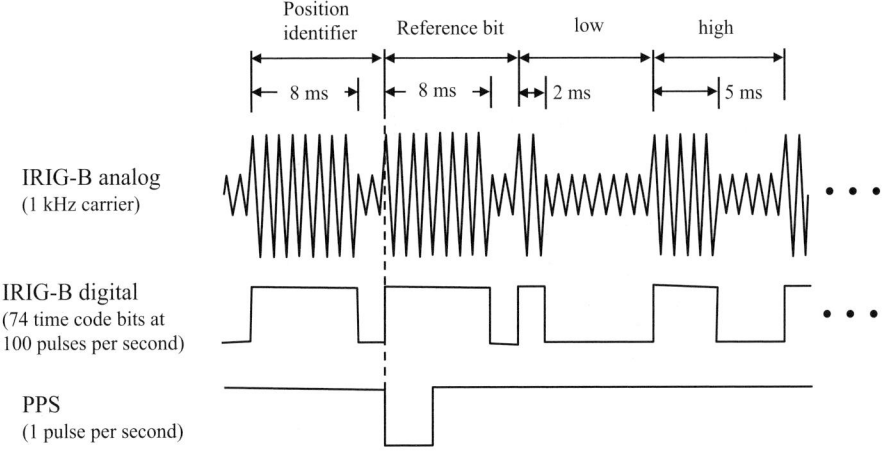

Figure 4.14 IRIG-B Synchronizing signals of a GPS receiver

Operating Characteristic

The fundamental equations and characteristics that were derived in section 3.2.4 apply also to the differential protection with phasor quantities. The operating and restraining quantities are however determined numerically by complex calculation. The differential algorithm may be repeated every quarter cycle (5 ms in case of 50 Hz systems) or in shorter intervals, depending on the processing power of the relay.

In case of line differential protection the real and imaginary part of the current phasors have to be exchanged via serial communication. The possible repetition rate of the differential algorithm therefore also depends on the bit rate of the data communication link.

The double inclined operating characteristic has generally been adopted for numerical relays. (Figure 4.15)

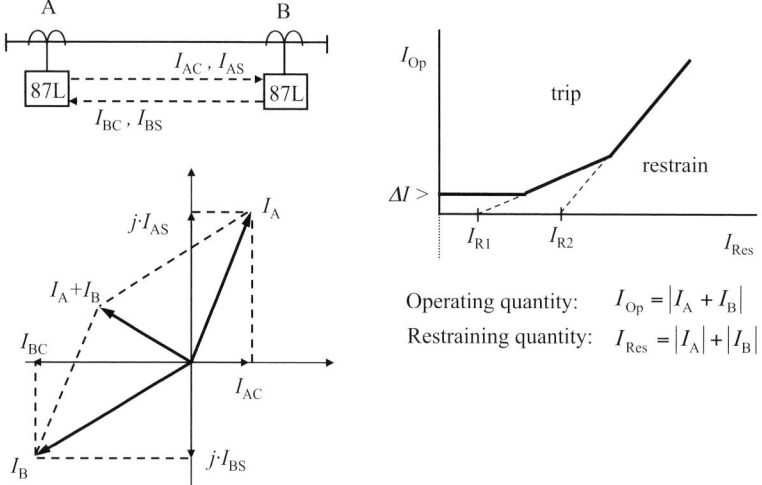

Figure 4.15 Differential protection with phasor quantities

4.2.4 Additional stabilisation with CT saturation

Stability of the differential protection in the event of CT saturation is very important for definition of the measuring technique. Modern protection devices include saturation detectors that detect saturation very quickly and activate an additional stabilisation. The principle of the saturation detector is essentially based on the fact that the CT initially, after fault inception, transforms the current correctly for a minimum duration and only goes into saturation later (Figure 4.16).

In the event of an external fault, the operating (differential) current I_{Op} (current vector sum) remains small (theoretically zero) during the first milliseconds while the restraint current (arithmetic sum) I_{Res} immediately shows a steep increase. In

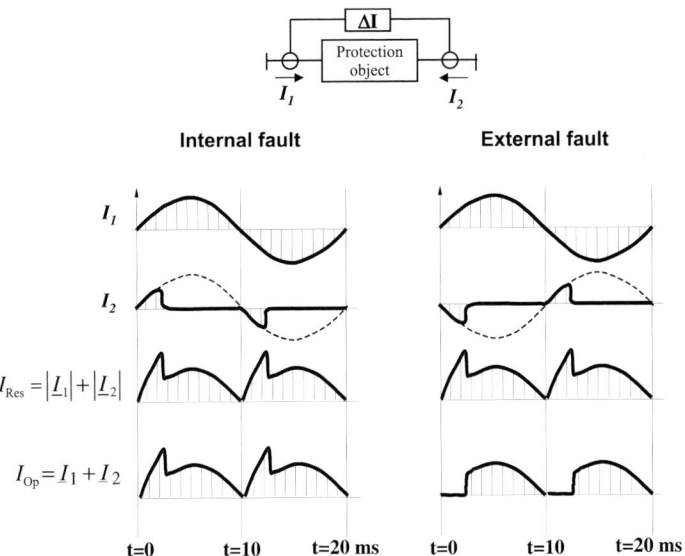

Figure 4.16 Differential protection with saturated currents

the event of an internal fault, both I_{Op} and I_{Res} immediately after fault inception exhibit a parallel rise.

This time delayed increase of differential current is therefore a clear indicator of CT saturation. In Figure 4.17 the course of the value pairs I_{Op} and I_{Res} in time are shown in the operating characteristic of the differential protection.

In the event of an internal fault, the value pairs I_{Op} and I_{Res} immediately have the same magnitude, in other words the locus of I_{Op}/I_{Res} progresses up and down the 45°-fault characteristic. In the event of an external fault with CT saturation, the value I_{Op} however initially is zero and the locus of I_{Op}/I_{Res} progresses along the hor-

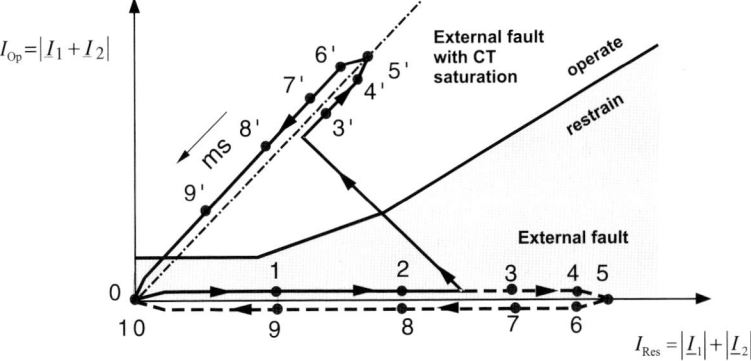

Figure 4.17 Locus of curves I_{Op}/I_{Res} during external faults without and with CT saturation according to Figure 4.16

izontal axis towards the right as the current increases and only jumps upwards into the fault characteristic when CT saturation ensues. This sequence is repeated in the subsequent half cycles. Intervals of unsaturated and saturated CT conditions alternate.

This difference in the response is used for the detection of CT saturation. If saturation is detected during an external fault, then a transient increase of the restraint or transient blocking of the trip output is initiated.

The faster the saturation detector responds, the earlier CT saturation may start, and therefore the current transformers may have a correspondingly smaller dimension. With busbar protection, the manufacturers indicate directly the minimum saturation free time that is required (e.g. 3 ms with 7SS52, refer to section 10.1.2). It corresponds with the CT transient dimensioning factor (refer to section 5.7). The protection stability can also be checked with calculated or measured current oscillograms, using the saturation free time.

As an example, the tripping logic with integrated saturation detector of the busbar protection 7SS5 is shown in Figure 4.18.

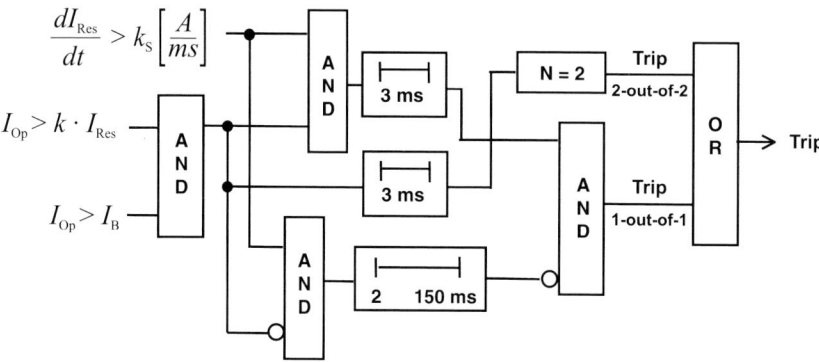

Figure 4.18
Tripping characteristic of the busbar protection 7SS5 with saturation detector (principle)

If, following an increase of the restraint current dI_{Res}/dt, there is no operating current I_{Op} within 3 ms, then an external fault is detected and the (1 out of 1) fast measurement is blocked for a period of 150 ms. The protection can then only trip with a 2-out- of-2 measurement (when it picks up in two consecutive half cycles). This ensures stability for the critical case where extreme CT saturation occurs with short circuit currents having large DC offset. This is so because the saturation only appears in the half cycles with positive DC offset, i.e. only in every second measurement (Refer to section 10.2). The blocking time of 150 ms covers the duration of the current flow until the external short circuit is switched off (a maximum of 100 ms for heavy current faults) with a security margin of approximately 50 ms.

CT saturation during symmetrical through flowing fault currents can be tolerated to a certain degree depending on the setting of the bias factor k (see section 10.2).

An increase of the restraint by switching to the 2-out-of-2 measurement would not make any sense in this case as the saturation arises in each half cycle. Security against extreme saturation would in this case require blocking of the tripping during the 150 ms. This is however not desired, as a sequential fault in the protected zone would only be cleared after a very long time. It is in any event advisable to select the CT dimension so that the largest through flowing fault current without offset does not cause extreme CT saturation. In general this will always apply as the demand on CT capacity (flux increase) in the event of symmetrical currents is substantially less (refer to section 10.2).

The saturation detector elaborated on was already applied successfully in analog static measuring systems (e.g. the busbar protection 7SS1).

Numerical technology also utilises harmonic analysis to determine CT saturation. Depending on the harmonic content of the short circuit current (deviation from the sinusoidal shape), the bias factor is automatically increased (adaptive measuring technique). This is applied in the feeder differential protections 7SD52/61 and 7SD84/86.

5 Current Transformers (CTs)

The role of the CT is particularly important with differential protection. The current comparison carried out by the differential protection only functions correctly when the primary currents are transformed to the secondary side with correct polarity and sufficient accuracy by the CTs. Swapped polarity or transformation errors result in erroneous differential currents that endanger the stability of the protection when external faults cause fault current to flow through the protected object. Particular note must be taken of CT saturation, as large error currents may be the result. A uniform dimension and design of the CTs should therefore always be strived for. This is particularly true if the protection does not contain any specific additional stabilisation against CT saturation. This applies to almost all electro-mechanical and most of the static analog relays.

Numerical relays on the other hand permit a relatively severe degree of saturation as the integrated saturation detector avoids incorrect operation.

5.1 Current Transformer equivalent circuits

In principle, the CT corresponds to a transformer that is current driven. During normal operation, its induction (flux density) is small in comparison with its saturation induction. The induction increases proportional to the increase of the primary current and correspondingly the voltage across the connected secondary burden. The CT in general is dimensioned such that a certain AC fault current can be transformed without saturation (the particular influence of the DC component in the fault current is elaborated on below).

From a protection point of view the leakage induction of the CT may be neglected. The simplified equivalent circuit shown in Figure 5.1 can therefore always be used.

During saturation free operation, the magnetising current may also be neglected. The Ampère's law states the following:

$$I_1 \cdot w_1 = I_2 \cdot w_2 \quad \text{or} \quad \frac{I_2}{I_1} = \frac{w_1}{w_2} = r_{CT} \tag{5-1}$$

The voltages at the secondary terminals of the CT correspond to the voltage drop across the connected burden $U_2 = I_2 \cdot R_B$[1].

The burden consists of the impedance of the CT secondary cables, the relay and if applied, interposing transformers and further devices. The reactive components

---

[1] With static and numerical relays, the reactive component of the burden (X_B) may in general be neglected. With electro-mechanical relays, it may be assumed that $X_B \approx R_B$.

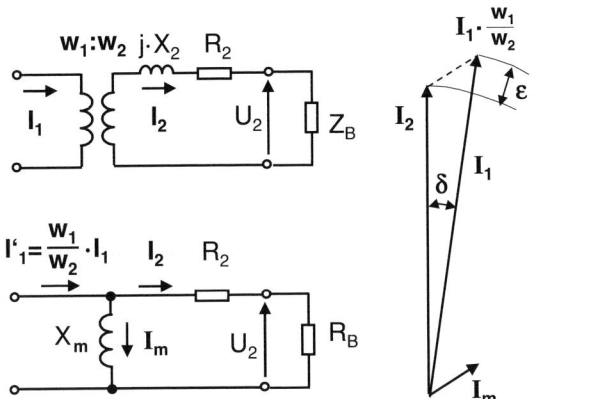

Figure 5.1
Equivalent CT circuit

may in general be neglected so that the pure resistive burden can be used for the calculation. If detailed information is not available, the indicated rating in VA is used as a purely resistive burden ($R \approx P[\text{VA}]/I[\text{A}]^2$). This corresponds to the worst case.

The power supplied by the CT is then:

$$P_2 = U_2 \cdot I_2 = I_2^2 \cdot R_\text{B}$$ (5-2)

The induction in the CT is proportional to the internal EMF

$$E_2 = I_2 \cdot (R_\text{CT} + R_\text{B})$$ (5-3)

The current transformer dimension is selected such that it provides a defined accuracy for fault currents[1] up to a threshold current (rated accuracy limit current) with the rated burden R_Bn connected.

The rated accuracy limit current is indicated as a multiple of the nominal current:

$$I_\text{al} = \text{ALF} \cdot I_\text{n}$$ (5-4)

The factor ALF is also known as the accuracy limit factor.

The internal EMF that arises when this rated accuracy limit current flows then corresponds to the saturation voltage of the CT.

$$E_\text{al} = \text{ALF} \cdot I_\text{2n} \cdot (R_\text{CT} + R_\text{Bn})$$ (5-5)

With Class PX according to IEC 60044-1 (formerly Class X to BS 3938) this voltage is designated as the knee-point voltage U_KN and is used to indicate the transformation capacity instead of the nominal burden and rated accuracy limit current. The accuracy is however defined somewhat differently. In general, the relation $U_\text{KN} \approx 0.8 \cdot E_\text{al}$ may be applied.

---

[1] Initially a pure AC current condition is considered. The influence of DC current components will be analysed later.

5.2 Specifications for the steady state response of current transformers

In the IEC, BS (British Standard) and ANSI/IEEE (US Standard) different definitions are used:

IEC 60044-1[1]

This standard applies for the steady state response of CTs. [5-1] The designation of the CTs for protection applications is done by means of the maximum total error (5 or 10%) at the rated accuracy limit current, followed by the letter P (for protection) as well as the accuracy limit factor (e.g. 5P20).

Two accuracy classes are defined (see Table 5.1).

Table 5.1 CT classes according to IEC 60044-1

Accuracy class	Current error at rated current I_n	Phase displacement δ at rated current I_n	Composite error at $\text{ALF} \cdot I_n^{**}$ ε_c (accuracy limit condition)
5P (5PR)*	±1%	±60 minutes	5%
10P (10PR)	±3%	–	10%
PX	Relates to low leakage reactance type CTs defined by knee-point voltage, magnetising current and secondary winding resistance (see below)		

* 5PR and 10PR classify low remanence type CTs (remanence factor $K_r \leq 10\%$)
** ALF is the accuracy limit factor of the CT

As a whole, a protection current transformer is defined by the following data:

Rated transformation ratio:	Ratio of rated primary to rated secondary current: $K_n = I_{pn}/I_{sn}$, e.g. 300/5 A
Rated power P_n:	Power provided by the CT on the secondary side at rated current and rated burden, e.g. 30 VA
Accuracy class:	5P or 10P
Accuracy limit factor:	This multiple of the rated current, without DC component, can be transformed by the CT with the defined accuracy class, if the connected burden equals the rated burden ($\cos \varphi = 0.8$ to 1.0). With larger current, the CT will saturate and distort the secondary current.
Secondary winding resistance:	R_{CT} in Ohm
Remanence factor: (for Class PR)	$K_r = 100 \cdot$ (Remanence flux/saturation flux), stated in %

---

[1] The new standard IEC 61869-2: Instrument Transformers, Part 2: "Current Transformers" is available as draft und will be published shortly (State 2011).

The different CT classes are applied in practice as follows:

Class 5P is preferred for differential and distance protection due to the higher accuracy.

Class 10P is typically applied with time overcurrent protection in distribution networks.

The PR classes are new in IEC 60044-1(2003) and used to a certain extent to avoid CT saturation problems caused by remanence of P Class CTs. (see discussion below)

Class PX is mainly used with high impedance differential protection.

Example 5-1: Real accuracy limit factor in operation (ALF')

Given: Protection current transformer: 400/1 A; 5P10; 30 VA; $R_{CT} = 6.2$ Ohm

              Connected burden: $R_B = R_L + R_R = 3.0 + 0.5 = 3.5$ Ohm

Task: Determine the real CT accuracy limit factor (ALF') under the given operation condition (actual burden).

Solution: The rated accuracy limit factor of the CT only applies with rated burden. If a smaller burden is connected, the exciting voltage E_2 (equation 5-3) is reduced and an increased operating accuracy limit factor ALF' results:

$$\text{ALF}' = \text{ALF} \cdot \frac{P_i + P_n}{P_i + P_B} = \text{ALF} \cdot \frac{R_{CT} + R_{Bn}}{R_{CT} + R_B} \qquad (5\text{-}6)$$

with: P_n = Rated CT burden:

         $P_i = R_{CT} \cdot I_{2n}^2$ = Internal burden of the CT

         $P_B = R_B \cdot I_{2n}^2$ = Actual connected burden

         R_{CT} = Secondary winding resistance:

         R_{Bn} = Rated burden resistance

         $R_B = R_L + R_R$ = burden resistance

         R_L = resistance of the connecting cable

         R_R = burden resistance of the relay

For the considered example we get:

$$\text{ALF}' = \text{ALF} \cdot \frac{6.2 + 30}{6.2 + 3.5} = 3.73 \cdot \text{ALF} = 37.3$$

The CT can therefore transmit 37.3 times rated current (a.c.) with an accuracy of 5%.

Class PX according to IEC 60044-1

This standard is used outside continental Europe primarily for differential protection.

It was formerly defined as Class X in BS3938 [5-2] but later (2000) adopted in IEC 60044-1 as Class PX. For high impedance protection only this type of CT is specified.

Current transformer dimension data used in this case is as follows:

- Ratio of rated primary to rated secondary current: $K_n = I_{pn}/I_{sn}$, e.g. 300/5 A
- Rated knee-point voltage (U_{KN})
- Magnetising current at the rated knee-point voltage or some other point, for example at $U_{KN}/2$ corresponding to the typical pickup voltage of a high impedance differential relay. (Refer to section 10.3)
- Resistance R_{CT} of the secondary winding (75°C)

The knee-point voltage U_{KN} is defined by the point on the magnetising curve at which a 10% increase in the voltage corresponds to a 50% increase in the magnetising current (Figure 5.2).

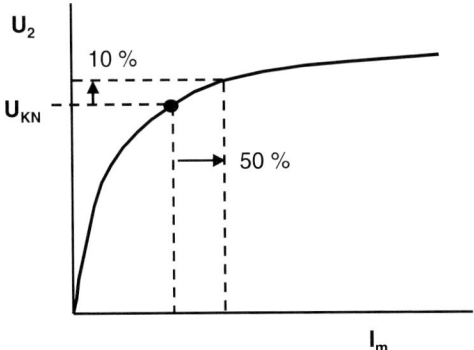

Figure 5.2
Definition of the Class PX
knee point voltage

The Class PX knee-point voltage corresponds to the accuracy limit voltage (E_{al}) calculated for the Class P with the equation (5-5) above. It is however not related to the total error current but to the slope of the magnetising characteristic. Due to the differences in the definitions a conversion from class P to class PX of IEC 60044-1, or vice versa, is only possible by approximation ($E_{al} \approx 1.25 \cdot U_{KN}$). More accurate values, according to the definition of the appropriate standard, must be determined by measurement.

Example 5-2: Class PX to Class P conversion of a protection CT

Given: Current transformer 500/1A, Class PX

$U_{KN} = 300$ V, $R_{CT} = 2$ Ohm

Magnetising current at I_m: 80 mA

Task: Determine the corresponding P Class of this CT.

Solution: $\quad E_{al} = 1.25 \cdot U_{KN} = (R_{CT} + R_{Bn}) \cdot ALF \cdot I_{2n}$

$\qquad ALF = 1.25 \cdot U_{KN}/[(R_{CT} + R_{Bn}) \cdot I_{2n}]$

The ALF value depends on the selected rated Burden.

Assuming a rated burden of 15 VA (15 Ohm for a 1 A CT), we get:

$ALF = 1.25 \cdot 300/[(2 + 15) \cdot 1] = 22.1$

The equivalent specification to Class P would then be approximately:
500/1 A, 5P20, 15 VA

Class C according to ANSI/IEEE C57.13

In this case the CT is dimensioned such that the transformation error does not exceed a value of 10% in the range from 1 to 20 times nominal secondary current (I_{2n}). [5-3] The class definition Cxxx states the secondary terminal voltage U_B at 20 times I_{2n} and the secondary winding resistance R_{CT}. This however only applies to 5 A CTs generally used in the US.

Example 5-3: ANSI Class C to IEC Class P conversion of a protection CT

Given: Current transformer 600/5 A, C200,

secondary winding resistance $R_{CT} = 0.2$ Ohm

Task: Determine the corresponding Class P specification of this CT.

Solution: C200 specifies a standard burden of $R_B = 2$ Ohm and a rated terminal voltage of $U_B = 20 \cdot 5$ A $\cdot 2$ Ohm $= 200$ V.

In accordance with IEC Class P, this would correspond to the specification: 10P, $P_n = I_{2n}^2 \cdot R_{Bn} = 5^2 \cdot 2 = 50$ VA

The comparable IEC specification would therefore be:
600/5 A, 5P20, 50 VA.

We further get the secondary excitation voltage to ANSI specification by adding the internal voltage drop: $E_S = U_B + R_{CT} \cdot 20 \cdot I_{2n}$ resulting in $E_S = 200 + 0.2 \cdot 20 \cdot 0.5 = 220$ V.

Again, a conversion to the accuracy limit voltage E_{al} of the IEC Class P can only be approximate as the definitions of the measuring error are not equivalent.

However, we can assume with good approximation:

E_{al} (to IEC) $\approx E_S$ (to ANSI).

Comment: As per standard, ANSI/IEEE C57.13 defines the rated burden with a 60° impedance angle. For simplicity, calculations are normally made with purely resistive burden. This provides some security margin and is also more realistic for static and numerical relays.

ANSI/IEEE C57.13 also defines a knee-point voltage as the point on the excitation curve, plotted on log-log axes with square decades, where the tangent to the curve makes a 45° angle with the abscissa. This ANSI knee-point voltage is about 50% of the excitation voltage E_S. (Figure 5.3)

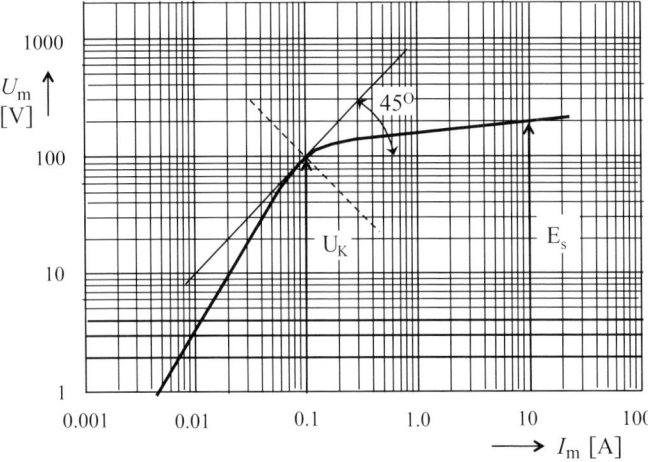

Figure 5.3 Typical magnetizing curve of a current transformer Class C to Standard IEEE C37.13 (C200, R_B = 2 Ohm)

Some older US application guides related the design of CTs to this knee-point voltage. [5-16]

The recent practice of CT design however is more precise and considers the real transient performance. [5-17 to 5-19] It is based on the ANSI defined saturation voltage E_S comparable to the accuracy limit voltage E_{al} of the IEC standard. (Refer to section 5.7 below.)

5.3 Transient response of the CT

The induction (flux density) of the CT is proportional to the integral of the secondary voltage across the CT magnetising inductance L_m:

$$B \text{ prop. } \int e_2(t)dt = (R_{CT} + R_B) \cdot \int i_2(t)dt$$

The induction B is therefore proportional to the area below the fault current.

The DC component in the fault current therefore results in the single-sided severe transient magnetisation of the CT which is several times larger than the AC component. The DC transient flux depends on the time constant of the fault loop impedance (system time constant T_N). (Figure 5.4) To accommodate this, the CTs must have a substantially increased dimension.

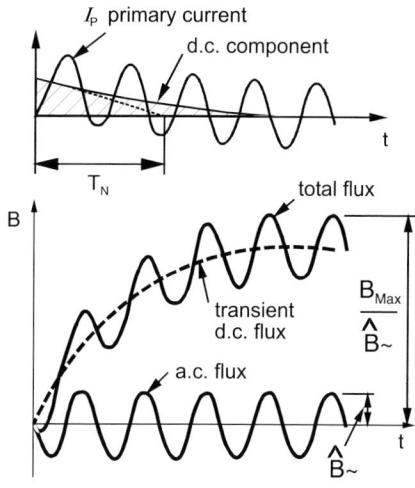

Figure 5.4
Course of CT induction for a
DC off-set short circuit current

According to IEC 60044-6, the ratio of total flux to the peak instantaneous value of the AC flux is defined as transient factor K_{TF}. [5-4 to 5-8]

For a fully offset short-circuit current (maximum DC component) we get the following relation:

$$K_{TF} = \frac{B(t)}{\hat{B}_\sim} = \frac{\omega \cdot T_N \cdot T_S}{T_N - T_S} \cdot \left(e^{-\frac{t}{T_N}} - e^{-\frac{t}{T_S}} \right) - \sin \omega \cdot t \qquad (5\text{-}7)$$

For practical reasons the formula (5-7) is simplified by setting $\sin \omega \cdot t = -1$. This simplified transient factor refers to the envelope curve of $B(t)$:

$$K_{TF} = 1 + \frac{\omega \cdot T_N \cdot T_S}{T_N - T_S} \cdot \left(e^{-\frac{t}{T_N}} - e^{-\frac{t}{T_S}} \right) \qquad (5\text{-}8)$$

Comment: The approximation of the CT induction by the envelope curve (as per IEC 60044-6 of 1992) is appropriate for operating times in the order of one cycle or more.

Modern differential relays however allow short time-to-saturation substantially below one half cycle. The envelope approach delivers in this case too high CT transient factors (K_{TF}).

The transient factor must therefore be calculated according to the real course of the induction including DC and AC components.

This is considered in more detail in the following paragraph.

The upcoming standard IEC 61869-2 for current transformers, replacing IEC 60044-1 and -6, specifies corresponding dimensioning factors for short accuracy limit times.

The transient factor as defined by (5-8) depends on the DC time constant of the primary current (network time constant) T_N and the secondary time constant T_S of the CT.

The maximum value is:

$$K_{TF} = 1 + \omega \cdot T_S \cdot \left(\frac{T_N}{T_S}\right)^{\frac{T_S}{T_S - T_N}} \tag{5-9}$$

It is reached after:

$$t_{B\,max} = \frac{T_N \cdot T_S}{T_S - T_N} \cdot \ln \frac{T_S}{T_N} \tag{5-10}$$

T_N depends on the X/R ratio of the applicable fault loop:

$$T_N = \frac{1}{\omega} \cdot \frac{\sum X}{\sum R}$$

T_S is determined by the main inductance L_m of the CT and the sum of the resistances in the secondary circuit:

$$T_S = \frac{L_m}{R_{CT} + R_B} = \frac{1}{\omega \cdot \tan \delta} \tag{5-11}$$

The secondary CT time constant therefore decreases as the error angle δ increases.

This is particularly true when the CT core is spilt by air gaps.

For 50 Hz (60 Hz) the following equations apply:

$$T_S = \frac{10900}{\delta_{[min]}} \; [ms] \qquad \left(T_S = \frac{9083}{\delta_{[min]}} \; [ms]\right) \tag{5-12}$$

It must be noted that the rated error angle indicated with the CT data only applies if the nominal burden is connected. When lower burdens are connected the error angle is reduced accordingly and the CT time constant increases.

The increase of induction according to equation (5-8) determines the required over-dimensioning of the CT to enable the transformation of a fully off-set fault current.

If the CT has to transform without saturation only up to the instant t_M, then $t = t_M$ must be applied in (5-7) and (5-8). If the total fault duration requires saturation-free transformation, then the CT must be dimensioned for a maximum flux according to 5-9.

IEC 60044-1 defines a so called transient dimensioning factor K_{TD} which considers the necessary saturation-free transmission time (permissible time to accuracy limit) and the duty cycle. An example is given further below with Figure 5.6 and equation (5-14).

Demagnetisation

Interruption of the fault current is done by the circuit breaker at the instant of current zero crossing. At this point the flux viz. induction is at a maximum. Furthermore it must be noted that following fast fault clearance, the CT may still be heavily magnetised due to the DC component at the instant of breaker opening.

The de-magnetisation takes place via a transient relaxation current in the secondary circuit of the CT. The induction is not reduced to zero as a result of this, but only down to the remanence induction B_R (Figure 5.5):

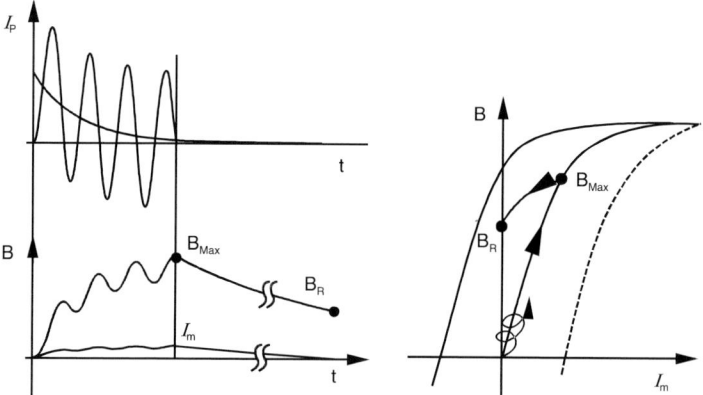

Figure 5.5 Magnetising and de-magnetising of a CT

$$B = B_R + (B_{max.} - B_R) \cdot e^{-\frac{t}{T_S}} \tag{5-13}$$

The remanence induction is reached after approximately $t = 3 \cdot T_S$.[1]

If re-closure onto a fault takes place at an earlier instant then the induction rises from an intermediate value according to (5-7).

Unsuccessful auto-reclosure

The most common case for rapid re-closure onto a fault is the unsuccessful auto-reclosure. In Figure 5.6, the course of induction for the entire fault cycle (C-O-C-O) is shown.

If fault inception and reclosure occur by chance at the same voltage angle (near positive or negative voltage zero-crossing), the induction of the CT increases twice in the same direction. The finally reached induction depends on the AR dead time and the secondary time constant of the CT. (see following paragraph)

---

[1] The remanence remains indefinitely and can only be removed by a forced de-magnetisation. [5-10, to 5-12]

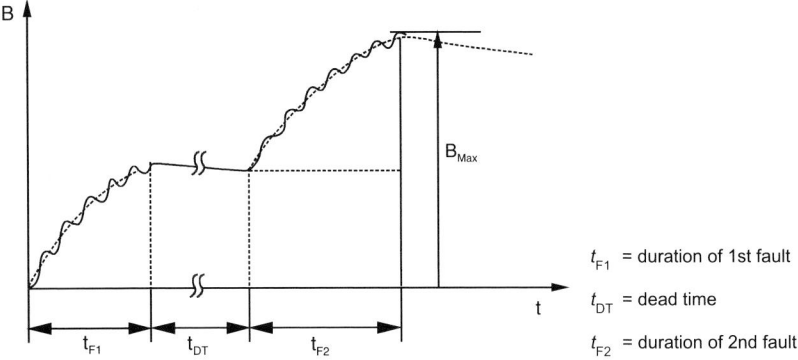

t_{F1} = duration of 1st fault

t_{DT} = dead time

t_{F2} = duration of 2nd fault

Figure 5.6 Course of induction on a CT during unsuccessful auto-reclosure

The successive increase of the induction requires the CT to be designed according to the following transient dimensioning factor:

$$K_{TD} = \frac{B_{max.}}{\hat{B}_{\sim}} = \left[1 + \frac{\omega \cdot T_N \cdot T_S}{T_N - T_S} \cdot \left(e^{-\frac{t_{F1}}{T_N}} - e^{-\frac{t_{F1}}{T_S}}\right)\right] \cdot e^{-\frac{t_{DT} + t_{F2}}{T_S}} +$$

$$+ \left[1 + \frac{\omega \cdot T_N \cdot T_S}{T_N - T_S} \cdot \left(e^{-\frac{t_{F2}}{T_N}} - e^{-\frac{t_{F2}}{T_S}}\right)\right]$$

(5-14)

5.4 TP Current transformer classes

The demands imposed on the CT with respect to transient response during off-set fault current transformation (**T**ransient **P**erformance Requirements) are defined in IEC 60044-6[1]. [5-9] This standard differentiates four classes depending on the construction of the CT core:

Class TPS:	Closed iron core with low leakage reactance
	Its transformation response is defined by the magnetising curve (knee-point voltage, magnetising current) and the secondary winding resistance.
	This Class has not been of importance in practice. Class PX of 60044-1 is used instead which corresponds to earlier Class X of BS3938. (see above)

---

[1] The new standard IEC 61869-2: Instrument Transformers, Part 2: "Current Transformer" exists as draft and shall be published in 2012. It shall replace IEC 60044-1 and IEC 60044-6.

Class TPX:	Closed iron core without limitation of the remanence
	The construction of these CTs is the same as the Class P CTs according to IEC 60044-1. TPX additionally specifies the transient response.
Class TPY:	CT with anti-remanence air-gap (Remanence ≤10%)
	Apart from this, the response is as with TPX.
Class TPZ:	CT with linear core (Remanence negligible)
	The indicated transformation accuracy only applies to the AC-current component.
	The DC component in the fault current is heavily attenuated.

The error limits of the individual classes are defined as follows in Table 5.2.

Table 5.2 CT classes according to IEC 60044-6

Class	Error at rated primary current:		Maximum peak instantaneous error at accuracy limit condition
	Ratio	Angle	
TPX	±0.5%	±30 Min	$\hat{\varepsilon} \leq 10\%$
TPY	±1.0%	±60 Min	$\hat{\varepsilon} \leq 10\%$
TPZ	±1.0%	±180 ±18 Min	$\hat{\varepsilon} \leq 10\%$ (AC current only)

Closed iron core CT

The closed iron core CTs transform DC and AC current components within the defined range with high accuracy. Their remanence is however very high, and may reach more than 80% (Figure 5.7). Almost the total flux reached after transformation of an offset fault current is trapped in the core. It can only be removed by forced de-magnetisation. [5-10 to 5-12]

The consequence of this is that the flux may accumulate according to the experienced fault history. In the worst case of an unsuccessful auto-reclose with the instant of closing at the most inopportune moment, the flux almost doubles. (Figure 5.8) The CT dimension for auto-reclose applications (AR) must therefore have twice the core cross section.[1]

Most of the protection CTs in the system are of class P. (In older plants, Class 1% and 3%, earlier designation of class 5P and 10P, can also be found). CT dimensioning, by consideration of the DC current component however, has only been done for approximately the last 40 years, ever since the transient CT response has been defined more accurately. [5-4 to 5-8]

---

[1] The permissible reduction of CT dimension resulting from the tolerated CT saturation is not considered here.

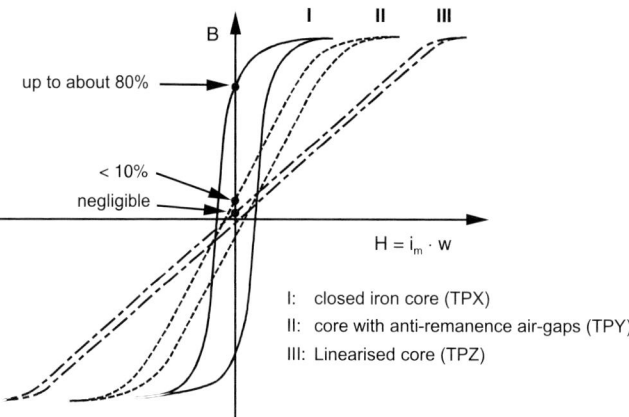

Figure 5.7 CT classes according to IEC 60044-6, magnetising curves

CTs with air gap

Air gaps in the CT core dramatically reduce the remanent flux. At the same time, the de-magnetising time is reduced to approximately 1 second or less.

The secondary time constant may however not be reduced too much, as the DC component of the current would then no longer be transformed correctly. 200 to 300 ms represent the approximate lower limit. Current transforners of the TPY Class (with air gaps in the millimeter range) therefore only partially de-magnetise during the dead time (Figure 5.8).

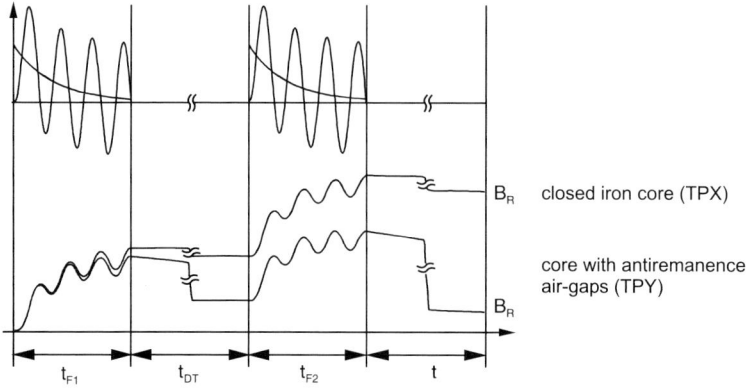

Figure 5.8 TPX and TPY CTs: magnetisation during unsuccessful AR

In the case of linearized TPZ cores with larger, centimeter sized airgaps, the DC component of the current is dampened so severely in the transformation that a much smaller flux increase is automatically achieved. Furthermore, the core de-magnetises itself completely in less than 200 ms, as the CT time constant is only

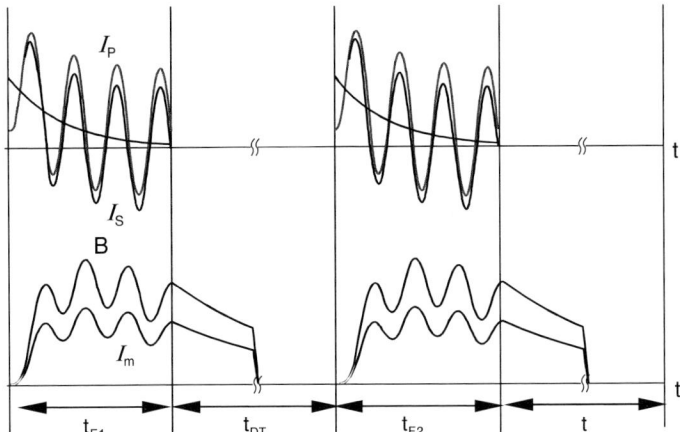

Figure 5.9 Linear CT core (TPZ): progress of the flux during unsuccessful AR cycle

approximately 60 ms. This means that even following the shortest auto-reclose dead time, the flux will have decayed to 0 prior to reclosure Figure 5.9).

Consequently the core cross-section may be substantially smaller. The segmented core however requires a more comprehensive mechanical fixation. A cost comparison must therefore be carried out under consideration of the application specific parameters (open air and GIS).

Linearized cores are mainly applied in some European countries. In Germany, they are used at the EHV level.

The TPZ core only transforms the AC current component with the specified accuracy. The DC component of the short circuit current is shortened due to the reduced magnetising inductance (short secondary CT time constant) and corresponds no longer to the primary system conditions.

It must also be noted that with the TPZ type CT a large de-magnetising (relaxation) current flows after the fault current is switched off. The initial value corresponds to the CT magnetising current at the instant that the primary current is interrupted. The decay is in accordance with the secondary CT time constant T_S.

This de-magnetising current may result in an increase of the reset time with conventional protection devices that have a sensitive current pick-up threshold setting (as in the case of circuit breaker failure protection). The numerical relays are not really affected by this as they almost entirely filter out the DC current component and check the current zero-crossing.

The mixture of TPZ class current transformers with closed core (P or TPX) types must be carefully considered when used with sensitive differential protection. The DC component of the through flowing fault current is in this case differently transmitted to the secondary side. Therefore a transient DC component appears as false differential current which could cause relay overfunction.

Electromechanical and analog static relays using DC average or RMS measurement were affected by this mixture of CT classes. Numerical relays based on fundamental component measurement are again less concerned.

It is however still good relaying practice to use the same or equivalent CT classes for differential protection.

5.5 Polarity of the CT

The polarity of the CT must be observed for the current comparison carried out by differential protection.

If the assumption is made that the primary and secondary windings are wound in the same direction around the CT core, then the conditions shown in Figure 5.10 apply.

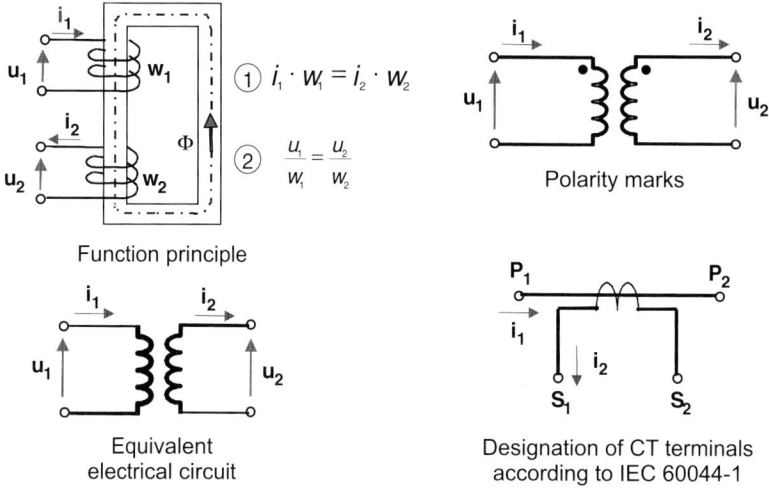

Function principle

Equivalent
electrical circuit

Polarity marks

Designation of CT terminals
according to IEC 60044-1

Figure 5.10 Polarity of current transformers

The currents i_1 and i_2 flow in opposite directions. The voltages across the windings have the same polarity (direction). This convention will also be used in this book.

This polarity sense marking is particularly useful when the windings are shown at separate positions in the circuit diagrams.

Figure 5.10 also shows the IEC 60044-1 designations for current transformers. According to this, the terminals P1 and S1 must have the same polarity at the same time. Therefore the secondary current must flow out at terminal S1 when the primary current flows in at terminal P1.

To avoid any uncertainty, polarity marks (dot, square or cross) may also be shown on the windings (Figure 5.10). The convention is then, that the secondary current flows out at the marked terminal when the primary current flows in at the corresponding marked terminal. (American practice)

5.6 CT errors

As long as the CT is operated in its linear range of the magnetising curve, the total error is small and its influence on the differential protection may be neglected. This is particularly true for the stabilised differential protection, where the pick-up sensitivity increases automatically as the measured current increases. The situation however becomes critical when the magnetic induction goes beyond the kneepoint of the magnetising curve and the magnetising current increases sharply, in other words when the CT goes into saturation. In this case large error currents arise, that threaten the protection stability.

CT saturation

It may arise due to pure AC current when its magnitude exceeds the accuracy limit current of the CT. (Figure 5.11, top). This only occurs with CTs having small dimension, where the operational accuracy limit factor ALF′ is lower than I_F/I_n.

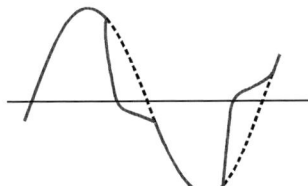

Saturation with steady-state current

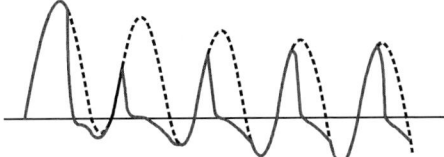

Saturation with offset current

Figure 5.11 CT saturation

CT saturation due to the DC current component of an offset fault current is the more likely cause for CT saturation Figure 5.11, bottom), as a multiple induction is required in this case (see section 5.3).

The missing parts of the secondary current correspond to the magnetising current of the CT. If it is assumed that the CT at the opposite side transforms its current without saturation, then in the case of an external fault with fault current flowing through the protected object, these current gaps appear as incorrect (false) tripping current in the differential protection. (Figure 5.12)

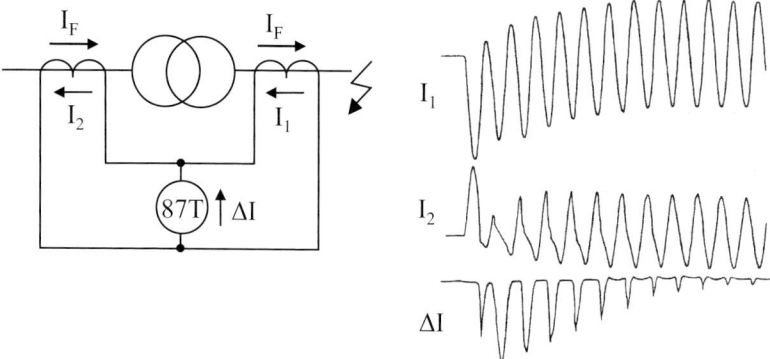

Figure 5.12 Differential protection: External fault with transient CT saturation

In Figure 5.13 the signals of a fault with CT saturation are shown in detail.

Due to the large fault current, the flux increase is very steep right from the beginning. When the saturation threshold is reached, the secondary current abruptly decreases to a smaller current that is phase shifted in the leading sense.

The primary current is only correctly transformed again when flux with a negative sense is required. A sequence of saturated and unsaturated intervals ensues. As the DC current component decays, the unsaturated intervals increase until there is no more saturation in the CT.

A small amount of saturation that only causes small error currents may be accommodated by the percentage restraint of the differential relay. Severe saturation

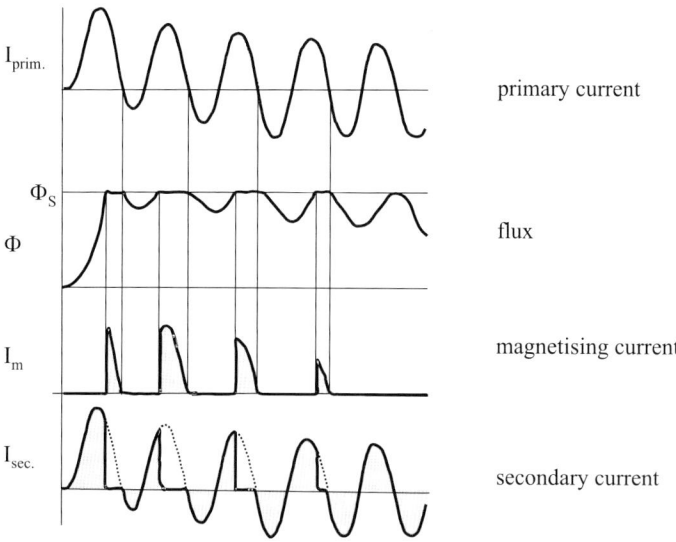

Figure 5.13 Transient CT saturation

however requires special measures (saturation detector with transient additional stabilising or blocking) to avoid an unwanted tripping by the protection.

Reduction of the DC current component

With closed iron CT cores the secondary time constant amounts to several seconds. The DC component of the fault current is therefore transformed correctly.

CTs with linearized core on the other hand have CT secondary time constants that are very short (60 ms). The DC current component is therefore dampened very quickly and even swings across to a negative value.

When connecting differential protection to CTs of different type, TPX and TPZ cores, the through flowing fault current would therefore result in a differential current consisting of the difference between the DC components of the two CT types. With a sensitive setting (e.g. generator or transformer differential protection) this may result in non-desired tripping if a corresponding blocking filter is not provided in the protection. This however does not present a problem with the Fourier filtering used in numerical protection.

5.7 Dimensioning of the CT

The required operating accuracy limit factor (ALF') may be derived with the following equation:

$$\text{ALF}' = \frac{I_\text{F}}{I_\text{n}} \cdot K_\text{TF} \cdot K_\text{Rem.} \tag{5-15}$$

The corresponding rated accuracy limit factor (ALF) is then:

$$\text{ALF} = \frac{R_\text{CT} + R_\text{B}}{R_\text{CT} + R_\text{n}} \cdot \text{ALF}' = \frac{P_\text{i} + P_\text{B}}{P_\text{i} + P_\text{n}} \cdot \text{ALF}' \tag{5-16}$$

The ratio I_F/I_n must consider the maximum fault current that can arise. Often, the rated short circuit current of the plant is used in this context.

K_TF is the transient factor that considers the single sided magnetising of the CT core due to the DC component in the fault current.

$K_\text{Rem.}$ is the over-dimensioning factor that considers the remanence. It is calculated using the remanence factor $K_\text{r} = B_\text{R}/B_\text{S}$:

$$K_\text{Rem.} = \frac{1}{(1 - K_\text{r})} \tag{5-17}$$

Transient factor K_{TF}

In most cases it is of primary importance for definition of the CT dimension.

Initially the common CTs with closed iron cores, 5P or 10P are considered. For these, the transient response is defined with Class TPX. The CT secondary time constant $(T_S = L_m/\Sigma R_2)$ in this case is always large in comparison to the system time constant T_N. Usually T_S is several seconds long, while T_N in the system only assumes a value above 100 ms in exceptional cases.

Equation (5-18) shows the simplified equation (5-8) for $T_S \gg T_N$:

$$K_{TF} = \frac{B_{max.}}{\hat{B}_\sim} = 1 + \omega \cdot T_N \tag{5-18}$$

To achieve saturation free transformation of the fully offset fault current for the entire duration of the fault, the CT must have a dimension calculated with the transient factor:

$$K_{TF} = \frac{B_{max.}}{\hat{B}_\sim} = 1 + \omega \cdot T_N = 1 + \frac{X_N}{R_N} = 1 + \frac{X_S + X_L}{R_S + R_L} \tag{5-19}$$

In the equation above, T_N is the DC time constant of the affected fault loop calculated with the values of the source impedance (X_S, R_S) and the line impedance (X_L, R_L).

The X/R ratio of the fault impedance close to generators and transformers and at the EHV level in general is very large so that large over-dimensioning factors of $K_{TF} = 30$ (at $T_N = 100$ ms) and more may appear.

CTs with such large dimensions can usually not be implemented due to the large costs involved. With encapsulated switchgear (e.g. GIS) where the CT must be fitted inside the encapsulation, space constrains prohibit such large CTs.

Most protective relays do not require saturation-free transformation for the total duration of the fault current, but permit saturation after a particular time. The dimension of the CT may therefore be reduced.

The minimum time to saturation must satisfy various criteria:

– With <u>internal faults</u>, a minimum saturation free time (a minimum non-distorted data window) is required to comply with the measurement process inside the protection. With distance protection this can assume a value of 25 ms especially if an exact fault location is required. Differential protection on the other hand can manage with shorter times of for example 5 ms.

– During <u>external faults</u>, protection stability is ensured if the CT transforms the through flowing fault current until the fault is switched off. In this case the relevant time is made up of the operating time of the protection and the circuit breaker. At the EHV level this equates to a time of approximately 100 ms.

For modern protection devices that contain a saturation detector, as described in section 4.2.4, the differential protection can cope with shorter times as the CT must only transform until the saturation is detected. In the case of for example the busbar protection 7SS52 a time of 3 ms is already sufficient.

The transient factor for time limited saturation free transformation K'_{TF} is calculated with equation (5-8), if the increase of flux is limited to the time t_M:

$$K'_{TF} = 1 + \frac{\omega \cdot T_N \cdot T_S}{T_N - T_S} \cdot \left(e^{-\frac{t_M}{T_N}} - e^{-\frac{t_M}{T_S}} \right) \qquad (5\text{-}20)$$

For closed iron CT cores, where $T_S \gg T_N$, the equation can be simplified as follows:

$$K'_{TF} = 1 + \omega \cdot T_N \cdot \left(1 - e^{-\frac{t_M}{T_N}} \right) \qquad (5\text{-}21)$$

Figure 5.14 provides a graphic representation of the equation.

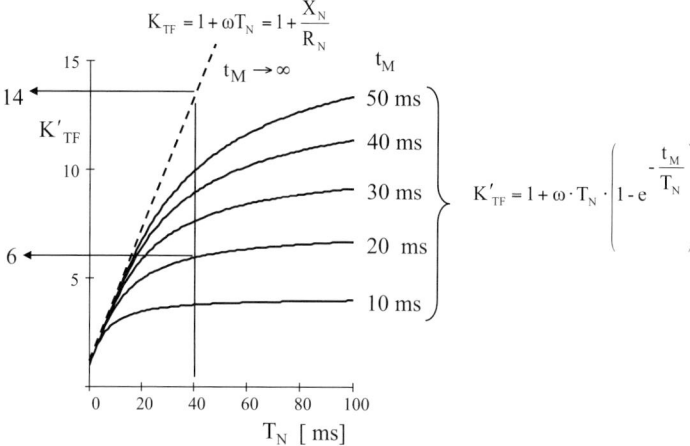

Figure 5.14 Transient factor for CTs with restricted saturation free transmission time

The required CT dimension may be substantially reduced if the saturation-free time t_M is much smaller than the system time constant. In the example at hand a reduction of $K_{TF} = 14$ to $K'_{TF} = 6$ is achieved.

In the equations (5-20) and (5-21) the AC component of the induction is assumed to have the peak value $(B_\sim / \hat{B}_\sim = 1)$, i.e. the envelope curve is taken for the transient increase of the induction. This is an acceptable approximation when current flow takes place for a longer time (several periods) as the DC induction in this case determines the transient over-dimensioning factor and large values for K_{TF} or K'_{TF} are obtained.

If the time to saturation however is short (below one period), the AC flux has the same order of magnitude as the DC induction and may even be larger if very short saturation-free times less than 5 ms are considered. The envelope based over-dimensioning factor delivers in this case much too high requirements.

Under these conditions the sinusoidal course of the AC induction component must be considered, i.e. the complete equation for the course of induction must the applied.

The transient increase of the CT induction $B(t)$ depends on the point-on-wave fault inception and the DC time constant of the short-circuit current. (Figure 5.15)

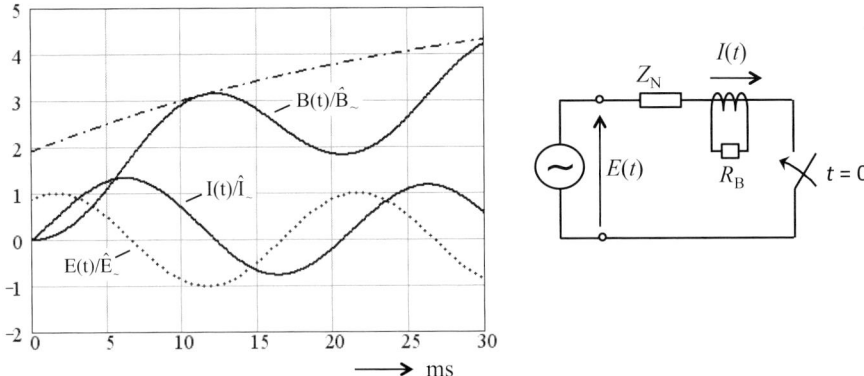

Figure 5.15 Transient CT induction $B(t)$ short time after fault inception

Instead of equation (5-7), we get here:

$$\frac{B(t)}{\hat{B}_\sim} = \frac{\omega \cdot T_N \cdot T_S}{T_N - T_S} \cdot \cos \Theta \cdot \left(e^{-\frac{t}{T_N}} - e^{-\frac{t}{T_S}} \right) +$$

$$+ \sin \Theta \cdot e^{-\frac{t}{T_S}} - \sin(\omega t + \Theta) \tag{5-22}$$

For simplicity we have introduced $\Theta = \Phi_{FI} - \Phi_{SC}$, which is the difference between Φ_{FI}, the point on wave fault inception angle and Φ_{SC}, the angle of the fault loop impedance.

The flux increase for a given time to saturation (t_M) is:

$$\frac{B(t)}{\hat{B}_\sim} = \frac{\omega \cdot T_N \cdot T_S}{T_N - T_S} \cdot \cos \Theta \cdot \left(e^{-\frac{t_M}{T_N}} - e^{-\frac{t_M}{T_S}} \right) +$$

$$+ \sin \Theta \cdot e^{-\frac{t_M}{T_S}} - \sin(\omega t_M + \Theta) \tag{5-23}$$

By differentiation and zero setting we can find the angle $\Theta(t_M, T_N)$ for the maximum induction increase. It depends on the required time to saturation and the DC time constant of the short-circuit current:

$$\Theta(t_M, T_N) = acos \left(\pm \sqrt{1 + \left(\cfrac{1 - \cos(\omega \cdot t_M)}{\omega \cdot \cfrac{T_N \cdot T_S}{T_S - T_N} \cdot \left(e^{-\frac{t_M}{T_S}} - e^{-\frac{t_M}{T_N}} \right) - \sin(\omega \cdot t_M)} \right)^2} \right)$$

(5-24)

The maximum induction corresponds to the transient factor for short time to saturation:

$$K_{TF}''(t_M, T_N) = \frac{\omega \cdot T_N \cdot T_S}{T_S - T_N} \cdot \cos \Theta(t_M, T_N) \cdot \left(e^{-\frac{t_M}{T_S}} - e^{-\frac{t_M}{T_N}} \right) +$$

$$+ \sin \Theta(t_M, T_N) \cdot e^{-\frac{t_M}{T_S}} - \sin(\omega \cdot t_M + \Theta(t_M, T_N))$$

(5-25)

The angle Θ is calculated by formula (5-24) (with+ sign of the square root) for the practical case when the time constant $T_N \geq 10$ ms or taken from the diagram in Figure 5.16. The secondary time constant of closed core CTs (Class P or TPX) can be assumed as $T_S = 5$ s, if not known.

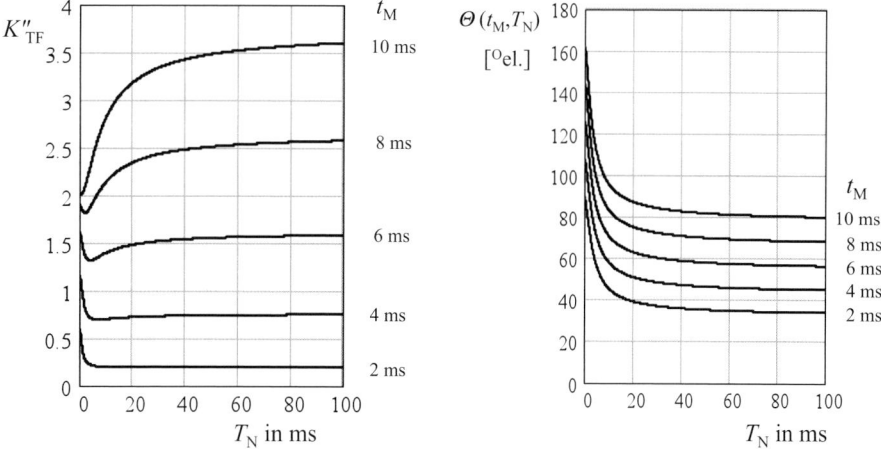

Figure 5.16 Transient factor K_{TF}'' and corresponding fault inception angle Θ for closed core CTs ($T_S \rightarrow \infty$)

With time constants $T_N < 10$ ms it would have to be checked in each case if the angle Θ, that corresponds to a negative sign of the square root of formula (5-24) does result in a larger K_{TF}''-factor (see Figure 5.16). This occurs when $\Theta > 90°$.

Comment: Formula (5-25) and the diagram in Figure 5.16 show that time to saturation and transient factor are directly interrelated. This means that the CT design requirement can be specified either as transient dimensioning factor or as time to saturation.

Both possibilities may be found in relay manuals.

The upcoming Standard IEC 61869-2[1]: "Instrument Transformers", Part 2: "Current Transformers" treats this subject in more detail and contains diagrams for the conversion of time to saturation into the corresponding transient factor.

[1] Presently (2011) this standard exists as draft of IEC subcommittee 38. It shall be published in 2012 and replace the Standards 60044-1 and 60044-6.

Example 5-4: Interrelation between K_{TF} factor and time to saturation

Given: A relay manufacturer specifies the CT requirement for a certain type line differential relay as follows:

Minimum time to saturation 4 ms.

Task: Determine the necessary transient factor K_{TF}.

Solution: From the left diagram in Figure 5.16 we take $K_{TF} \geq 0.8$.

(From formula (5-25) we would get $K_{TF} \geq 0.74$ for $T_N = 50$ ms)

Equivalent time constant

If several sources feed onto the fault with different DC time constants, the CT dimension calculation may be carried out with the sum of the fault currents and the following equivalent time constant [5-13]:

$$T_{equivalent} = \frac{I_{F1} \cdot T_1 + I_{F2} \cdot T_2 + \ldots + I_{Fn} \cdot T_n}{I_{F1} + I_{F2} + \ldots + I_{Fn}} \qquad (5\text{-}26)$$

It produces an equivalent area for the DC current component (same flux rise and therefore same CT dimension requirement) as the sum of the individual currents would have produced. Equation (5-26) in actual fact is only applicable to the area of the total fault ($t \rightarrow \infty$). If saturation may occur after a short time, the equivalent time constant should be calculated for this short interval. With the given equation, the resultant CT dimension is a little larger and therefore the result on the safe side. In example 8-4, section 8.5 the equation is applied.

Over-dimensioning factor for remanence $K_{Rem.}$

The remanence may be as much as 80% ($K_r = 0.8$) in closed iron core CTs. Consequently only 20% flux increase up to saturation would remain (see Figure 5.7) and the CT dimension would have to be increased by the remanance over-dimensioning factor $K_{Rem.} = 1/(1 - K_r) = 5$.

CT dimensioning in practice

Only a small number of statistical analyses exist regarding the frequency and magnitude of DC current components and CT remanence.

A full off-set of the fault current (fault inception close to the zero crossing of the voltage signal) seldom occurs (low probability during lighting strikes exists).

In general the DC off-set is smaller than 70%. [5-14]

With regard to remanance of CTs in the system, only an old Canadian investigation is available. According to this, large remanance values of 60-79% were observed in 27% of 141 CTs. [5-10]

The worst case condition, when fully off-set fault current and maximum remanence coincide, should in practice hardly ever occur. If the closed iron core (P or TPX) CT is dimensioned for this case, it would result in an extremely large and expensive CT. This is not commonly applied in practice. If at all, remanence is only considered at the EHV level. To take into account re-closure onto a permanent fault in the course of an unsuccessful AR cycle (refer Figure 5.8), the CT dimension is calculated to allow for a C-O-C-O switching sequence. For this purpose, CTs with an air gap TPY or TPZ are applied. The dimension (transient dimensioning factor K_{TD}) is based on the course of flux shown in Figure 5.6 and calculated with equation (5-14). IEC60044-6 provides a detailed guideline with calculation examples for this purpose.

The basis for calculating the CT dimension is the recommendations provided by the manufacturers for each relay. These either state a minimum transient dimensioning factor K_{TD} or a corresponding knee-point voltage for the maximum through flowing fault current and a minimum saturation free time after fault inception. In the case of feeder differential protection, an additional requirement often is that the operational accuracy limit factors ALF' at the two line ends may not deviate substantially.

Stability in the presence of through flowing currents (external fault)

This criterion is normally used for the dimension of CTs applied with differential protection. In general, a minimum operational accuracy limit factor is required which is dependent on the maximum through flowing fault current:

$$ALF' \geq K_{TD} \cdot \frac{I_{F\text{-max.-through fault}}}{I_{n\text{-CT}}} \tag{5-27}$$

For most Siemens differential relays, only $K_{TD} > 1$ is specified, i.e. the CTs must transform the maximum non-offset through fault current without saturation. (Table 5.3) In the presence of DC offset, a substantial amount of saturation is therefore permissible.

Exceptions are generator and transformer differential protection with sensitive (10 to 20% I_N) pick-up setting and low percentage restraint. In this case it must be assured that DC saturation does not occur already at very low currents ($< 2 \cdot I_N$) where the saturation detector cannot operate.

With busbar protection, the situation is particularly difficult, as the fault currents of several in-feeds flow into the protected object and exit via the faulted feeder as the sum total of all fault current in-feeds. The busbar protection must therefore allow a particularly large degree of CT saturation. The Siemens busbar protection type 7SS52 for example only demands 3 ms of non-saturated current transformation to reach a secure stabilising. This corresponds to $K_{TD} = K_{TF}'' \geq 0.5$, which means that the CT must only be capable of transforming half the maximum sum total fault current of the busbar.

Secure trip decision during internal faults

For this purpose a minimum saturation free time is necessary so that non-delayed tripping can be achieved. In some cases this may even be the critical CT dimension criterion.

The transformer protection is an example. In the event of an internal fault close to the terminals, a very large fault current will result while the current in case of external faults is much smaller due to the limiting transformer short circuit impedance.

Modern relays have fast high set measuring elements which use only a few current samples for the trip decision and therefore can allow very short time to saturation.

More detailed advice for the CT dimension calculations may be found in the product documentation or engineering guides provided by the protection manufacturers. [5-15]

The following table contains the recommendations stated by Siemens for their differential protection relays:

Table 5.3 CT requirements for digital differential protection

Type of protection	Requirements for external faults (through-fault condition)	Additional requirements
Generator- and transformer differential protection 7UM6, 7UT6	$K_{TD} \geq 4$	None
Feeder differential protection with digital communication 7SD52/53/61/84 to 87	$K_{TD} \geq 1.2$	None
7SD80	$K_{TD} \geq 1.2$	$ALF' \geq 30$
Feeder differential protection with pilot wires 7SD600	$K_{TD} \geq 1$	$\dfrac{3}{4} \leq \dfrac{ALF_1'}{ALF_2'} \leq \dfrac{4}{3}$
Busbar differential protection 7SS50/51 and 7SS52	$K_{TD} \geq 0.5$ (corresponds to 3 ms saturation free time)	$ALF' \leq 100$
Busbar differential protection 7SS600	$K_{TD} \geq 0.75$ (corresponds to 4 ms saturation free time)	$ALF' \leq 100$

Example 5-5: Practical dimension of the CTs

Task: The CTs must be dimensioned for the application shown in Figure 5.17.

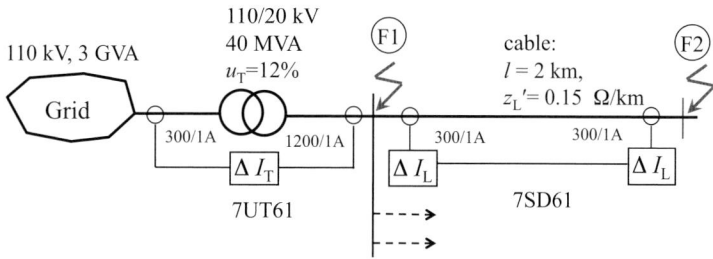

Figure 5.17 Single line diagram and data for example 5-5: CT dimensioning

Solution: Initially the short circuit impedances are calculated.

Impedances based on 110 kV:

$$\text{Network: } Z_N = \frac{U_n^2 [\text{kV}^2]}{\text{SCC}''[\text{MVA}]} = \frac{110^2}{3000} = 4.03 \ \Omega$$

$$\text{Transformer: } Z_T = \frac{U_n^2 [\text{kV}^2]}{P_{n\text{-}T}[\text{MVA}]} \cdot \frac{u_T[\%]}{100} = \frac{110^2}{40} \cdot \frac{12\%}{100} = 36.3 \ \Omega$$

Impedances based on 20 kV:

$$\text{Network: } Z_N = \frac{U_n^2 [\text{kV}^2]}{\text{SCC}''[\text{MVA}]} = \frac{20^2}{3000} = 0.13 \ \Omega$$

$$\text{Transformer: } Z_T = \frac{U_n^2 [\text{kV}^2]}{P_{n\text{-}T}[\text{MVA}]} \cdot \frac{u_T[\%]}{100} = \frac{20^2}{40} \cdot \frac{12\%}{100} = 1.2 \ \Omega$$

Line: $Z_L = l \,[\text{km}] \cdot z_L' \,[\Omega/\text{km}] = 2 \cdot 0.15 = 0.3 \ \Omega$

For the fault locations F1 and F2 the following fault currents are calculated:

Maximum transformer through fault current (related to 110 kV):

$$I_{F1\text{-}110kV} = \frac{1.1 \cdot U_n /\sqrt{3}}{Z_N + Z_T} = \frac{1.1 \cdot 110 \,\text{kV}/\sqrt{3}}{4.03 \ \Omega + 36.03 \ \Omega} = 1.73 \ \text{kA}$$

Maximum transformer through fault current (related to 20 kV):

$$I_{F1\text{-}20kV} = \frac{110}{20} \cdot 1.73 = 9.5 \ \text{kA}$$

Maximum line through fault current (20 kV):

$$I_{F2\text{-}20kV} = \frac{1.1 \cdot U_n /\sqrt{3}}{Z_N + Z_T + Z_L} = \frac{1.1 \cdot 20 \,\text{kV}/\sqrt{3}}{0.13 \ \Omega + 1.2 \ \Omega + 0.3 \ \Omega} = 7.8 \ \text{kA}$$

Dimensioning of the CTs for the transformer differential protection:

CTs on 110 kV-side:

From Table 5.3 the requirement for the relay 7UT61 are extracted:

Dimension criterion: $K_{TD} \geq 4$ for maximum through fault current.

The corresponding accuracy limit factor in operation therefore is:

$$\text{ALF}' = K_{TD} \cdot \frac{I_{F1\text{-}110kV}}{I_{n\text{-}CT}} = 4 \cdot \frac{1730}{300} = 23$$

Initially the following CT design data are assumed:

5P?, 300/1 A, 10 VA, internal burden 2 VA.

The connected burden (CT secondary cables and relay) is approx. 2.5 VA.

The required rated accuracy limit factor is then:

$$\text{ALF} \geq \frac{P_i + P_{\text{actual}}}{P_i + P_{\text{rated}}} \cdot \text{ALF}' = \frac{2 + 2.5}{2 + 10} \cdot 23 = 8.6$$

The following CT type is selected:

300/1 A, 5P10, 10 VA, $R_{CT} \leq 2$ Ohm ($P_i \leq 2$ VA)

CTs on 20 kV-side:

$$\text{ALF}' = K_{TD} \cdot \frac{I_{F1\text{-}20kV}}{I_{n\text{-}CT}} = 4 \cdot \frac{9550}{1200} = 31.7$$

A symmetrical dimensioning of the CTs is planned on both sides of the transformer as a basis for the best stability in the event of external faults:

We choose a higher rated power of 15 VA and assume again about 20% internal burden ($P_i = 3$ VA)

Then we get:

$$\text{ALF} \geq \frac{P_i + P_{\text{actual}}}{P_i + P_{\text{rated}}} \cdot \text{ALF}' = \frac{3 + 2.5}{3 + 15} \cdot 31.7 = 9.7$$

The following CT type is selected:

1200/1 A, 5P10, 15 VA, $R_{CT} \leq 3$ Ohm ($P_i \leq 3$ VA)

We can now finally check the symmetry of the CT design:

The actual ALF' factors on both sides of the transformer are:

$$\text{ALF}'_{110\text{-}kV} \geq \frac{10}{8.6} \cdot 23 = 27.74 \quad \text{and} \quad \text{ALF}'_{20\text{-}kV} \geq \frac{10}{9.7} \cdot 31.7 = 32.68$$

Related to the through-flowing current we get

$$\frac{\text{ALF}_{110\text{-}kV} \cdot I_{n\text{-}CT\text{-}110\text{-}kV}}{I_{n\text{-}Transformer\text{-}110\text{-}kV}} = \frac{27.74 \cdot 300}{40000/(110 \cdot \sqrt{3})} = \frac{27.74 \cdot 300}{210} = 39.6$$

$$\frac{\text{ALF}_{20\text{-kV}} \cdot I_{n\text{-CT-20-kV}}}{I_{n\text{-Transformer-20-kV}}} = \frac{32.68 \cdot 1200}{40000/(20 \cdot \sqrt{3})} = \frac{32.68 \cdot 1200}{1155} = 34$$

We see that the CT dimension at high and low voltage side of the transformer are reasonably balanced.

Note: Numerical relays are less sensitive against CT unbalance due to the integrated saturation detectors. However it is still good relaying practice to balance the CT dimension for differential relays as far as possible.

Dimensioning of the CTs for the feeder differential protection:

We take the CT requirements for relay 7SD61 again from Table 5.3:
Dimension criterion: $K_{TD} \geq 1.2$ for maximum through fault current.

The required accuracy limit factor in operation therefore is:

$$\text{ALF}' = K_{TD} \cdot \frac{I_{F2\text{-20kV}}}{I_n} = 1.2 \cdot \frac{7800}{300} = 31.2$$

A CT type 300/1 A, 5 VA, 5P?, internal burden $\leq$ 1 VA is planned.

The connected burden is assumed to be 0.5 VA, as the numerical protection is located very close to the switchgear (short CT connecting cables).

The required nominal accuracy limit factor is then:

$$\text{ALF} \geq \frac{P_i + P_{actual}}{P_i + P_{rated}} \cdot \text{ALF}' = \frac{1 + 0.5}{1 + 5} \cdot 31.2 = 7.8$$

CTs with the next greater standard accuracy limit factor (ALF = 10) are finally chosen at both sides of the cable:

300/1 A, 5P10, 5 VA, $R_{CT} \leq 1$ Ohm ($P_i \leq 1$ VA)

5.8 Interposing CTs

With conventional protection, interposing CTs were often used to compensate unequal transformation ratios and the vector group of transformers. They were also applied to obtain a galvanic separation of current transformer circuits.

Numerical protection devices contain a numeric ratio and vector group correction inside the relays so that interposing CTs are generally not required. Under special circumstances they may still be required when the numeric compensation is not sufficient. Furthermore, interposing CTs are still used for special applications.

The following discussion of the traditional interposing circuits will also be helpful for the understanding of the equivalent numerical algorithms of digital relays.

Current ratio correction

The current is transformed proportional to the inverse turns ratio according to the Ampère's law (Figure 5.18):

$$\underline{I}_1 \cdot w_1 = \underline{I}_2 \cdot w_2 \qquad \text{or} \qquad \underline{I}_2 = \frac{w_1}{w_2} \cdot \underline{I}_1 \qquad\qquad (5\text{-}28)$$

By suitable selection of the terminals, the current may be adapted without reversing the phase or with phase reversal.

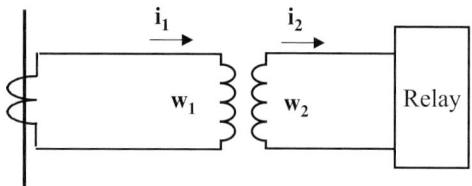

Figure 5.18
Current matching with interposing CTs: separate winding type

If galvanic separation is not required then the interposing transformer may also be an auto-transformer. In this manner the number of turns may be reduced whilst maintaining the same voltage across the winding. The losses (internal burden) of the interposing transformer are also reduced (Figure 5.19). The balance of the ampere-turns results in the following equation:

$$w_a \cdot \underline{I}_2 = w_b \cdot (\underline{I}_1 - \underline{I}_2) \qquad \text{or} \qquad \underline{I}_2 = \frac{w_b}{w_a + w_b} \cdot \underline{I}_1 \qquad\qquad (5\text{-}29)$$

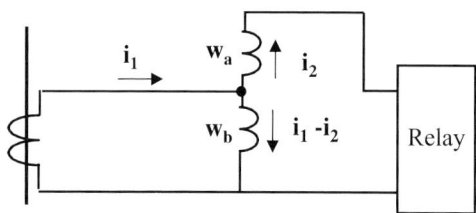

Figure 5.19
Current matching with interposing CTs: auto transformer type

For a given current transformation ratio the turns ratio may be calculated with the following equation:

$$\frac{w_a}{w_b} = \frac{(\underline{I}_1 - \underline{I}_2)}{\underline{I}_2} \qquad\qquad (5\text{-}30)$$

The stated equations apply for the case when I_2 is smaller than I_1. If I_2 is greater than I_1, then I_1 and I_2 must be swapped in the equation (reverse connection of the interposing CT).

Example 5-6: Comparison of the separate winding and auto-transformer connected interposing CT

Task:
The CT transformation ratio in a feeder must be adapted from 600/1 to 400/1 for a differential protection.

The standard interposing CT 4AM5170-7AA must be employed (for data refer to Figure 5.35 below)

The burden of the protection including CT connecting cables is 1 Ohm.

Which CT connection is best? (Galvanic separation is not necessary.)

Figure 5.20 shows the two alternative connections.

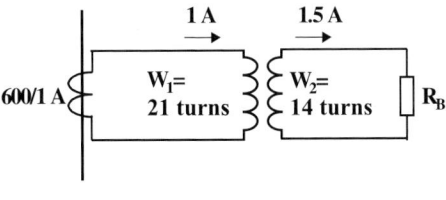

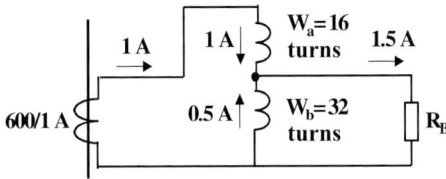

Figure 5.20 Interposing CT: Equivalent circuit for calculation example

Solution:
Separate winding connection:

The adaptation from 600/1 A to 400/1 A implies that the secondary current must be transformed up by a factor of 1.5. This corresponds to a turn's ratio of 3 to 2. The 2 times 16 turns of the interposing CT are only designed for 1 A and may therefore only be applied on the primary side.

By trial and error the following transformation of $21(1+2+2+16)$ to $14(7+7)$ is found to be suitable.

The resistance of the primary winding is:

$$R_{1i} = 0.013 + 0.025 + 0.025 + 0.75 = 0.813\ \Omega$$

The resistance of the secondary winding is then:

$$R_{2i} = 0.08 + 0.08 = 0.16\ \Omega$$

The internal voltage across the winding (magnetising voltage at nominal CT current $I_N = 600$ A) is:

$$e_{2\text{-wndg.}} = E_2/w_2 = I_1 \cdot \frac{w_1}{w_2} \cdot (R_{2i} + R_B)/w_2 =$$

$$= 1\ \text{A} \cdot \frac{21}{14} \cdot (0.16\ \Omega + 1\ \Omega)/14 = 0.124\ \text{V}$$

The internal losses are:

$$P_i = R_{1i} \cdot i_1^2 + R_{2i} \cdot i_2^2 = 0.813 \cdot 1^2 + 0.16 \cdot 1.5^2 = 1.17 \text{ VA}$$

Auto-transformer connection:

The required transformation ratio for the secondary current from 1 A to 1.5 A also applies here.

Therefore according to (5-30) and by exchanging I_1 with I_2 the following is obtained:

$$\frac{w_a}{w_b} = \frac{I_2 - I_1}{I_1} = \frac{1.5 - 1.0}{1.0} = 0.5$$

In this case all sub-windings may be used as the current in the windings will not be greater than 1 A when nominal current flows on the primary side.

The ratio $w_a/w_b = 16(16)$ to $32(2+7+7+16)$ uses nearly all of the available sub-windings. The winding resistance of w_b then is $R_b = 0.025 + 0.08 + 0.08 + 0.75 = 0.935 \ \Omega$.

The internal voltage $e_{b\text{-wdg.}}$ results from the voltage drop across the burden and the internal voltage drop across R_b of the winding w_b.

$$e_{b\text{-wndg.}} = E_b/w_b = (I_2 \cdot R_B + I_b \cdot R_b)/w_b =$$
$$= (1.5 \text{ A} \cdot 1 \ \Omega + 0.5 \text{ A} \cdot 0.935)/32 = 0.062 \text{ V}$$

The internal losses are:

$$P_i = R_a \cdot I_1^2 + R_b \cdot (I_2 - I_1)^2 = 0.75 \cdot 1^2 + 0.935 \cdot 0.5^2 = 0.98 \text{ VA}$$

Result:

The transformation response of the auto-transformer is better.

The magnetising is only half as high. This corresponds to a doubled transient dimensioning factor K_{TD}.

Elimination of zero sequence currents (zero sequence current blocking)

This circuit is required when the zero sequence current may not reach the measuring relay. For this purpose a magnetically coupled delta connected winding, which shunts the zero sequence current is required. (Figure 5.21)

The co-phasal zero sequence currents flowing in the three phases encounter an opposing flux as a result of the current circulating in the delta winding, i.e. the zero sequence currents are short circuited by the delta winding. On the relay side, the star- point may however not be connected as this would constitute an alternate path for the zero sequence current.

The positive and negative sequence system is transformed to the secondary side in accordance with the transformation ratio w_1/w_2.

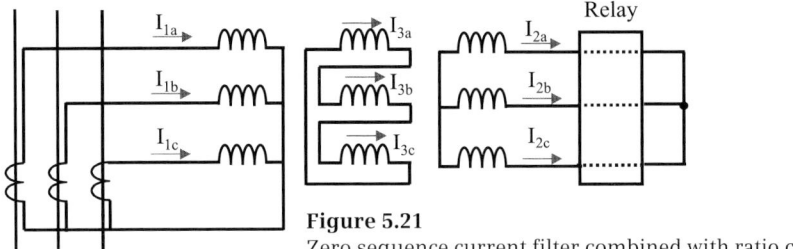

Figure 5.21
Zero sequence current filter combined with ratio correction

From Figure 5.21 the following equations are obtained when assuming the indicated current directions:

$$\underline{I}_{1a} \cdot w_1 - \underline{I}_{2a} \cdot w_2 - \underline{I}_{3a} \cdot w_3 = 0 \tag{5-31}$$

$$\underline{I}_{1b} \cdot w_1 - \underline{I}_{2b} \cdot w_2 - \underline{I}_{3b} \cdot w_3 = 0 \tag{5-32}$$

$$\underline{I}_{1c} \cdot w_1 - \underline{I}_{2c} \cdot w_2 - \underline{I}_{2c} \cdot w_3 = 0 \tag{5-33}$$

$$\text{with} \quad \underline{I}_{3a} = \underline{I}_{3b} = \underline{I}_{3c} \tag{5-34} \quad \text{and} \quad \underline{I}_{2a} + \underline{I}_{2b} + \underline{I}_{2c} = 0 \tag{5-35}$$

From (5-31) to (5-35) the following is obtained:

$$\underline{I}_{3a} = \underline{I}_{3b} = \underline{I}_{3c} = \frac{w_1}{w_3} \cdot \frac{\underline{I}_{1a} + \underline{I}_{1b} + \underline{I}_{1c}}{3} \tag{5-36}$$

$$\text{and} \quad \underline{I}_{2a} = \frac{w_1}{w_2} \cdot \left(\underline{I}_{1a} - \frac{\underline{I}_{1a} + \underline{I}_{1b} + \underline{I}_{1c}}{3} \right) \tag{5-37}$$

$$\underline{I}_{2b} = \frac{w_1}{w_2} \cdot \left(\underline{I}_{1b} - \frac{\underline{I}_{1a} + \underline{I}_{1b} + \underline{I}_{1c}}{3} \right) \tag{5-38}$$

$$\underline{I}_{2c} = \frac{w_1}{w_2} \cdot \left(\underline{I}_{1c} - \frac{\underline{I}_{1a} + \underline{I}_{1b} + \underline{I}_{1c}}{3} \right) \tag{5-39}$$

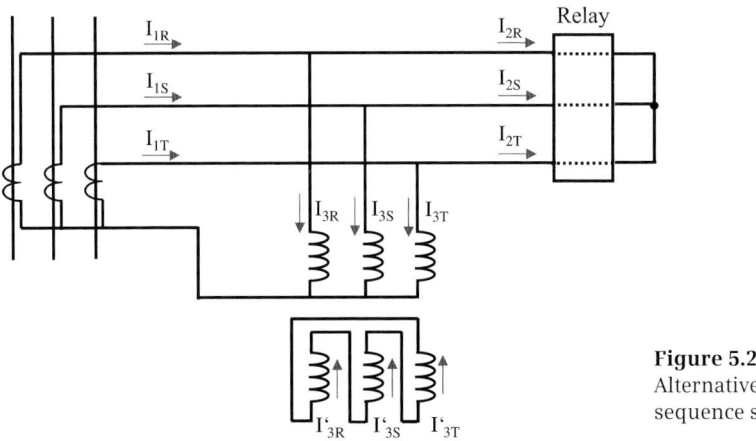

Figure 5.22
Alternative zero sequence shunt

101

An alternative connection for the zero sequence current elimination without ratio correction is shown in Figure 5.22.

Example 5-7: Elimination of the zero sequence current in transformer differential protection

Figure 5.23 initially shows that current flows via the differential path if no zero sequence shunt is applied. In the event of an external fault this would cause tripping. The difference between the currents is equal to the zero sequence component of the fault current which is short circuited in the delta winding of the power transformer and therefore not transformed to the high voltage side.

This condition may be rectified by application of a zero sequence current shunt (delta winding) in the CT circuit of the low voltage side. (Figure 5.24)

The star point of the interposing transformers on the relay side may <u>not</u> be connected

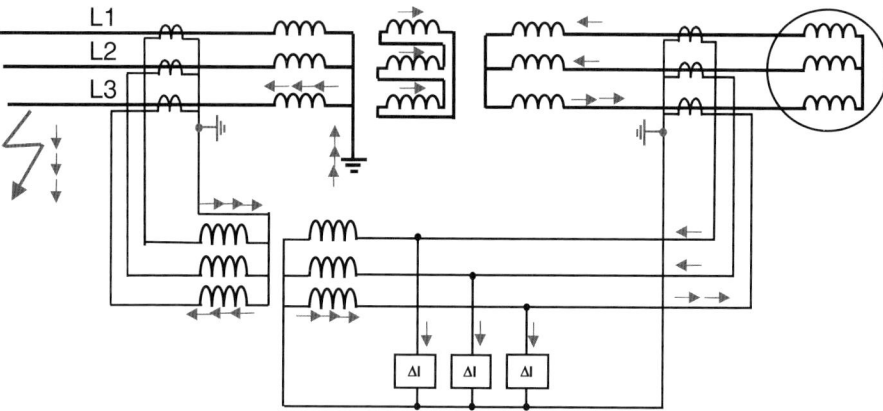

Figure 5.23 External fault, current distribution **without** zero sequence current filter

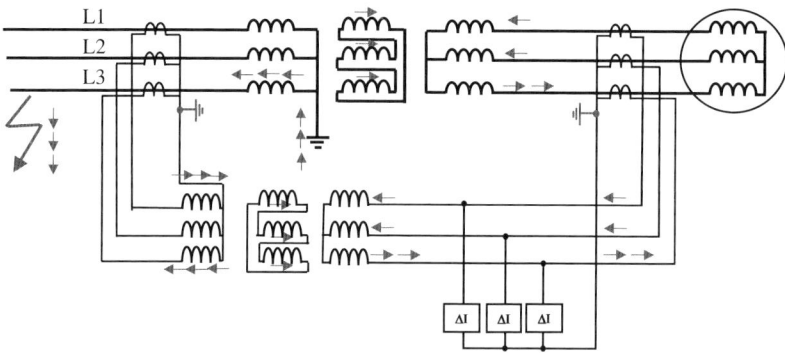

Figure 5.24 External fault, current distribution **with** zero sequence current filter

with the star point at the relay side, as this would otherwise constitute a connection alternate to the delta connection. The current distribution during an external single pole fault shows that current now no longer flows via the differential path of the relay.

In the event of an internal earth fault, the zero sequence current is however also filtered out. The pick-up sensitivity of the protection is therefore reduced. For single pole faults this reduction corresponds to 1/3rd. Furthermore, a tripping current corresponding to the zero-sequence system appears in the healthy phases so that the protection indication is no longer phase selective. This loss of sensitivity and selectivity may again be compensated by introduction of a CT in the star point of the earthed transformer winding. This is illustrated by means of the following example:

Example 5-8: Zero sequence current correction

This technique may always be applied when a CT is available in the star point of the earthed transformer winding (Figure 5.25 and Figure 5.26).

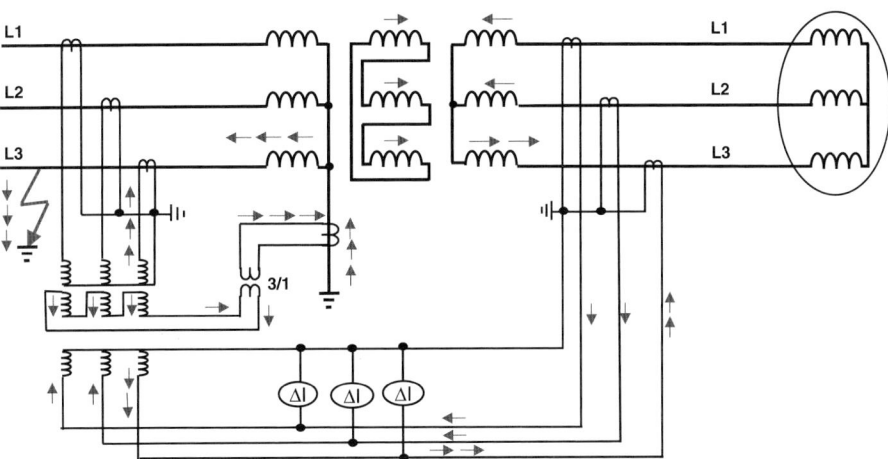

Figure 5.25 Zero sequence correction during external earth fault

The zero sequence current in this case is not short circuited by means of a delta winding when an external fault occurs, but rather is sucked off by the star point CT (Figure 5.25). During an internal fault, the zero sequence current is fed back onto the delta connection (Figure 5.26), so that the differential protection maintains full pick-up sensitivity. It must be noted that the current in the delta winding of the interposing CT equals the zero sequence current of the symmetrical components (it flows in the same direction in all three phases), while the current flowing via the star point earth connection is the earth current ($I_E = 3 \cdot I_0$). An adaptation of 3 to 1 is therefore required. For the purposes of simplification, a separate interposing CT is indicated in Figure 5.25 and Figure 5.26. In practice however, the number of turns of the delta winding of the interposing CT would be reduced by a factor 3.

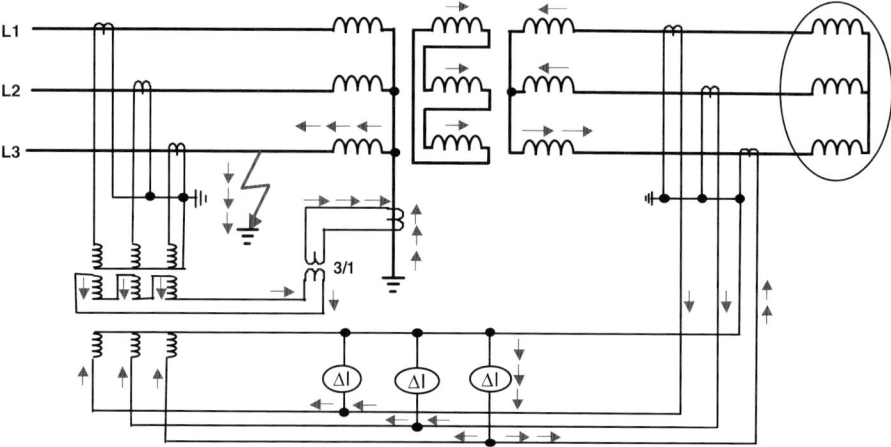

Figure 5.26 Zero sequence current correction with internal fault

From Figure 5.26 it is apparent that 3 current arrows flow via the measuring system in the faulted phase which means that the full fault current is effective. The measuring systems of the other phases contain no current so that indication of the fault type is clear.

Matching CTs in star delta connection

This type of connection for the matching CTs is often applied with conventional transformer differential protection. Three functions are provided in this manner:

- Matching of the transformation ratio
- Matching of the transformer vector group
- Isolation of the zero sequence current from the differential protection (zero sequence shunt) as described above.

In Figure 5.27 an example of matching is shown.

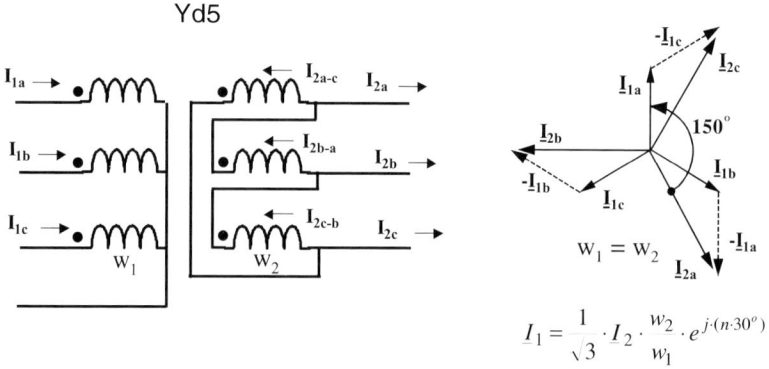

Figure 5.27 Interposing CTs in Star-Delta-connection

The current transformation is defined by the following equations:

Currents in the delta winding:

$$\underline{I}_{2\Delta a\text{-}c} = \frac{w_1}{w_2} \cdot \underline{I}_{1a}, \quad \underline{I}_{2\Delta b\text{-}a} = \frac{w_1}{w_2} \cdot \underline{I}_{1b} \text{ and } \underline{I}_{2\Delta c\text{-}b} = \frac{w_1}{w_2} \cdot \underline{I}_{1c} \tag{5-40}$$

Currents on the secondary side:

$$\underline{I}_{2a} = \underline{I}_{2\Delta b\text{-}a} - \underline{I}_{2\Delta a\text{-}c} = \frac{w_1}{w_2} \cdot (\underline{I}_{1b} - \underline{I}_{1a}) \tag{5-41}$$

$$\underline{I}_{2b} = \underline{I}_{2\Delta c\text{-}b} - \underline{I}_{2\Delta b\text{-}a} = \frac{w_1}{w_2} \cdot (\underline{I}_{1c} - \underline{I}_{1b}) \tag{5-42}$$

$$\underline{I}_{2c} = \underline{I}_{2\Delta a\text{-}c} - \underline{I}_{2\Delta c\text{-}b} = \frac{w_1}{w_2} \cdot (\underline{I}_{1a} - \underline{I}_{1c}) \tag{5-43}$$

The corresponding phasor diagram is shown for $w_1 = w_2$ in Figure 5.27. It can be seen that the secondary side currents lag the primary side currents by $150° = 5 \cdot 30°$. This is in accordance with the transformer vector group Yd5. (The vector group and manner of counting are referred to below)

In most cases the interposing CTs are applied so they have the same vector group as the primary transformer (image). In this way a simple reverse transformation (vector adaptation) of the secondary currents is obtained for comparison in the differential protection.

The star points of the CTs on both sides of the power transformer must in this case be the same: both towards the transformer or towards the system (refer to the following example).

From equations (5-40) to (5-43) it may also be seen that the zero sequence current that flows via the star connected winding is not flowing on the secondary side. As the zero-sequence current component in I_{1a}, I_{1b} and I_{1c} has the same magnitude and phase angle, it is cancelled out on the phase outputs, which means it is shunted by the delta connection.

Example 5-9: Calculation of the interposing CT ratio

Given: Transformer with tap changer 10 MVA, 110 kV ±16% / 6.3 kV,

Vector group Yd5

CT: 110 kV-side: 75/1 A

CT: 6 kV-side: 1200/5 A

Nominal current of the differential relay: 5 A

Procedure: Initially the circuit diagram of the transformer and differential protection is drawn (Figure 5.28).

The Yd5 transformer connection diagram may be derived from technical literature. Alternatively it may be constructed using the guidelines provided below.

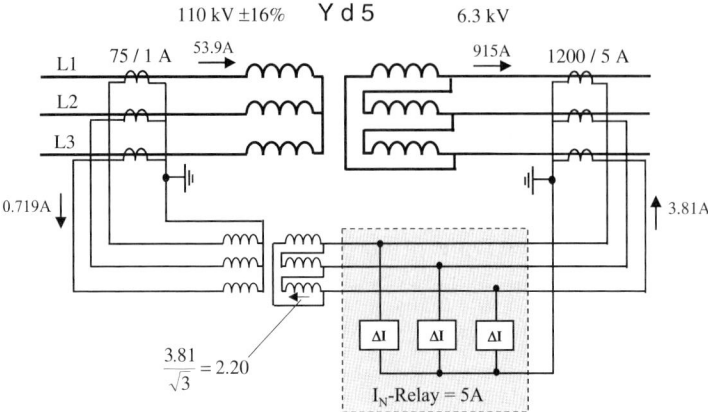

Figure 5.28 Current matching with interposing CTs (example)

The transformer star point is not earthed so that a zero sequence shunt is not required on the star point side. In principle, the interposing CT could therefore be positioned between the high voltage CTs and the relay or between the low voltage CTs and the relay. As the relay has a nominal current of 5 A, the second alternative is chosen.

The interposing CT must in any event be connected in Y-Δ to provide adaptation to the transformer vector group Yd5. In this case, the star point side must be towards the CT side so that the star point of the CT can be connected, providing a return path for the residual current. It is assumed that the CT neutrals are towards the transformer in both cases. (This in general depends on the common practice applied by the user.) For the interposing CTs the same connection circuit as for the transformer itself results. If the CT neutrals are not similar (one side towards the power system and the other side towards the transformer) then the connection of the star connected interposing CT windings would have to be reversed to provide correct polarity adaptation. (A guideline to obtain the proper connection circuits is provided in the following section).

Calculation of the interposing CT ratio:

The transformer nominal current is used as through flowing reference current:

<u>110 kV-side:</u>

The average current based on the highest and lowest tap changer setting is calculated:

$$I_1 = \frac{10{,}000\ \text{kVA}}{\sqrt{3} \cdot (110\ \text{kV} + 16\%)} = 45.2\ \text{A}$$

$$I_1' = \frac{10{,}000\ \text{kVA}}{\sqrt{3} \cdot (110\ \text{kV} - 16\%)} = 62.5\ \text{A}$$

$$I_{1\text{-mean}} = \frac{45.2 + 62.5}{2} = 53.9\ \text{A}$$

6 kV-side:

$$I_2 = \frac{10{,}000 \text{ kVA}}{\sqrt{3} \cdot 6.3 \text{ kV}} = 915 \text{ A}$$

The corresponding secondary side currents then are:

$$i_1 = 53.9 \cdot \frac{1}{75} = 0.719 \text{ A} \quad \text{and} \quad i_2 = 915 \cdot \frac{5}{1200} = 3.813 \text{ A}.$$

The current in the primary winding of the interposing CT is i_1.

The current in the delta connected secondary winding is $i_2/\sqrt{3}$.

Correspondingly the transformation ratio of the interposing CTs must be:

$$\frac{w_1}{w_2} = \frac{i_2/\sqrt{3}}{i_1} = \frac{3.813/\sqrt{3}}{0.719} = \frac{2.202}{0.719} = 3.06$$

Vector group designation of the transformer according to the clock-wise notation

This designation is defined in IEC 60076-1.

The connection diagram always shows the high voltage winding at the top and the low voltage winding at the bottom (Figure 5.29). The direction of the induced voltages is designated by means of dot polarity marks.

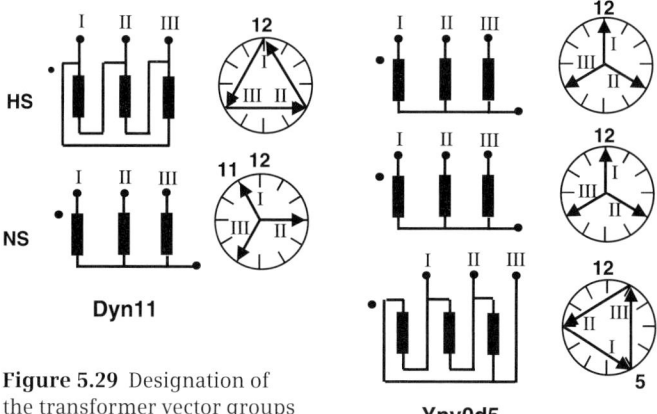

Figure 5.29 Designation of the transformer vector groups

Dyn11

Yny0d5

The vector diagram of the high voltage winding is positioned such that the voltage phase L1 is at 12 o'clock. The vector of the L1 phase voltage on the low winding side is entered according to the position of the induced voltage direction. Increments of 30° are counted in the lagging sense from the high voltage side to the low voltage side. In the left hand example, 330° lagging are thus obtained, i.e. the vector group is Dy11. (The high voltage side is always indicated by means of the capital letter.) Other common vector groups are shown in IEC 60076-1.

The application of the clock-wise designation is shown in Figure 5.30.

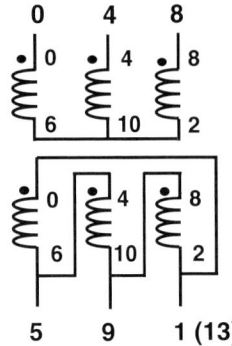

Figure 5.30
Clock-wise notation principle for
determining the vector group

Example 5-10: Transformer vector groups

Given: Transformer connection diagram (Figure 5.30)

Wanted: Vector group

Solution: The following steps must be followed:

- Starting on the high voltage winding, the phase connection terminals are numbered with 0, 4, 8 (always $4 \cdot 30° = 120°$ phase shift)
- The opposite end of each winding is labelled with a number incre-
  mented by 6 relative to the phase connection (180°)
- The secondary windings are numbered the same. In this context it
  is assumed that the polarity of the windings is the same in the diagram. (If in doubt, polarity marks may also be applied)
- The phase connection is labelled with the average value of the corresponding terminal designations belonging to the winding terminals connected to this phase terminal, e.g. $(6+4)/2 = 5$.
- The difference between the high and low voltage side terminal
  numbers of same phases corresponds each with the vector group
  number, being Yd5 in this case.

Example 5-11: Transformer winding connection

Given: Vector group

Wanted: Connection circuit

Solution: Proceed as follows:

- Draw the star winding completely. On the delta side only draw the
  windings without the interconnections.
- Number the primary and secondary windings as in the example
  above.

- At the phase L1 terminal of the delta winding enter the number which is equal to the star point terminal number plus the vector group number. This applies if the low voltage winding is connected in delta. If the high voltage winding is connected in delta, the reverse applies. The other phase terminals must be numbered clock-wise with incremental steps of 4 (corresponds to a phase shift of 120° in each case).

- Connect the two winding terminals that are designated with numbers that neighbour the vector group number (in this case 4 and 6). Carry on with the other windings in an analogous manner.

- The difference between the numbers at the high and low voltage side phase terminals must be the same in each phase and correspond with the vector group number.

Application of the clock-wise method

Using this method, the measuring circuit of the differential protection can be checked with regard to its termination Figure 5.31).

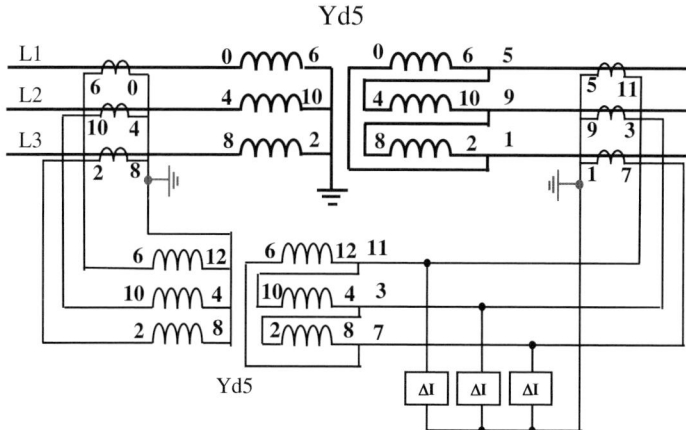

Figure 5.31 Checking the connection of the transformer differential protection using the clock-wise notation method

Initially the clockwise notation numbers must be entered on the transformer circuit terminals corresponding with the vector group as described above. Starting on one side of the transformer, the terminals of the CTs and the devices are numbered in sequence. Circuit positions that must obtain a number and that are directly connected are provided with the same number, as they have the same potential. For all CTs and transformers it is assumed that the voltage in the windings is always induced in the same direction (polarity). The numbering sequence must form a continuous loop similar to the numbering sequence of a clock.

To determine the interposing CT, the numbering of the terminals may be carried out from each side up to the terminals of the interposing CT. The connections at the interposing CT itself can then be done according to the rules listed above.

Checking with the arrow method

The arrow method facilitates a simple technique for checking the polarity and connection of the measuring circuits. For this purpose external faults on both sides of the transformer are assumed and the current flow is indicated by means of directional arrows. The currents must form a closed loop with their flow direction and no current component may flow via the differential path if the connections are correct. It is sufficient to carry out this check for a single-phase earth fault in L1 and a two-phase fault without earth between the phases L2 and L3. With star-delta transformers, a particular current distribution will result: the single pole fault appears as a two-phase and the two- phase fault as a non symmetrical three-phase fault on the other side of the transformer (Figure 5.32).

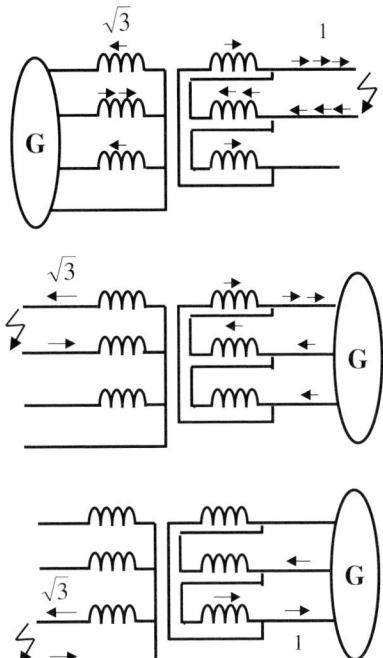

Figure 5.32
Current distribution for
faults at a Y-Δ-transformer

The flow of current may for example be determined analytically by application of the symmetrical component analysis technique. In practice, the only correct current flow can be easily found with the trial and error method by simply noting that the arrows must be directed such that the summated ampere-turns on each winding of a core is equal to zero. With a little practice the common current distribution patterns are quickly memorised so that one can easily recognise whether one should start with one, two or three arrows.

To check the turns ratio of the interposing CTs, the weighting of the arrows may also be entered. In this context it must be noted that the currents inside the delta windings are always a factor $\sqrt{3}$ smaller than the currents in the star connected windings due to the fact that the number of turns is greater by the factor $\sqrt{3}$ when referenced to the same voltage.

Application of the arrow method is indicated by means of two-phase and single-phase fault examples. In the event of an external fault and correct connection, no current (arrows) may flow via the differential relay in any of the three phases Figure 5.33 and Figure 5.34).

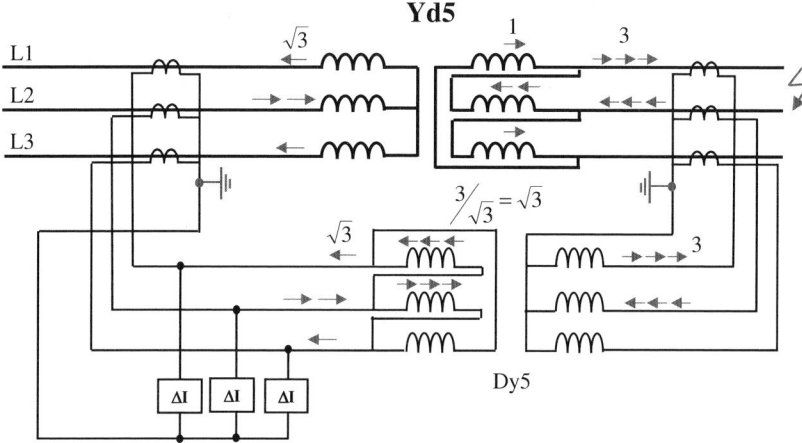

Figure 5.33 Checking the connections of a transformer differential protection using the arrow method (two-phase short circuit)

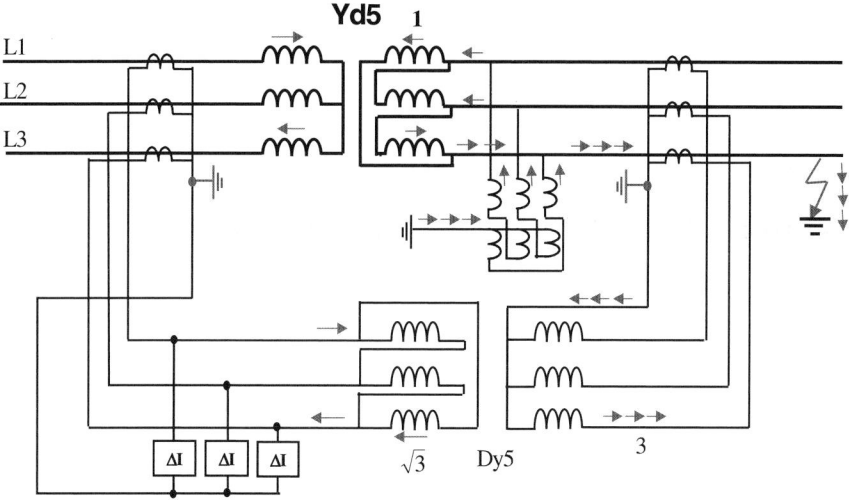

Figure 5.34 Checking the connection of a transformer differential protection using the arrow method (single-phase earth fault)

As a cross check, the current distribution can also be indicated with arrows in the event of an internal fault to check that secure operation is achieved.

With the single-phase earth fault example (Figure 5.34) a slightly more complicated case with earthing transformer was deliberately chosen to illustrate the current distribution in the zero sequence system.

To ensure that the connections of all phases and the zero-sequence system (star point connection and delta connection) are checked it is recommended to carry out the check with a single-phase earth fault in one phase (e.g. L1-E) and the two-phase fault between the other two phases (e.g. L2-L3).

Specification of the interposing current transformers

In principle the interposing CTs must be specified the same as normal CTs. If they are mounted in the protection cubicle or very close to the protection relay, which is generally the case, only the relay burden must be considered so that a small rating is sufficient.

If the interposing CTs are installed close to the main CTs, for example to reduce the current level on long interconnections to the protection and thereby reduce the burden imposed on the main CT, a higher rating of the interposing CT will be necessary.

It is common practice to use a standard interposing CT that may be configured by jumper connections of tapped windings to obtain the ratio required for varying applications.

Figure 5.35 shows the corresponding Siemens CT 4AM5170 that has been applied for a number of decades with only minor modification. It has 8 sub-windings which may be combined by addition (series connection with same polarity) and subtraction (series connection with opposing polarity) to obtain the total winding.

During the configuration it must be observed that the saturation voltage and current rating of the sub-windings are not exceeded.

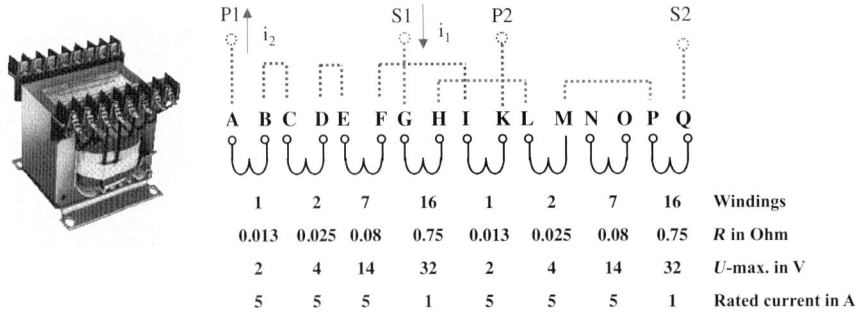

1	2	7	16	1	2	7	16	Windings
0.013	0.025	0.08	0.75	0.013	0.025	0.08	0.75	R in Ohm
2	4	14	32	2	4	14	32	U-max. in V
5	5	5	1	5	5	5	1	Rated current in A

Figure 5.35 Link selectable interposing CT (Siemens type 4AM5170)

Example 5-12: Application of the interposing CT 4AM5170

Task: The matching ratio calculated in example 5-9

$$\frac{w_1}{w_2} = \frac{i_2}{i_1} = \frac{2.202 \text{ A}}{0.719 \text{ A}} = 3.06 \text{ shall be produced.}$$

Solution: By trial and error it becomes apparent that the ratio 34/11 = 3.09 windings is sufficiently accurate.

The available sub-windings of the 4AM CT must now be combined in such a manner that the required number of turns is obtained. This is achieved with the following arrangement: $w_1 = 6+2+16 = 34$ to $w_2 = 1+2+7+1 = 11$ (refer to Figure 5.35)

Required operational accuracy limit factor for interposing CTs

In section 5.7 it was stated that, for saturation free transformation of fault currents with DC component, the current transformer must be over-dimensioned by the factor $K_{TF} = 1 + \omega \cdot T_N$.

The minimum operational accuracy limit factor resulting from this consideration applies to the main CT (M-CT for short) as well as the interposing CT (IP-CT for short):

$$\text{ALF}'_{\text{IP-CT}} \geq K_{TD} \cdot \frac{I_{\text{F-max.}}}{I_N} \qquad (5\text{-}44)$$

As such CT dimensions would imply unacceptably high cost; the protection must permit a larger degree of CT saturation.

Even the static analog protection was provided with additional stabilising against CT saturation effects. With numerical protection, the demands placed on the CT could be further reduced by intelligent numerical algorithms (saturation detector). The principle was described in section 4.2.4. (Refer to the examples in chapters 7 to 10.)

The manufacturer normally states the CT dimensioning requirements for each type of relay. (Refer to Table 5.3).

In general the interposing CTs should be dimensioned at least to the same K_{TD} as the main CTs.

How to determine the accuracy limit factor of the interposing CT 4AM 5170

The principle is the same as applied with main CTs. The ALF of this link selectable interposing CT (IP-CT for short) however is variable dependent on the selected ratio and the used sub-windings.

The magnetising characteristic of the IP-CT is therefore provided in general form with volts-per-turn and ampere-turns as scales. (Figure 5.36)

The magnetising voltage of the IP-CT is calculated as follows:

$$E_2 = I_2 \cdot (R_{w2} + R_B)$$
(5-45)

R_{w2} is the total resistance of the series connected secondary sub-windings (Figure 5.34) and R_B is the connected burden resistance.

We convert formula (5-45) into a more convenient form and get the corresponding volts-per-turn dependent on the magnetising ampere-turns:

$$\frac{E_2}{w_2} = (I_2 \cdot w_2) \cdot (R_{w2} + R_B)/w_2^2$$
(5-46)

Herewith, the necessary ampere-turns $I_{2m} \cdot w_2$ can be taken from the magnetising curve for given volts per second E_2/w_2. (Figure 5.36)

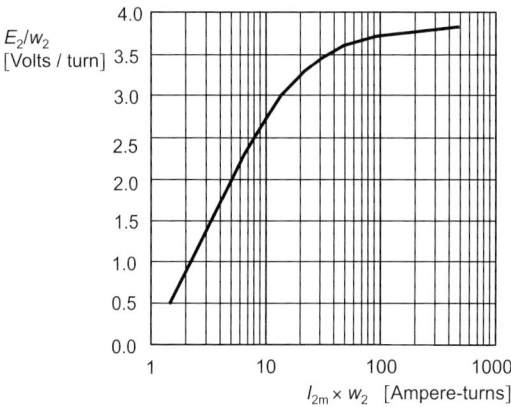

Figure 5.36 Magnetizing characteristic of an interposing CT (Siemens type 7AM5170)

To find the accuracy limit, we must assume increasing current values and use the try and error method. The 5P (10P) accuracy limit is reached when the magnetising ampere-turns exceed 5% (10%) of the ampere-turns of the assumed fault current.

The operational accuracy limit factor is then:

$$ALF'_{(IP\text{-}CT)} = \frac{\text{accuracy limit current}}{\text{rated current}}$$
(5-47)

Example 5-13: Checking the interposing CT configuration

Task: An existing protection device must have its sensitivity increased by increasing the current by a factor 2 by means of an interposing CT 4AM5170 connected in front of the relay.

It must be checked whether the interposing CT is sufficient for this purpose.

Given:
- Technical data of the interposing CT as shown in Figure 5.35 and magnetising curve according to Figure 5.36
- Burden of the protection including connecting cables: 0.2 Ω
- Main CT: 400/1 A
- Maximum fault current 10 kA
- The protection data sheet demands a transient dimensioning factor $K_{TD} \geq 2$ for non-delayed operation.

Solution: The interposing CT has a total of 52 turns.

To achieve as high a saturation voltage as possible, as many as possible turns must be applied. With a ratio of $w_1/w_2 = 32/16$ turns, i.e. 48 turns in total, the interposing CT is almost fully utilised.

To reduce the internal voltage drop across the secondary winding to a minimum, the sub-windings with the lowest resistances are used for the secondary side.

We select $w_2 = $ (C-D)+(E-F)+(N-O) $= 2+7+7 = 16$ turns

Thus: $R_{w2} = 0.025 + 0.08 + 0.08 = 0.185 \ \Omega$

Using the equation (5-46) derived above, we get:

$$\left(\frac{E_2}{16}\right) = (I_2 \cdot 16) \cdot (0.185 + 0.2)/16^2 = (I_2 \cdot 16) \cdot 0.0015$$

Due to the non-linearity of the magnetising curve, we must find the accuracy limit by trial and error.

For this purpose, we take as first estimation the given maximum fault current of 10 kA. The IP-CT secondary current is $I_2 = 10,000$ A/ $(400/1) \cdot (2/1) = 50$ A corresponding to $16 \cdot 50 = 800$ ampere-turns.

The secondary excitation voltage is:

$$\left(\frac{E_2}{16}\right) = (50 \cdot 16) \cdot 0.0015 = 800 \cdot 0.0015 = 1.2 \text{ Volts per turn}$$

From the magnetising curve we read $I_{2m} \cdot w_2 = 3$ ampere-turns.

The operating point at maximum AC fault current is in the linear range of the magnetising curve, the error is therefore negligible: $(3/800) \cdot 100 = 0.4\%$.

To reach the accuracy limit of the IP-CT, we must increase the fault current considerably.

By iteration we find that the 5% accuracy limit is reached at $I_{2m} \cdot w_2 = 2500$ ampere-turns corresponding to $I_{2\text{-AL}} = 156$ A.

The accuracy limit voltage per turn is about $E_{2\text{-AL}} = 3.7$ V, and the magnetising ampere-turns are $I_{2m} \cdot w_2 = 125$.

The accuracy limit factor in operation is herewith:

$$\text{ALF}'_{(\text{IP-CT})} = \frac{I_{2\text{-AL}}}{I_{2n}} = \frac{156}{2} = 78$$

The transient dimensioning factor is

$$K_{TD(IP\text{-}CT)} = \frac{ALF'_{(IP\text{-}CT)}}{I_{F\text{-max.}}/I_n} = \frac{78}{25} \approx 3, \qquad \text{i.e. greater than the required value 2.}$$

The interposing CT therefore has a sufficient dimension for this application.

Comment: For a rule of thumb calculation, a constant accuracy limit voltage could be assumed. Including some security margin, this would be about 3.5 volts per turn for the interposing CT type of the example.

The detailed calculation is only necessary in the borderline case.

6 Communications

For differential protection on cables and overhead lines measured values must be transmitted to the opposite line end over large distances. With conventional technology, this was done with low power analog signals (50 Hz or voice frequency). With state of the art technology, digital communication signals are employed (binary coded information). [6-16]

The signal transmission system consists of communication devices (this function may also be integrated in the protection) at each line terminal and the communication path (medium). With conventional differential protection, the only available communication medium is metallic pilot wires. Fiber optic cables (FO) and directional radio links have been increasingly employed in the last number of years. In this context, digital communication techniques are employed. This allows for the transmission of large data volumes with multiplexing. The dedicated point to point communication applied with conventional protection systems is increasingly being replaced by transmission channels of digital communication networks.

Apart from FO connections, the signal transmission paths are subjected to electromagnetic influence. On the one hand this is the system frequency based influence resulting from ground fault currents and on the other hand the influence from high frequency transient currents and voltages resulting from switching operation or lightning strikes. This noise must be considered when selecting and configuring the communication link. Screening and isolation (barriers) must be applied where necessary.

6.1 Transmission channels

Initially the features of the various transmission media are considered.

6.1.1 Pilot wires

Wire connections have a limited transmission bandwidth (3.4 kHz in the case of telephone twisted pairs). In the past they were used for differential protection with analog (50 Hz) measured value comparison and for phase comparison protection by implementation of modulated voice frequency signal transmission.

Only a few Kbit/s can be transmitted digitally in the base-band. Modern modulation techniques (multiple level instead of simple binary coding, phase modulation) allow for higher transmission rates of above 100 Kbit/s. The application of differential protection with digital communication via pilot wires of up to approx. 20 km is therefore now possible (refer to section 9.3).

The choice of pilot wires depends on the type of protection, the distance that must be spanned and the level of interference voltage that is expected.

On short links up to approx. 2 km, where the level of system frequency (50 Hz) interference voltage is not expected to be very large, normal control cables (e.g. NYY) with 2 kV rated voltage are suitable. They are often used with current differential protection (3 core differential protection).

On longer line lengths, pilot wire cables with symmetrically twisted cores must be used. The voltage influence on the cores due to earth currents must be minimised by positioning the pilot wire cable away from the power cable or by means of metallic conducting screens [6-2]. Wherever possible, high voltage isolated communication cables (special protection class cables) should be applied. This is particularly so when the cable runs parallel to a high voltage cable or overhead line. Additional isolation is also required at the termination of the pilot wire cable in the high voltage substation to prevent breakdown due to the transverse voltage induced in the pilot wire pair by earth short circuit currents (see below).[1]

The pilot wires are usually isolated against ground ("floating"). The terminal devices (protection devices or communication equipment) are thereby symmetrically connected. In Figure 6.1 an example of a differential protection for two pilot wire cores (telephone twisted pair) is shown. The cable screen is earthed on both sides so that it effectively eliminates inductive interference. [6-3]

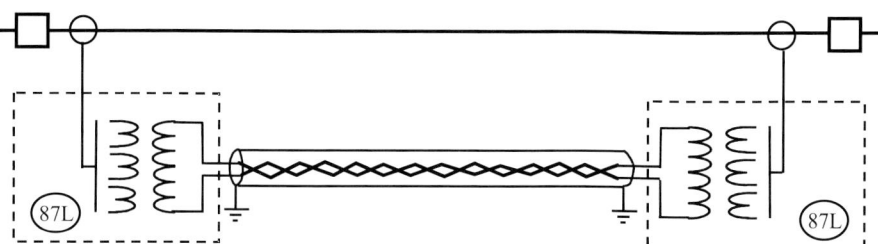

Figure 6.1 Line differential protection with pilot wires

The induced transverse voltage drives a charging current via the earth capacitance of the cable cores so that the transverse voltage to earth, shown in Figure 6.2, arises. The areas F_1 and F_2 above and below a potential of zero correspond to the product of line length and voltage to earth at the corresponding location. On non-earthed cables they are always the same.

Assuming a constant level of interference along the entire pilot wire the potential to ground is distributed symmetrically at both cable ends. Therefore only half the induced voltage between the terminals of the cable appears as shunt voltage to the screen or earth.

---

[1] Earth short circuit currents occur in earthed systems during single and two phase earth short circuits, in isolated or resonant grounded systems only during double earth fault (cross-country fault).

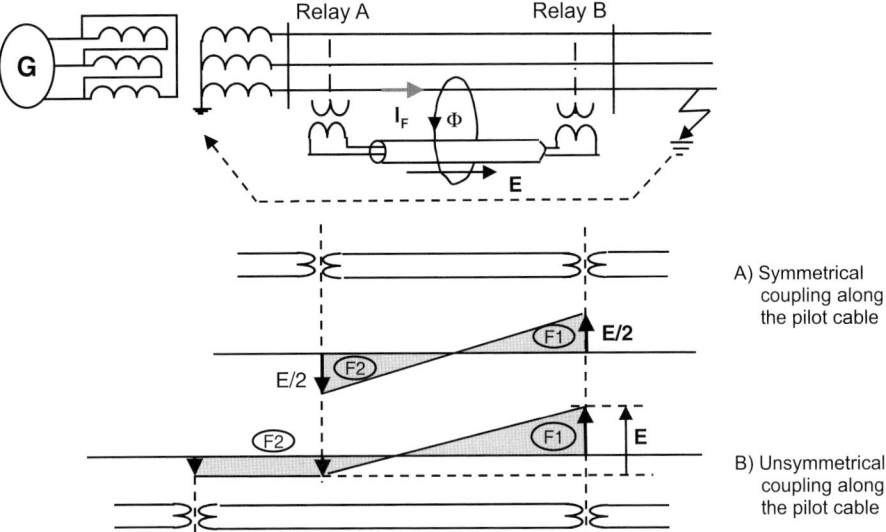

A) Symmetrical coupling along the pilot cable

B) Unsymmetrical coupling along the pilot cable

Figure 6.2 Inductive interference of wire pilot cables

For an uneven distribution, for example if at one end the pilot wire continues, a non-symmetrical distribution of the potential will result.

Network protection cable

Special network protection cables with different numbers of twisted pairs and triplets can be supplied (Figure 6.3).

Example: Type AD-2YF(L)2Y $\dfrac{3 \times 3 \times 1.4/2.8}{3 \times 2 \times 0.8P(M60)}$

The pairs and triplets are separately twisted and in PE insulation. The pilot cores are enclosed with a screen consisting of plastic coated aluminium tape and an external PE sleeve.

Figure 6.3 Construction of a protection cable (cross section and side view)

Table 6.1 shows the typical parameters.

119

Table 6.1 Network protection cable, PE-insulated with copper core triplets and pairs for direct installation in the ground or pulling into pipes.

Pilot core-diameter	Maximum Pilot wire resistance		Maximum operating capacity	Rated voltage (RMS)				
mm	Core Ω/km	Loop Ω/km	nF/km	Core-Core kV	Core-Screen kV	Triplet-Triplet kV	Pair-Pair kV	Pair-Triplet kV
1.4	11.9	---	---	2.5	8	8		8
0.8	---	73.2	60	2	2		2	

The permissible temporary induced voltage according to VDE 0228, Part 1[1] is 60% of the rated AC voltage between core and screen, i.e. for the cable described above, a voltage of 4.8 kV for the triplet cores and 1.2 kV for the twisted pairs. If the voltage between the terminals is larger, the cable must be split and barrier transformers have to be applied.

In the example in question (Table 6.1) one of the triplet cores with increased isolation is designated for the current differential protection. Other core combinations, for example with high voltage isolated twisted pairs can however also be provided by the manufacturer.

The protection devices themselves are usually designed with an AC rated voltage of 2 kV so that barrier transformers must in any event be applied at the cable terminals if the induced voltage across the cable exceeds 1.2 kV. Common isolation voltages of the barrier transformers for this purpose are 5 or 15 kV.

In the event of non-symmetry, the voltage induced on the pilot wire core differs, whereby the voltage difference appears directly in the pilot core loop and thus influences the protection measurement. Relay specific limits may not be exceeded here.

By careful stranding a symmetry factor of 10^{-4} (80 dB) at 50/60 Hz or 10^{-3} (60 dB) at 800/1000 Hz is obtained in network protection cables. This corresponds with the specification of telephone cables (cross talk rejection better than 80 dB). Therefore, protection and telephone channels or multiple protection systems may be operated on parallel cable cores or triplets in the same pilot cable. The influence on the measurement (transverse voltage) is kept small ($T_{ransverse} \leq 10^{-4} \cdot U_{lateral}$), even in the presence of large induced lateral voltages. The screen consisting of plastic coated aluminium tape however only has limited effectiveness (reduction factor $k_r = 0.9$).

The pilot wire pairs (telephone twisted pair) are suitable for the transmission of analog signals with frequencies ranging from low frequency up to approximately 4 kHz (differential protection, phase comparison protection, voice frequency inter-

---

[1] German Standard VDE 0228, Part 1 (December 1987): Proceedings in the case of interference on telecommunication installations by power installations, General. See [6-1].

trip devices). They can only carry limited current levels (some tens of mA), so that differential protection can only be implemented using the voltage comparison protection (two core pilot wire differential protection) technique.

For digital signal transmission, modems must be applied. In the past they were capable of transferring only at rates of some tens of Kbit/s. With a newly developed technique it is even possible to transmit 128 Kbit/s up to a distance of approximately 20 km. (See section 9.3).

The triplet (three core pilot cable) with increased cross-section is intended for the current differential protection (three core pilot wire differential protection). The current in the pilot wire cores in this case is approximately 100 mA at CT nominal current.

Leased telephone lines

Twisted pairs can also be leased from telephone companies. In some countries this has been common practice. This may have the advantage that the pilot wire connection is not in the vicinity of the high voltage overhead lines or power cables and is therefore not significantly influenced by power system interference. On the other hand reduced security against manipulation of the pilot wires must be accepted.

Two aspects in respect of pilot wire cores with low level insulation deserve special attention:

1. Normal telephone cores are only designed for voltage ratings ≤ 2000 V and are usually provided with arrestors for protection against over-voltages. [6-4] For this purpose the arrestors are connected at each terminal between the core and earth. When two arrestors at one terminal conduct at the same time a short circuit of the pilot wires may be induced. This is not allowed for protection applications, as even a very brief short circuit of the pilot wire cores can result in unwanted tripping of the protection. It must therefore be ensured that the cores applied for protection purposes do not have any arrestors connected to them[1]. Cores with higher voltage ratings may have to be applied.

   Over-voltages that originate in the substation must be kept away from the telephone cores by connecting barrier transformers.[2]

2. In Germany, the operating voltage (maximum transverse voltage) must according to VDE 0816, Part 1 [6-5] be restricted to a peak value of 225 V[3]. The differential protection must therefore be designed such that the voltage between the two cores remains below this value when the largest short circuit current flows or, alternatively, a varistor must be applied to limit the voltage.

---

[1] Twisted pairs used for telephony that are in parallel in the pilot may be protected with arrestors. Unused cores may be earthed at both ends as they then act as reduction circuits.

[2] In the UK for example, where leased signal channels are often used, all components that are connected to public telephone cables have to be isolated from the substation equipment (CT's, battery circuits, etc.) with 15 kV barrier transformers.

[3] In the UK a restriction of the voltage to 130 V (peak value) is specified. The current may not exceed 60 mA (rms).

Interference induced by the earth short circuit current

The pilot wire circuits are mainly affected by the large short circuit currents flowing via earth. [6-1 to 6-3] In this context there are two methods by which the overvoltages are induced:

- by the potential difference (ohmic coupling) via the station earthing
- by magnetically induced lateral voltages

Ohmic voltage coupling

Figure 6.4 shows a "voltage funnel" that arises in the vicinity of a substation during an earth short circuit.

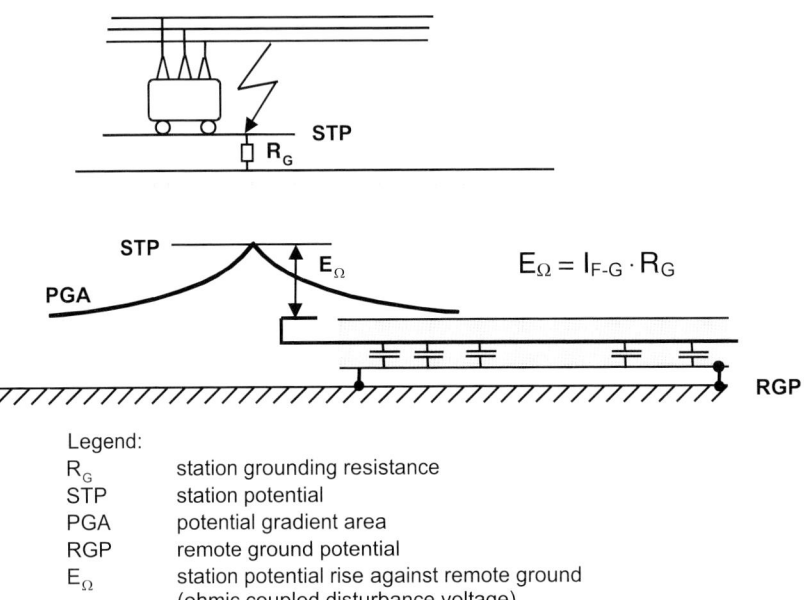

$$E_\Omega = I_{F\text{-}G} \cdot R_G$$

Legend:
R_G	station grounding resistance
STP	station potential
PGA	potential gradient area
RGP	remote ground potential
E_Ω	station potential rise against remote ground (ohmic coupled disturbance voltage)

Figure 6.4 Risk of damaging the pilot cables due to the "voltage funnel" resulting from an earthing resistance of the substation that is too large

Due to the voltage drop across the station earth a difference between the earth potential in the station (device earth, screen) and the remote end (core potential) arises. For example, with a station earth resistance of $R_G = 0.5\ \Omega$ and an earth short circuit current of $I_{F\text{-}G} = 10\ \text{kA}$ a potential difference $E_\Omega = 10000\ \text{A} \cdot 0.5\ \Omega = 5000\ \text{V}$ would arise. The pilot wire core, acting like a sensor, brings the potential zero from a large distance right into the centre of the "voltage funnel". A pilot wire cable with 2 kV insulation level between the core and screen, or a protection relay with 2 kV isolation, would in this case already experience flash-over. The station earthing would have to be improved, or a barrier transformer would have to be applied for isolation of the high voltage.

A conducting sheath (good conductor) can also be fitted to the pilot cable in the vicinity of the station. The field of the short circuit current flowing in the screen induces a lateral voltage in the cores that reduces the voltage difference against the station earth or between the cores and the cable screen. In the USA neutralizing reactors are used at the point of entry to the substation Figure 6.5.

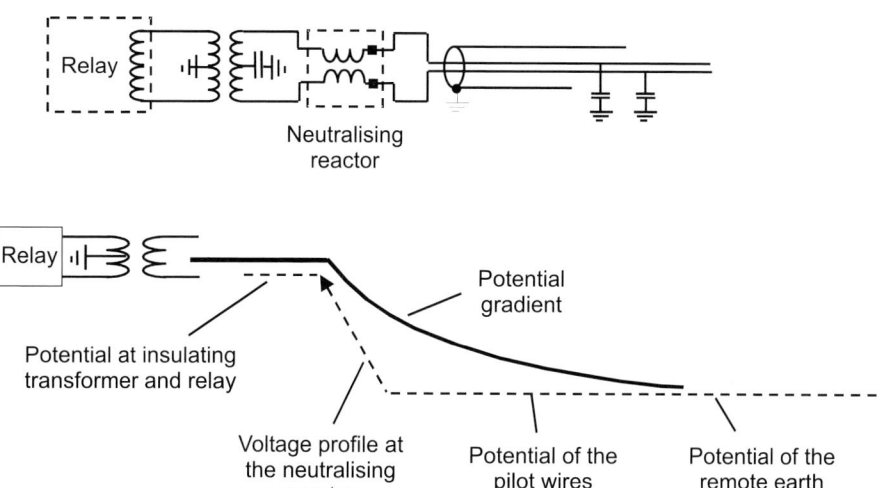

Figure 6.5 Application of a neutralising reactor for potential balancing

They are connected in series with the pilot cores and serve to compensate the potential difference. [6-6 to 6-8]

In this case the voltage difference at the "voltage funnel" causes currents to flow in the same direction through both windings of the reactor, returning via the capacitance to earth of the pilot cable cores and the additional capacitance at the centre tap of the barrier transformer. The voltage drop across the large unloaded inductance of the reactors then bridges the voltage difference between the remote earth (pilot cores) and the station earth

The current in the pilot wire cores of the differential protection is not affected by the neutralizing reactors as it flows through the windings in opposite directions and therefore only encounters the small short circuit reactance.

Inductive voltage coupling

If the pilot cable is close to a high voltage overhead line or a high voltage cable, lateral voltages are induced in the cores by earth short-circuit currents. These can also be several kV (see Figure 6.2). Guidelines for the computations of these guidelines are given in the German standard VDE 0228, Part 1. [6-1]

For detailed computation this standard should be referred to. Only the general equation is given here:

$$E_i = 2\pi \cdot f \cdot w \cdot r \cdot I_{FE} \cdot M_{LL\text{-}E} \cdot l \cdot 10^{-6} \qquad\qquad (6\text{-}1)$$

Whereby:

E_i Induced lateral EMF in V

f Nominal frequency of the HV plant in Hz

I_{FE} Inducing earth short-circuit current ($= 3 \cdot I_{F\text{-}0}$) in A

w Probability factor for I_{FE} (considers the fact that for the computation of I_{FE} the most unfavourable conditions are assumed, the coincidence of which factors is highly unlikely. For protection application of the pilot cables the value $w = 1$ must be applied.)

$M_{LL\text{-}E}$ Referenced coupling inductance between two individual conductors with return via earth in µH/km (see Figure 6.6)

l Effective length with this spacing in km (for different distance to the influencing feeder, the computation must be carried out in several separate sections)

r Resulting reduction factor: $r = r_K \cdot r_S \cdot r_E \cdot r_X$ for cable armouring, tracks, earth wires and other neighbouring metallic conductors

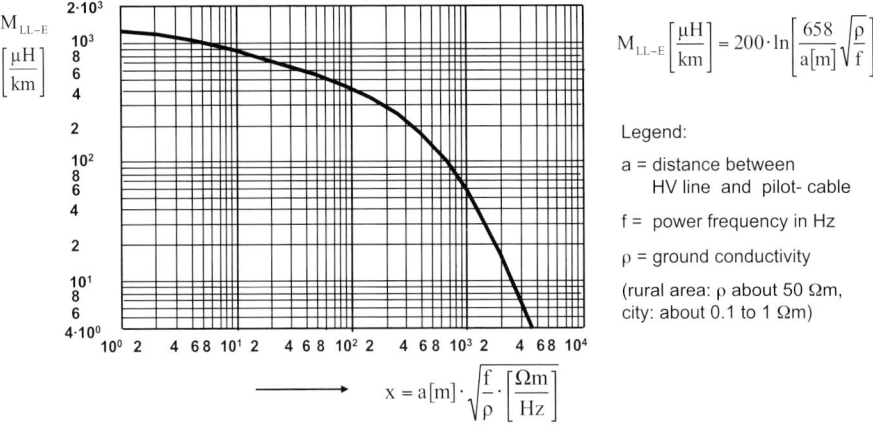

$$M_{LL\text{-}E}\left[\frac{\mu H}{km}\right] = 200 \cdot \ln\left[\frac{658}{a[m]}\sqrt{\frac{\rho}{f}}\right]$$

Legend:

a = distance between
      HV line and pilot-cable

f = power frequency in Hz

ρ = ground conductivity

(rural area: ρ about 50 Ωm,
city: about 0.1 to 1 Ωm)

$$x = a[m] \cdot \sqrt{\frac{f}{\rho}} \cdot \left[\frac{\Omega m}{Hz}\right]$$

Figure 6.6 Coupling inductance between two simple conductors with return via earth

The influencing voltage is therefore proportional to the earth fault current, the coupling inductance (proximity of the pilot cable to the HV-feeder) and the length over which this proximity of pilot cable and HV-feeder exists. A reduction of the influencing voltage is obtained by short-circuit loops located in parallel. A current that partially compensates the interfering field is induced in these. This is expressed by the reduction factors mentioned above (Figure 6.7).

The level of reduction introduced by overhead line earth wires and the conducting screens (armouring) of high voltage cables depends on their ohmic resistance (material, cross section) and lies in the range from 0.9 (steel earth wire) to 0.2 (lead cable screen with steel armouring). Railway tracks can introduce an additional reduction factor of between 0.8 and 0.2.

On the communication cable which is the influenced circuit, the influencing voltage may be further reduced by earthing the cable screen at both ends.

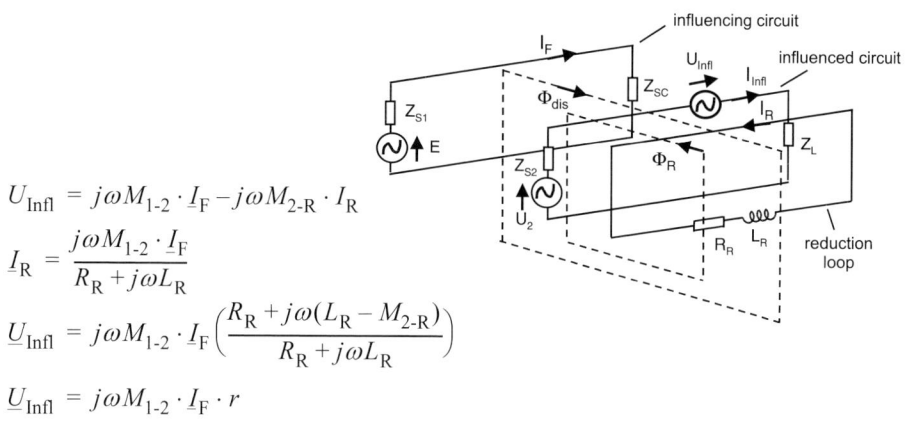

$$U_{Infl} = j\omega M_{1\text{-}2} \cdot I_F - j\omega M_{2\text{-}R} \cdot I_R$$

$$I_R = \frac{j\omega M_{1\text{-}2} \cdot I_F}{R_R + j\omega L_R}$$

$$U_{Infl} = j\omega M_{1\text{-}2} \cdot I_F \left(\frac{R_R + j\omega(L_R - M_{2\text{-}R})}{R_R + j\omega L_R} \right)$$

$$U_{Infl} = j\omega M_{1\text{-}2} \cdot I_F \cdot r$$

Figure 6.7 Reduction of the influencing voltage by screens (earthed on both sides) and reduction conductors (r = reduction factor)

On plastic cables (see above) the effectiveness of the screen is small ($r = 0.9$). However on communication cables with conducting screen (e.g. aluminium) and steel tape armouring reduction factor values that are similar to those of high voltage cables, down to approximately $r = 0.2$ are obtained. [6-9]

Example 6-1: Pilot cables for pilot wire differential protection

Given: 20-kV-three core single sheath cable NEKBA

Earth short circuit current 4 kA

Reduction factor of the cable: $r_K = 0.7$

Length of the parallel course: 6 km

As it involves a built up area an additional reduction factor of approx. $r_X = 0.5$ may be included for pipes, tracks etc.

Protection cable according to Table 6.1 (Reduction factor $r_K = 0.9$)

Distance of power cable – protection cable: 1 m

Wanted: Influencing voltage on the pilot wires

Solution: According to Figure 6.6 the mutual inductance is $M_{LL\text{-}E} = 1300\ \mu H/km$

With equation (6-1) the following is then obtained:

$$E_i = 2\pi \cdot 50\frac{1}{s} \cdot 1 \cdot 0.7 \cdot 0.5 \cdot 0.9 \cdot 4000\ A \cdot 1300 \cdot 10^{-6}\frac{H}{km} \cdot 6.0\ km =$$
$$= 3086\ V$$

Sum of the influencing voltages

Depending on the substation and system conditions, the magnetically induced lateral voltage E_i (Figure 6.2) and the influencing voltage due to ohmic effects E_Ω (Figure 6.4) may appear separately or together. As the voltages have a phase displacement of approximately 90° with respect to each other, the addition should be carried out by vector summation:

$$E_\Sigma = \sqrt{E_\Omega^2 + E_i^2} \tag{6-2}$$

In the pilot cable, the insulation of the core to the screen must be selected for this voltage sum. If the voltage exceeds 1.2 kV (60% of 2 kV) then the protection and inter-tripping devices must be higher insulated (e.g. 5 kV) or connected via barrier transformers to the pilot cores.

Measures against voltage interference

The following influencing factors with positive effect must be included in the planning:

a. Attempt to achieve the lowest possible earthing resistance in the HV substations.

b. Ensure that the pilot cables in the proximity of HV substations are fitted with a metal sheath having very good conducting capabilities. (The field of the current flowing in the sheath induces a voltage on the core that reduces the potential difference between the core and the sheath in the "voltage funnel".[1])

c. Maintain as large a separation as possible between the HV equipment and the pilot cables.

d. Try to make the reduction factors as small as possible on the influencing side (cable metal armouring, cable screens, earth wires).

e. Small reduction factors on the influenced side are obtained by metal armouring and conducting screens on pilot cables, earthing of unused cores at both ends, additional reduction conductors (co-ordinated with f.).

f. Use pilot cables with high insulation levels (insulation between protection core-screen and protection core-parallel core) (co-ordinated with e.).

g. Ensure that the pilot cores are symmetrical (twisted).

h. Apply barrier transformers if needed.

Monitoring of the pilot cores

Pilot cores can be damaged for example by earth moving. In particular on long distances a continuous monitoring is therefore important.

Older pilot wire differential protection devices applied superimposed DC currents of a few mA for this purpose. This however had the disadvantage that a galvanic connection to the pilot wires had to be present. Barrier transformers could not be applied.

Modern digital relays achieve the monitoring with superimposed AC current signals in the voice frequency spectrum so that barrier transformers can still be used. In the case of phase comparison protection or with the signal communication device that operates in the voice frequency spectrum, the keyed measuring signal or the quiescent frequency may be directly used for the monitoring. For signal

---

[1] In the USA neutralising reactors connected in series with the pilot wire cores are applied to compensate the potential difference in the „voltage funnel". [6-6 to 6-8]

transfer with DC voltage, a circuit with current flow in the quiescent state should be applied.

With numerical signal transfer, the monitoring is contained in the protection devices.

6.1.2 Fiber Optic Cables

Fiber optic cables (FO) have found wide application as a communication medium since the nineties.[1] They provide a practically interference free signal transmission with very large band-width. Data transmission is in digital form with data rates ranging up to several gigabit/s.

Electric utilities apply FO-cables as buried cables, self-supporting aerial cables or in the earth wire of overhead transmission lines. The FO cable can also be strapped on to the earth or phase conductor of an overhead line, so that it does not have to provide its own mechanical support (latch cable). The cables may contain up to 96 individual FO fibers. [6-10]

In general the FO cables are used in a multiplexed mode within a communication network, providing several services. Only in exceptional cases are dedicated fibers for the protection application available.

Firstly the protection is connected via a standardised electrical interface (X.21 or G.703) to the multiplexer. The data transfer is then executed by the communication authority.

Secondly, a dedicated FO interface is available at the protection device for direct FO connection. For transmission over larger distances, a special protection data communication device may be inserted in the FO channel.

FO data networks

The 64 Kbit/s voice channel is the basis for the digital communication in data communication networks. This results from the sampling of an analog voice frequency signal at a sampling rate of 8 kHz (i.e. every 125 µs) and a resolution of 8 bits.

Based on this base channel of $8 \times 8 = 64$ Kb/s a hierarchy of multiplexing systems is available starting with the PCM-30 system which provides thirty 64 Kb/s channels (+2 auxiliary channels) in time multiplex form via a 2.048 Kb/s link. This is commonly referred to as transmission at 2 Mb/s.

By further multiplexing, systems with a larger number of channels at 8, 34, 140 or 565 Mb/s have been standardised in the legacy PDH (Plesiochronous-Digital-Hierarchy) data networks.

Modern SDH (Synchronous-Data-Hierarchy) networks now allow data transmission up to 10 Gab/s. [6-11 to 6-13, 6-21]

---

[1] The first FO link for the Deutsche Telekom AG was installed by Siemens in 1977.

FO Basics

The FO consists of a core that is surrounded by a sleeve (cladding) and a coating that provides mechanical protection and increases the tensile strength. The core is the central part of the FO cable and provides the channel for guiding the light signal. The light signal is only restricted to the core because the refraction index of the core n_1 is larger than the refraction index of the cladding n_2. The light waves are subjected to total reflection at the boundary between core and cladding, thereby propagating inside the core with several modes. For transmission of protection signals between the line ends, only glass fibers can be used as these have a sufficiently small attenuation of the light signal. Plastic fibers are only suitable for short distances (smaller than 100 m).

Principally, two types of FO cables are differentiated:

– *Multi-mode fibers*
  In this case, a large number of discrete light waves propagate which in sum total provide the signal transmission.

– *Mono-mode fibers*
  In this case only one light wave is capable of progressing.

Multi-mode fibers may have a stepped index profile having an abrupt change of the optical density between core and cladding, or a graded profile index which has a continuous reduction of the core density proportional to the radius.

In practice the following types are applied:

Graded index fibers
for short distances < 1.5 km (connections within the substation e.g. between protection and data terminal equipment)

Mono-mode fibers
for greater distances (up to approx. 100 km) from one line end to the other i.e. also for the line differential protection.

Graded index fibers

The refraction index profile, which has a parabolic shape causes the propagation path of the individual light waves to be sinusoidal (Figure 6.8, A). The propagation time difference of the light wave along the individual paths therefore is only 0.1 ns based on an overall propagation time of 5 μs/km. The typical dimension of a FO with graded index profile is 50 μm for the core and 125 μm for the cladding diameter.

Mono-mode fibers

The light wave propagation is along the axis of the fiber (Figure 6.8, B). A pre-requisite for this is that the core diameter is not much larger than the wave length of the light signal that is used. For example not greater than 9-10 μm for light waves with 1300 nm wave length. Mono-mode fibers always have a stepped index. The mono-mode fiber provides excellent transmission characteristics for the light signal (very small attenuation and negligible dispersion). Due to the small core diam-

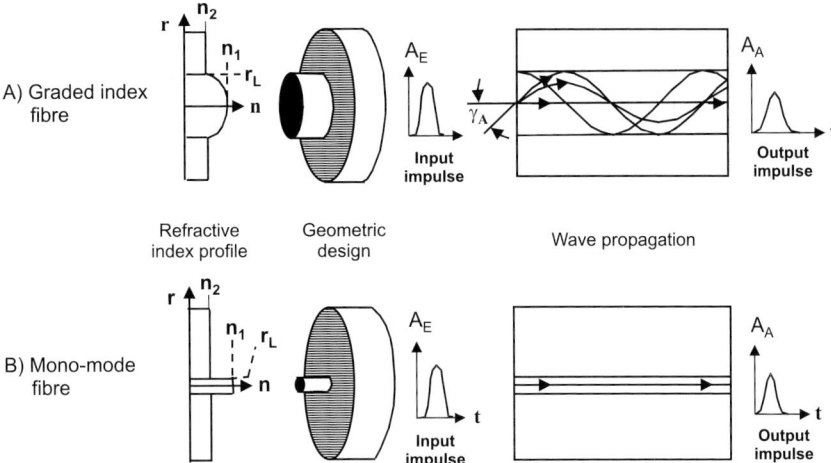

Figure 6.8 Wave propagation in FO fibres (principle)

eter, the application is more complex. Coupling the required signal level at the sending end is more difficult and requires the application of a laser diode instead of the less complex and less costly LEDs.

FO-attenuation (damping)

The loss of light signal strength in the FO is indicated in db/km. The optical damping is mainly caused by:

- Dispersion resulting from non-homogeneous propagation (Rayleigh-dispersion),

- Absorption due to contamination of material.

While the Rayleigh-dispersion is unavoidable due to the nature of things, the absorption losses are an indication for the quality of the FO and can be reduced by improved manufacturing techniques.

Both attenuation causes are dependent on the optical wave length of the applied signal. The Rayleigh-dispersion is inversely proportional to the wave length to the power of 4, while the absorption losses increase sharply (resonance losses) at certain wave lengths (Figure 6.9). As shown in the diagram, there are 3 wave length windows with particularly small attenuation losses at 850 nm, 1300 nm and 1550 nm. The common transmission systems used today operate at these wave lengths. The 850 nm systems utilise graded index fibers, while the 1300 and 1550 nm systems apply mono-mode fibers. In particular the 1550 nm applications currently have an attenuation of approx. 0.3 db/km which approaches the theoretical minimum of 0.12 db/km of silicone glass.

Larger wave lengths are not practical for silicone glass as the infra-red absorption causes a sharp increase of the attenuation.

Larger losses due to bending (creasing) during the installation of the FO are not expected. Only severe bending by approx. 1 mm would lead to a significant increase of the attenuation.

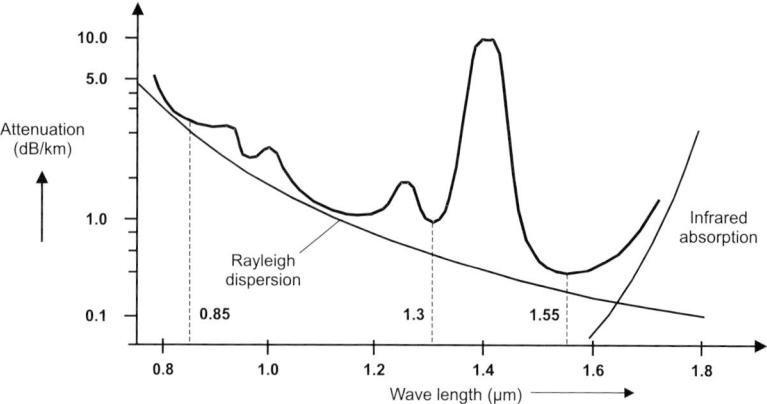

Figure 6.9 Attenuation of the mono-mode fibre depending on the wave length

FO transmitter and receiver

Light transmitters may be implemented with light emitting diodes (LEDs) and laser diodes (LDs). The transmitter must have a large radiation density and the emitting surface should be smaller than the FO core, if possible. The rated emission of an LED is below 5 mW, while laser diodes emit approximately 3 times more.

Receivers (opto-electric converters) are photo diodes of various types. Pin diodes or the much more sensitive avalanche diodes are common.

Planning a FO link

A FO cable installation consists of the laid and spliced together cable sections and ends at the data communication device plugs at the two cable ends.

Only the attenuation is considered here. Due to the slow data rates used by the protection, this is the decisive factor. With higher optical modulation rates in the megabit range, which is common with telecommunication application, the bandwidth of the optical fiber must also be considered.

Reference literature is recommended in this regard. [6-10]

Attenuation plan

The attenuation α_{Total} of a cable installation comprises the cable length l with a damping co-efficient α_{FO} and the number n of splice attenuations α_{SPL}. The attenuation of the plug connectors α_{PC} must be included in the above.

The number of splices should be kept to a minimum by making sure that the individual sections are as long as possible.

As cable installations are planned with a long life expectancy, allowance must be made for additional repair splices α_{RES}. These may be critical when the cable is damaged during building or earth moving measures, or when the cable is re-routed by inserting new cable sections. The margin allowed must be determined by the end user.

The following guidelines may be used for rough estimation:

Table 6.2 Light attenuation, rough estimation

Component of the FO system:		FO attenuation:
Mono-mode-fibre	at 1300 nm	$\alpha_{FO} = 0.45$ dB/km
	at 1550 nm	$\alpha_{FO} = 0.30$ dB/km
Graded index fibre	at 850 nm	$\alpha_{FO} = 2.5$ to 3.5 dB/km
	at 1300 nm	$\alpha_{FO} = 0.7$ to 1.0 dB/km
Per splice		$\alpha_{SPL} = 0.1$ dB
Per plug connector	ST	$\alpha_{PC} = 0.2$ to 0.4 dB
	LC	$\alpha_{PC} = 0.1$ to 0.3 dB
Reserve		$\alpha_{RES} = 0.1$ to 0.4 dB/km

The total attenuation of the cable system is obtained as follows:

$$\alpha_{Total} = l \cdot \alpha_{FO} + n \cdot \alpha_{SPL} + 2 \cdot \alpha_{PC} + l \cdot \alpha_{RES} \qquad (6\text{-}3)$$

Example 6-2: Computation of the maximum distance that can be bridged by the line differential protection 7SD61

Given: Laser type fiber optic sender/receiver interface with an optic budget (maximum difference between sender and receiver) of 29 dB

Optical wave length: 1550 nm

The FO cable is supplied in 2 km segments.

A reserve margin of 0.2 dB/km was selected.

Wanted: Maximum distance that can be spanned

Solution: The cable length is: $l = x \cdot 2$ km

The number of splices is: $n = x - 1$

The permissible attenuation (optical budget) is 29 dB

From formula (6-3) the following is then obtained:

$$29 \text{ dB} = x \cdot 2 \text{ km} \cdot 0.3 \frac{\text{dB}}{\text{km}} + (x - 1) \cdot 0.1 \text{ dB} + 2 \cdot 0.5 \text{ dB} +$$
$$+ x \cdot 2 \text{ km} \cdot 0.2 \frac{\text{dB}}{\text{km}}$$

which results in: $x \approx 25$ equivalent to: $l \approx 50$ km

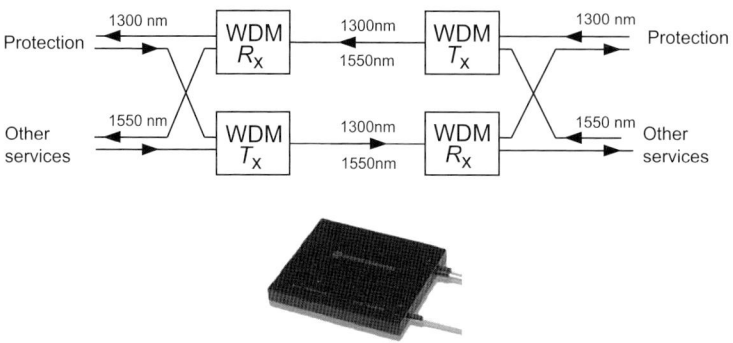

Figure 6.10 Wavelength Division Multiplexing: Principle and module sample

Wavelength division multiplex

The capacity of a fiber optic pair can be multiplied by wavelength division multiplexing (WDM).This method assigns services to different light wavelengths in the same way as traditional frequency division multiplexing. Light sensitive filters combine light wave lengths at the sending end and separate them at the receiving end. (Figure 6.10)

The WDM modules are passive and need no external supply. The operation principle is based on the prisma effect.

The WDM filter introduces some additional margin which reduces the path length to the wave length with the highest losses. Transmission ranges of up to about 60 km can be achieved without regenerators.

It is good practice to design light wave systems with enough gain to allow the future use of WDM for extension of the transmission capacity.

6.1.3 Line of sight radio links

Directional radio links operate with wave lengths in the centimetre band (1 to 20 GHz) and implement fixed transmitters and receivers. A high directionality is achieved by using parabolic antennae with a diameter of 1 to 2 m that focus the transmission signal. A line of sight link (approx. 50 km) is required, as the directional radio waves do not follow the curvature of the earth. If the required link distance is greater than line of sight, then intermediate repeaters are required. (Figure 6.11)

The advantage of directional radio links is that their course is independent of that of the HV feeder, and that they are therefore not influenced by faults or switching operations. On the other hand, atmospheric interference (heavy rain) and reflections may cause fading.

It is common practice to use digital radio systems that utilise PCM, similar to FO transmission. These broad-band microwave radio links are usually part of the utility private digital network. The protection utilises individual 64 Kb/s-channels. The connection is usually made via an interface having the X.21 or G.703 standard.

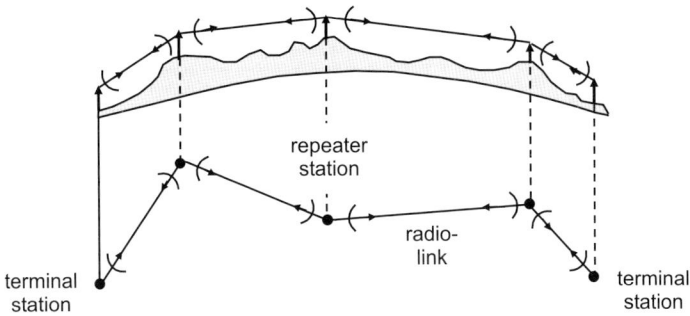

Figure 6.11 Radio (micro-wave) link

6.2 Digital protection communication

Modern digital protection devices provide a large number of communication options [6-16]:

Serial interface for local servicing via PC

This interface is usually located on the front of the relay and has not been standardised. Siemens relays utilise the interfaces V.24/V.28 (up to SIPROTEC 4) and newly USB (SIPROTEC 5). The protocol structure complies with IEC 870-5-103.

Service interface for remote access

In this case, a RS 485 daisy chain or an optical interface is used. The relays of a station may be coupled via star coupler to a central modem with telephone link for remote operation. The protocol is the same as that of the front service interface.

System interface

This interface provides the connection of the protection relay to a control system (substation or network control device). A number of non-compatible, legacy protocols have been used, including IEC 870-5-103 or Profibus in Europe and DNP3.0 or Modbus in the USA. In 2005 the standard IEC 61850 became valid. It is now the world-wide accepted single standard for substation communication.

Protection data interface

It is intended for the communication with the protection at the remote terminal. No international standard exists for this. The transfer of measured values by the differential protection is in any event specific to each vendor.
This connection between two protection devices must be very carefully planned as the reliability of the protection system directly depends thereon. This is particularly so when the dedicated transmission link is subjected to interference and when transmission through public data networks is used. Telegram structures and

transmission techniques must be designed with particular consideration being taken of interference immunity and availability.

Data transmission techniques

Data transmission follows a fixed step by step procedure. [6-11, 6-12]

The information that needs to be transferred may be visualised as a sequence of bits (digitised measured values, coded control signals, text in ASCII code etc.) In this regard 8 information bits are combined to a byte (Octet). A complete message is generally made up of several bytes[1].

The sequence of execution during serial transfer is shown in Figure 6.12.

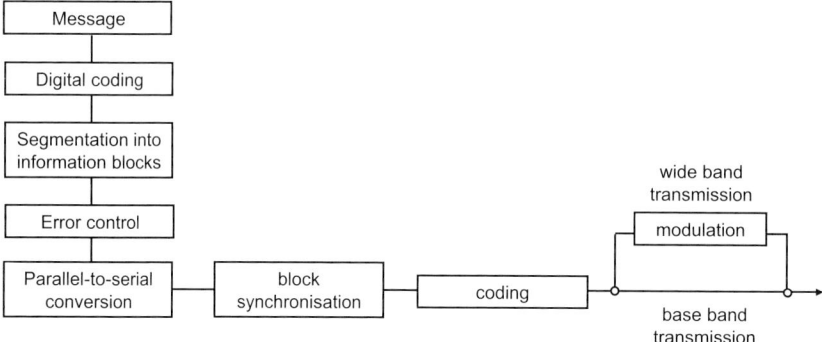

Figure 6.12 Sequence of execution during serial data transfer (sending end)

The message must initially be split into information blocks, which may also be of variable length but have a defined maximum length. The information blocks are then supplemented with test bits for error detection (redundant coding) which allow for the detection of transfer errors at the receiving end.

Synchronisation of the blocks within the bit stream is necessary to ensure that the beginning of the block and the end of the block are clearly identified at the receiving end. Thereby the individual elements of the block (address, data, etc.) which are referenced to the beginning of the block can be decoded correctly. (Figure 6.13)

The subsequent coding and modulation are required for the physical signal conditioning necessary in transmitting the signal across the communication channel.

At the receiving end, corresponding stages (demodulation, decoding and serial/parallel conversion) are applied to recover the message.

---

[1] The smallest information element is a bit. Several bits make up a character. A character with 8 bit length is designated as a byte. Several bytes are combined to make up information or broadcast block.

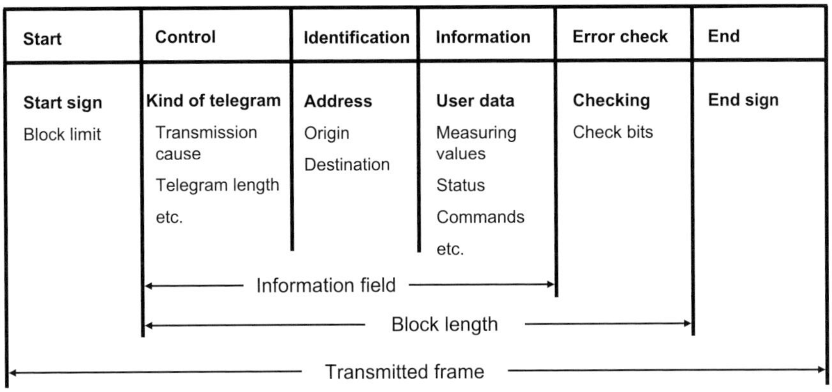

Start	Control	Identification	Information	Error check	End
Start sign	Kind of telegram	Address	User data	Checking	End sign
Block limit	Transmission cause	Origin	Measuring values	Check bits	
	Telegram length	Destination	Status		
	etc.		Commands		
			etc.		

Information field

Block length

Transmitted frame

Figure 6.13 Format of a telecontrol message

Signal coding

For digital information transfer, the logic states "0" and "1" must be represented by electrical signal states, i.e. pulse code modulation (PCM). The smallest unit of a digital signal is called a code element. In general several definitive states may exist. A code element with two states is binary. The duration T of a code element determines the signal speed:

$$v_S = \frac{1}{T} \, [\text{Baud}] \tag{6-4}$$

If the code element contains several definitive states, that are associated with individual information bits (logic state "1" or "0"), then the resulting transmission speed (equivalent bit rate in bps[1]) may be a multiple of the signal speed:

$$v_T = v_S \cdot \lg_2(n) \, [\text{Baud}] \tag{6-5}$$

With binary codes, having only two definitive states, the bit rate equals the signal speed.

The (non-modulated) baseband transmission through wire bound interfaces and lines (cables) requires special coding of the binary data. The code must in most applications be d.c. free to allow transmission through transmitters (I/O-transformers) and it should contain enough signal changes to allow clock regeneration at the receiver.

The following line codes are common (Figure 6.14):

The NRZ-Code (Non-Return-to-Zero) has the simplest form. The pulse length of the rectangular impulses corresponds to a code length. With this code, a sequence of logical "1" produces a continuous signal with a fixed polarity. The signal therefore

---

[1] Bits per second

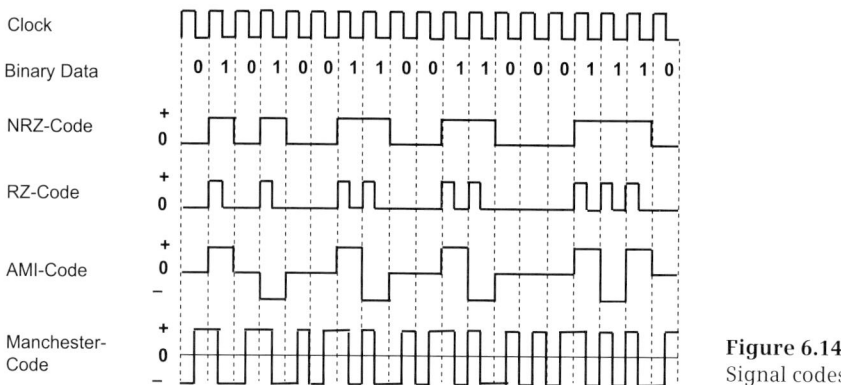

Figure 6.14
Signal codes

contains a DC component and also does not facilitate the clock recovery at the receiving line end.

The <u>RZ-Code</u> (Return-to-Zero) has similar properties, however during a sequence of logical "1"-states, the clock signal is transmitted.

The <u>AMI-Code</u> (Alternate Mark Inversion) has three definitive states, which are however only used to represent two discrete values. By alternating the polarity of the logical state "1" the signal does not contain a DC component. The clock information is only transferred with the signal state "1". This code is for example used at the G.703 interface of data networks.

The <u>Manchester-Code</u> signal consists of two halves which are displaced by 180°. The clock frequency is therefore twice the signal speed, so that twice the bit rate or band width of the communication channel is required for the information transfer. The signal is DC free and can therefore be transferred via twisted pairs with galvanic isolation.

Modulated transmission

Here, in contrary to base band transmission, the source data are shifted to an advantageous frequency band by means of a carrier frequency. This allows more effective use of the transmission medium. Examples for this are micro-wave or optic fiber transmission.

Possible signal combinations

0000
0001
0010
0011
0100
0101
0110
0111
1000
1001
1010
1011
1100
1101
1110
1111

Phase-space diagram with possible signal combinations

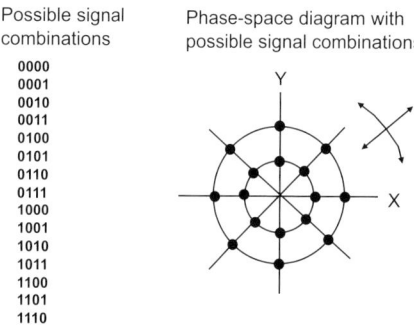

Figure 6.15
16-bit QAM (Quadrature Amplitude-Modulated) signal constellations

For transmission over voice-grade channels (pilot wires), a modem modulates the pulses into a combination of analog tones with amplitude and phase changes that fits within the pass band of the channel.

Complex modulation methods are used for encoding multiple bits in one transmission step. (Refer to equation (6-5) above)

The quadrature amplitude modulation (QAM) of high speed modems is an example for this. (Figure 6.15)

It allows transmitting data for line differential protection with 128 Kb/s over pilot wires by combined phase and amplitude modulation.

Asynchronous and synchronous transmission

The receiver of the digital message must be controlled so that the sampling takes place as close to the centre of the signal as possible.

Asynchronous transmission is the simplest technical solution. Each character (byte) is framed by a start and a stop bit. The bit sequence in a character is in fixed time slots. Each character is synchronized individually, that is, the sampling is re-synchronised by the start bit of each received character. Long pauses may occur between the characters. The bits of two separate sequential characters have no fixed timing with respect to each other. As the clocks of two physically separate devices can never be exactly the same, and as communication channel time delays will occur, a character may not be longer than approximately one byte and the data rate may also not be very large (< 64 Kb/s) with this technique. The performance of asynchronous systems is lower than that of the synchronous techniques described hereafter. It can however be implemented simply and cost effectively. It is particularly suited to lower data rates and spontaneous transmission.

A typical example of asynchronous data transmission is the telegram format FT 1.2 according to IEC 60870-5-1 [6-17], which was developed for telecontrol systems and was also used to implement protection data communication (Figure 6.16).

It provides the basis for the standard (IEC 870-5-103) for protection data interfaces and the protection signalling interface with small data rates in conjunction with comparison protection (e.g. Siemens current comparison protection 7SD51 with 19.2 Kb/s).

With *synchronous transmission* on the other hand, all bits are in fixed time slots and synchronism exists between the two data terminals. The common clock may be transferred via a separate clock link; alternatively self clocking signal coding may be implemented. With these, the clock signal is recovered along with the transmitted data from the received signal. In principal, any signal edge on the receiving end may be used for synchronising. It must only be ensured that signal edges are available with sufficient frequency, i.e. long continuous signal states as would occur with a sequence of "1" or "0" must be avoided. This may be achieved with scramblers on the transmitting end and equivalent de-scramblers (mirrored) at the receiving end.

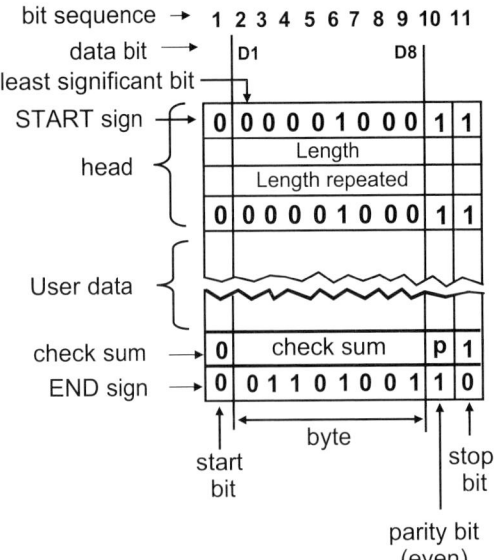

Figure 6.16
Telegram format acc. to
IEC 60870-5-1, Format class
FT 1.2 (variable telegram length)

The beginning and end of a data block is indicated with fixed coded opening and closing flags (block synchronising).

As a large amount of user data can be transmitted in a block, and as the synchronisation may be maintained with high accuracy by crystal controlled clocks, the synchronous data transfer has a substantially higher performance and efficiency in comparison to the asynchronous technique. The complexity and cost is however greater.

Whereas asynchronous data terminals can communicate with each other if the speed, code, and error checking conventions are identical, synchronous terminals require protocol compatibility and intelligence in the data terminal.

Synchronous data communication is for example applied with line differential protection where information is continuously transferred with a high data rate.

Synchronous data transmission is now generally used for protection either through dedicated optic fibers, digital micro-wave or data communication networks.

Figure 6.17 shows the HDLC (High-level data link control) protocol which is recommended by ITU and widely applied, also in line differential protection [6-18].

Opening Flag	Address Field	Control Field	Information Field	Frame Check FCS	Closing Flag
01111110	8 or more bits	8 or 16 bits	any length	16 or 32 bits	01111110

Figure 6.17 HDLC-Protocol: Frame structure according to ISO 3309

Error handling

The occurrence of errors with data transmission cannot be excluded completely. The main task of error handling is the detection of transmission errors and ensuring that the faulty telegrams are rejected or re-transmitted. A direct error correction of faulty telegrams is in principle possible, but is however hardly applied as too much redundant information would have to be transmitted for this purpose.

The effort that has to be invested in the increased transmission security depends on the bit error probability of the transmission channel and the importance of the transmitted information.

The expected bit error rate (BER)[1] of the transmission channel depends on the transmission medium that is used. The following typical values for data communication apply [6-11]:

- 10^{-5} for telephone circuits
- 10^{-6} to 10^{-7} for digital data networks of German Telecom
- 10^{-9} for coaxial cables in local areas
- 10^{-12} for FO

The stated values apply for normal operating conditions. Unusual conditions which cause higher error rates such as for example bit slipping, synchronisation errors, connection failures, device faults or fading for directional radio links, must also be considered for the protection application.

According to a 2001 CIGRE report [6-13] regarding protection with tele-communication the following bit error rates are applicable:

- 10^{-6} for FO communication
- 10^{-6} for data networks (PDH, SDH, ATM)
- 10^{-3} for directional radio
- ---- no information is provided for wire bound communication

The published data regarding bit error rates differs considerably and is heavily dependent on the local conditions and operating environment. Directional radio in particular has a bad reputation due to the known fading during adverse weather conditions, although certain users have indicated good performance [6-14, 6-15].

In the CIGRE report referred to [6-13], the following bit error rate requirements for data communication systems applied with differential protection (analog comparison systems) are mentioned:

- 10^{-6} during normal operating conditions
- 10^{-5} during faults in the high voltage system

Based on the definition for availability in ITU-T G.821 the following recommendation is provided:

---

[1] BER: The *Bit Error Rate* is the number of bit errors in relation to the total number of transferred bits.

In general, the protection or tele-control should still function correctly with bit error rates of up to 10^{-3} whereby it is accepted that the performance (availability, tripping times) are worse if the bit error rate exceeds the value of 10^{-6}.

The specification according to IEC 60834-2 of 1993 has a similar formulation [6-19].

The availability of the telecommunication service should be better than 99.99%.

Measures for secure data transfer

Redundant coding is applied to achieve security against bit errors. The transmitter supplements the telegram with additional information (test bits) which is derived from the service data (transmitted data content). The receiver computes the test bits with the same procedure from the received service data and compares the received test bits with the computed test bits. A measure for the effectiveness of this test is the "Hamming distance". It indicates how many bits of a data block must be incorrect before the coding rules would indicate that the block code is again valid. A code with Hamming distance $d = 4$ is therefore secure even with up to 3 simultaneous bit errors in a single telegram.

Transverse and longitudinal parity

Although a single parity bit (corresponds with $d = 2$) reduces the block error rate (rate of not detected erroneous telegrams) by two orders of magnitude, it is not sufficient for an acceptable data integrity.

By monitoring the transverse parity of each character and the arithmetic bit sum of all service data in each block, a Hamming distance of $d = 4$ can be obtained thereby reducing the block error rate by a factor 10^{-6}. In general this is sufficient for tele-control and protection. It is however a pre-requisite that the bit error rate of the transmission channel does not exceed 10^{-4}.

Cyclic redundancy check

The protocols used for synchronous data transmission (e.g. HDLC) have random bit sequence. In this case a cyclic block checking is applied. It does not require that the information is structured in bytes or other units. 16 or 32 test bits in the form of CRC (Cyclic Redundancy Check) or FCS (Frame Check Sequence) are supplemented to the chain of bits (block) that is transmitted. The test bits of the CRC field are computed from all bits in the data frame using a theoretical procedure with Modulus 2 arithmetic (complex polynomial). The corresponding circuitry for generating the test codes consists of a combination of shift registers and EXOR gates [6-11]

At the receiving end, the test bits are calculated in the same way from the received telegram. An error is detected if the locally calculated test bits do not match the test bits in the received CRC field.

The rate of undetected block errors can be reduced by a factor 10^{-5} (CRC-16 technique) or 10^{-10} (CRC-32 technique).

The probability of an undetected error with CRC is so slight that the checked and acknowledged telegrams can be considered as practically error free.

Communication standards

The standardisation of protocols and interfaces must facilitate compatible and open communication.

In general, the OSI 7-layer model (Open Systems Interconnection) of the ISO is applied [6-17, 6-20]. The protocol is divided into 7 layers, whereby each layer provides a service for the layer above it. For tele-control and protection tasks the reduced EPA model which consists of only the layers 1, 2 and 7 can be used. This 3-layer model has better transmission efficiency and is more suited for time critical information transfer in particular with limited channel capacity.

For permanently connected end to end links the middle layers with functions such as packet-switching or routing are not required. With transmission via digital broadband networks, these functions are provided by the digital signal transmission units.

Asynchronous transmission standards

In this case, protocols according to IEC 60870-5 are predominantly used in Europe, as they can be implemented cost-effectively [6-17]. The corresponding standard in the USA is DNP 3.0.

These protocols have originally been developed for telecontrol purposes and applied as point to point connections with slow speech band channels (< 64 Kb/s).

With the introduction of digital substation automation (around 1990), the protocol IEC 60870-5-103 was developed (now widely replaced by the standard IEC 61850) for open communication between relays and control centres.

For teleprotection, the asynchronous transmission has only been used in older devices with simpler communication. The digital phase comparison protection 7SD51, which only transmits binary direction signals and therefore demands only low data rate (19.2 Kb/s) is an example for this. The application layer in this case is specific to each manufacturer.

Synchronous transmission standards

With modern relays (SIPROTEC 4 and 5), the communication between the line ends is done with synchronous transmission having n times 64 Kb/s. It corresponds with the standards of digital data networks. The HDLC protocol, applied in the second layer, thereby guarantees high transmission security [6-18].

For the access to PCM multiplexers and transmission through data networks, the CCITT interface standards G.703 or X.21 are generally used.

6.3 Digital communication networks

Currently, most utilities have access to digital communication networks. These utility owned or private networks have a hierarchical structure and combine as a rule different legacy (PDH) and modern (SDH) communication standards. (Figure 6.18)

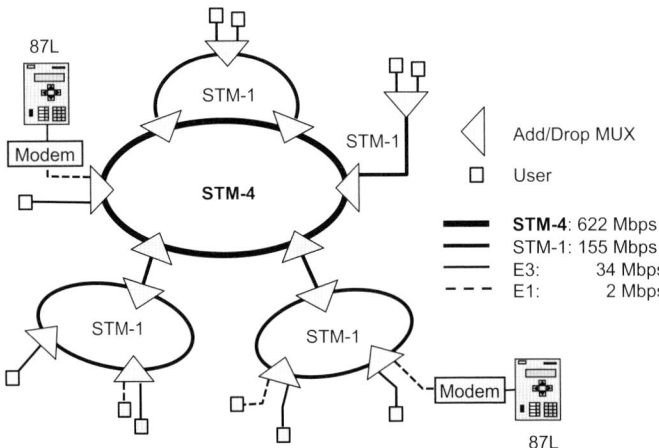

Figure 6.18 Modern data communication network

The user has access to the network via multiplexers at network hubs according to the interface standards G.703 and X.21.

Both are wire-bound interfaces and therefore require a conversion from FO (protection device output) to the data network input.

The relay manufacturers offer corresponding modems to adapt the relay communication interface to these standards.

An IEEE Standard C37.94 [6-31] for optical interfaces between relays and multiplexers was published in 2002 to dispense with the modems and the EMC critical wiring.

It shall provide plug-and-play transparent communications between different manufacturers' teleprotection and multiplexer devices using multimode optical fiber.

Devices to this standard are already available (e.g. 7SD8 of the SIPROTEC5 series).

If there is no network connecting point in the substation, an additional spur line (via optic fiber, micro-wave or pilot wire) may be necessary to connect the relay terminal with a next network terminal (multiplexer).

Nearly synchronous and synchronous networks

Data networks use the multiplexer technology to bundle data streams for the common transmission with higher bit rates. This allows the effective exploitation of modern communication media (up to some 10 Gb/s in case of optic fiber links).

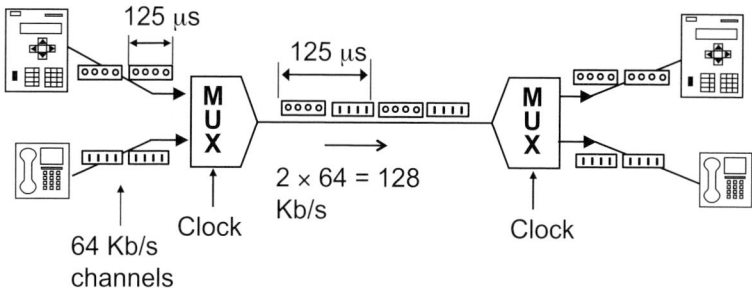

Figure 6.19 Time Division Multiplex (TDM), principle

The Time-Division Multiplexing (TDM) is used to transmit data of different telecom services in a time sharing procedure. (Figure 6.19) [6-21]

The multiplexing is done by interleaving a fixed number of bits (frame) per channel. One channel is dropped by taking out the corresponding bit frame. Another signal can be added in the same way occupying the bit frame freed by the dropped signal.

The bit rate at the line is proportional to the number of multiplexed channels.

By connecting multiplexers in series the bit rates increase correspondingly.

The basic channel bit-rate is 64 Kb/s which was originally designed for the transmission of a PCM coded speech band (8 bit resolution times 8 bit sampling rate = 64 Kbits per second).

The sequence of bits is divided in eight bit long coded words (bytes) corresponding to 125 µs long pulse frames. This basic channel hierarchy level is called E0 in Europe and DS-0 in North America.

The digital line differential relays of all manufacturers need at least this base rate.

On the next higher European hierarchy level E1, 30 base channels of 64 Kb/s (together with two additional control channels) are multiplexed up to the PCM 30 system resulting in a bit rate of 32×64 Kb/s = 2.048 Mb/s. (The corresponding North American T1 level consists of 24 base channels and uses a bit rate of 1.544 Mb/s).

Direct access to this E1(T1) level allows much faster telegram transmission (and shorter relay operating times) by connecting base channels in parallel. For example, the differential protection 7SD52/61 can be operated with the bit rate of 8 base channels (8×64 = 512 Kb/s) resulting in a reduction of the operating time by nearly 10 ms.

Many utilities already use this possibility.

Plesiochronous networks

PDH (Plesiochronous Digital Hierarchy) networks have been used since about 1970, at first with pilot wires and coaxial cables, and later with optic fiber links.

The name PDH refers to the fact that the multiplexers distributed in the network have independent clocks which are only nearly synchronous. The bit streams arriving from different network areas must therefore be adapted by bit stuffing. The multiplexers contain buffer memories which store the incoming frames until they can be inserted in the multiplex-frame. [A-19]

The PDH hierarchy is built up from PCM 30 systems. (Figure 6.20)

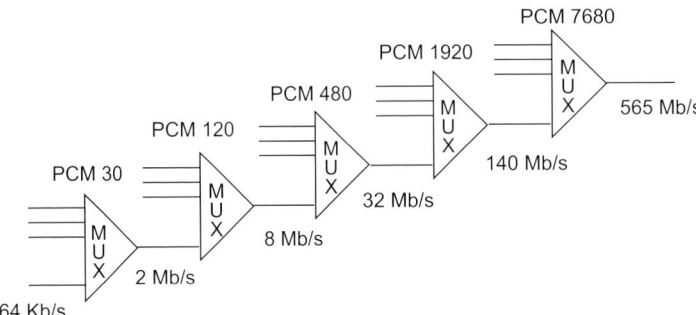

Figure 6.20 Plesiochronous Digital Hierarchy (Europe)

Four subsystems are each bundled by a multiplexer to the next highest level.

The bit rates were fixed by CCITT (ITU-T) in 1972. The base rate (Hierarchy level 0) of 64 Kb/s is world-wide the same, but the higher levels deviate regionally. (Table 6.3).

The disadvantage of the now legacy PDH networks is that access to the data of lower level channels is only possible by complete down multiplexing through the complete hierarchy range. This causes corresponding transmission time delay (125 µs per multiplexing step).

Existing PDH networks are however further used as local data networks.

Table 6.3 Hierarchy levels of PDH networks in Kb/s

Hierarchy level	North America		Europe		Japan
0	DS-0	64	E0	64	64
1	DS-1 (T1)	1544	E1	2048	1544
2	DS-2 (T2)	6312	E2	8048	6312
3	DS-3 (T3)	44736	E3	34368	32064
4	DS-4 (T4)	274176	E4	139264	97728
5			E5	564992	

Synchronous networks

The Synchronous Digital Hierarchy (SDH) was proposed by ITU-T (formerly CCITT)

in 1987. It was designed for high bit rates and in particular for data transmission over optic fiber cables. It should further harmonise the formerly incompatible systems of Europe, Japan and North America.

In North America the ITU-T compatible system is called SONET (Synchronous Optical Network) as only optical systems are applied for the transmission.

In this case, the basic element is the STM-1 transport module with a bit rate of 155.52 Mb/s. It brings together the existing PCM multiplex systems of lower bit rate. The further hierarchy levels are shown in Figure 6.21.

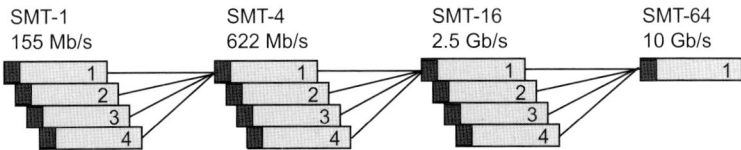

SMT-1 155 Mb/s   SMT-4 622 Mb/s   SMT-16 2.5 Gb/s   SMT-64 10 Gb/s

Figure 6.21 SDH hierarchy

In SDH systems the multiplexers of the total network are exactly synchronised by a central clock.

SDH/SONET is like an E1(T1) carrier because it is in the same way byte synchronous up to the highest hierarchy level.

As a result, byte-wise multiplexing is possible on all network levels. That is, individual E0 (T0) channels (n times 64 Kb/s) can be extracted from the bit stream with relatively simple equipment (Add-Drop multiplexers).

All multiplexers are controlled by a higher order network management center.

This offers centralised end-to-end network management and performance monitoring.

Modern fiber optic SDH networks are designed in a self-healing double ring structure (Figure 6.22). A path protection mechanism is implemented to switch the communication services at the time of communication failure to a standby path in the ring.

There are two main modes of switching: uni-directional and bi-directional.

Uni-directional mode is where only the end that detects the failure switches to the standby mode, whereas in bi-directional mode both ends switch to the standby path.

In the bi-directional mode transmit and receive path are both switched to the standby loop. The transmission times in both directions change by the same value.

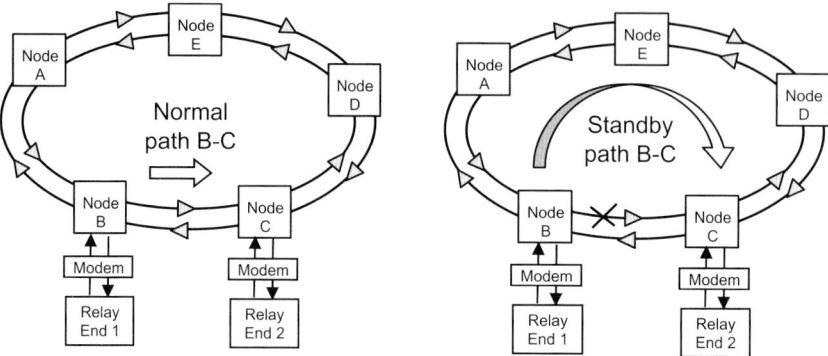

Figure 6.22 Self-healing ring structure of modern SDH networks

For large rings this may be some milliseconds. Only a short transient unequal propagation delay occurs.

The uni-directional switching however results in a permanent path split with consequential permanent differential delay (unequal transmit and receive times).

The described path switching and change of data transmission times may affect the sampling synchronisation of line differential protection.

In case of bi-directional switching, transmit and receive times change by the same value. This symmetrical time change causes no problems as the relays measure the transmission time (loop delay divided by two) continuously and adapt automatically to changes.

The uni-directional switching however causes a measuring error (false differential current) as the "loop delay divided by two" principle only functions properly with symmetrical propagation times. (Refer to section 4.2)

ISDN networks

The ISDN (Integrated Services Digital Network) offers bit-transparent communication over two independent base channels B with 64 Kb/s each, fully duplex up to the end user. In addition there is a 16 Kb/s signalling channel. The channel structure at the S0 interface is therefore 2B+D = 144 Kb/s.

The two base channels can be used for transmitting differential protection data with a bit rate of 128 Kb/s. Corresponding modems are offered by the relay manufacturers.

Application aspects

For permissive tripping schemes or differential protection the communication from station to station is time critical and demands high security and availability.

A fixed, non-switched main and standby connection with symmetrical transmission times would be the best solution. This however is not usual in digital networks as it would require special provisioning procedures and routing constraints.

As a rule, switched channels n×64 Kb/s compatible to X.21 or G.703 are therefore applied in case of PDH and SDH networks.

Routine channel switching should however be minimised as the switchover time (some 10 ms) cause resynchronisation (which may take seconds) and loss of protection during this time.

Virtual point to point connections provided by an exchange via an X.21 interface may be used in networks with ISDN technology.

The following delays must be reckoned with:

- FO: 5 μs/km
- Directional radio: 3.3 μs/km
- Multiplexing of 64 Kb/s to n Mb/s: 500 to 1000 μs (PDH-networks)

For signal transmission via a number of sub-links in the network, total transmission times to the order of 5 ms will arise.

Furthermore, the low bit error rate of 10^{-12} stated above only applies to the FO link itself. Communication via a network may result in the loss of clock or synchronism, whereby the de-multiplexer outputs undefined signals for a short time (some ms), from the instant of error detection up to the output of the AIS signal (alarm indication signal) as a sequence of logical "1 s". The protection must detect the transmission failure and block the teleprotection.

Special demands are placed on the data network by the line differential protection:

Changing or non-symmetrical signal transmission times are not critical for permissive transfer trip teleprotection schemes. A variation up to 5 or even 10 ms can normally be accepted. (Blocking schemes, which are also time critical, are not recommended for use with signal transmission via data networks.)

The differential protection must however continuously measure and monitor the signal transmission time so that it will automatically adapt the transmission time compensation when the channel is switched over.

Synchronisation of the sampled measured values (current phasors) is done via the data channel by the modern devices. (Refer to section 4.2.3)

The generally applied technique assumes that the telegram transmission time for the transmit and receive path is the same. In case of deviations, as a result of channel switching to different paths, an angle error in the comparison of the current phasors will result. (Figure 6.23) It can be calculated from the channel delay time difference as follows:

$$\Delta \varphi = \frac{t_{\text{transmit}} - t_{\text{receive}}}{2} \text{ [ms]} \cdot \frac{18°}{1 \text{ ms}} \quad \text{for } f_{\text{n}} = 50 \text{ Hz} \qquad (6\text{-}6)$$

The false differential current $\Delta I = 2 \cdot I_{\text{thru}} \cdot \sin(\Delta\varphi/2)$, corresponds to this angle error, whereby $I_1 = I_2 = I_{\text{thru}}$ is the current flowing through the feeder. In the event of 1 ms channel delay time difference, a false operating current ΔI of 16% I_{thru} and a ratio $\Delta I/\Sigma I \ (= I_{\text{OP}}/I_{\text{Res}})$ of 8% will already result.

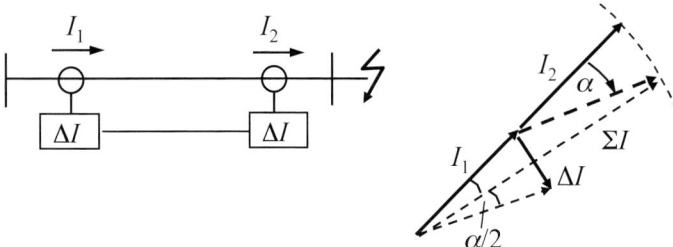

Figure 6.23 Impact of unsymmetrical data transmission times on line differential protection

The Dutch national utility (previously SEP) formulated the following requirements for the communication network in 1995 [6-22]:

- Channel delay time < 6 ms
- Channel delay difference: < 1 ms
- Bit error rate (BER) $< 10^{-5}$

A Japanese operational experience report [6-26] indicates a user specified requirement of maximum synchronisation error equal to 4% I_{thru}. This corresponds to a maximum channel delay difference of 200 to 240 µs. This accuracy could not be achieved by network operators. Direct channel allocation with bit stuffing was therefore used to achieve a synchronisation accuracy of several µs.

As a rule, it should be tried to keep the propagation time difference below about 0.1 to 0.25 ms. This would limit the false differential current to 2 to 5%.

If this requirement cannot be met by the network operator, an external synchronisation via GPS must be applied [6-23, 6-24]. The information exchange is then carried out with exact time stamps having micro-second accuracy so that channel delay differences and changes are no longer relevant. Modern line differential protection relays provide this option (refer to section 9.3).

7 Generator/Motor Differential Protection

Electrical generators/motors are complex equipment that requires protection against a number of fault types (stator earth faults and short circuits, unbalanced load, rotor earth faults, etc.).

With conventional technology, each fault type had a special protection relay applied for it. Consequently, the protection of large generators/motors often necessitated 10 to 20 devices depending on the size of the plant and redundancy requirements. [7-1, 7-2, A18]. With modern numerical technology, these functions can now be integrated in multifunctional devices so that a redundant generator/motor protection system can be constructed with as little as two devices. (Figure 7.1)

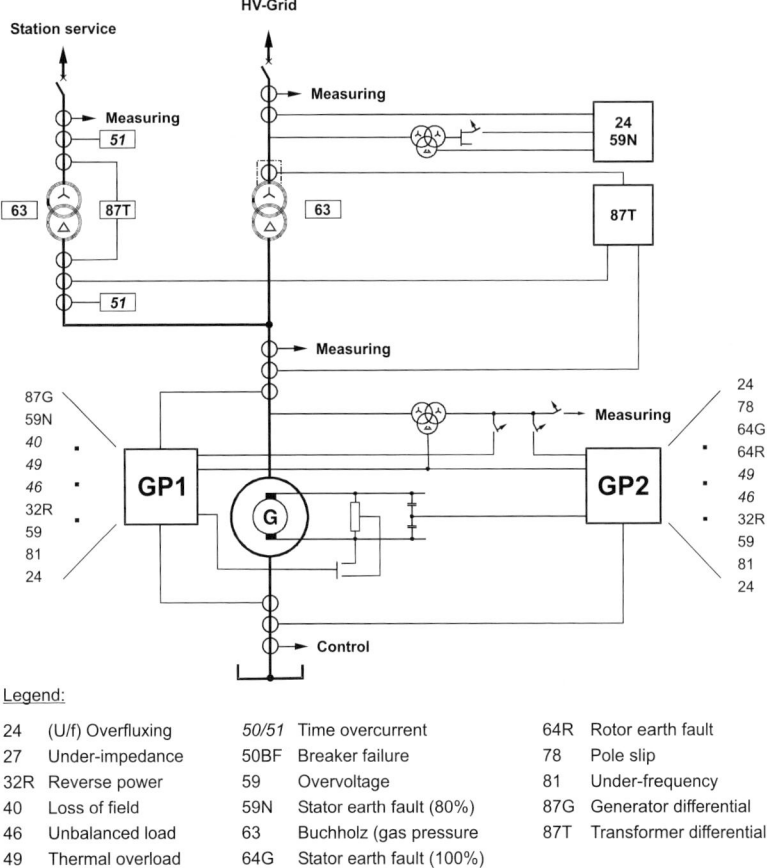

Legend:

24	(U/f) Overfluxing	50/51	Time overcurrent	64R	Rotor earth fault
27	Under-impedance	50BF	Breaker failure	78	Pole slip
32R	Reverse power	59	Overvoltage	81	Under-frequency
40	Loss of field	59N	Stator earth fault (80%)	87G	Generator differential
46	Unbalanced load	63	Buchholz (gas pressure	87T	Transformer differential
49	Thermal overload	64G	Stator earth fault (100%)		

Figure 7.1 Numerical generator/motor unit protection

On generators/motors that are larger than 1 MVA, differential protection is some-times applied for fast clearance of two and three phase short circuit faults. Above 5 MVA, it is always applied. On generators/motors with low impedance neutral earthing, or direct connection to an earthed system, single phase earth faults are also covered to some extent. On generators/motors with high impedance neutral earthing, or connection to a system with high impedance earthing, a separate earth fault protection must be applied. [7-3]

7.1 Generator differential protection

On generators, winding short circuits mainly occur on the winding overhang where the phases cross over. The initial (sub-transient) short circuit current pro-vided by the generator itself is approximately 3 to 10 times the generator rated cur-rent, depending on the magnitude of the sub-transient reactance. The short circuit current from the system must be added thereto, it is at least of the same order of magnitude. To prevent severe damage non-delayed fault clearance is of the essence. Differential protection is ideally suited for this purpose.

Inter-turn faults are not detected by differential protection. The circulating cur-rents in the short circuited turns are very large; the change in the current flowing in and out of the terminals however is not significantly different.

For the detection of earth faults, a differentiation must be made between direct connection to the system (busbar connection) and connection via a dedicated transformer (unit transformer) as shown in Figure 7.2.

With the *unit connection* the generator neutral is usually earthed via a high imped-ance so that in the event of an earth fault in the generator, only small currents

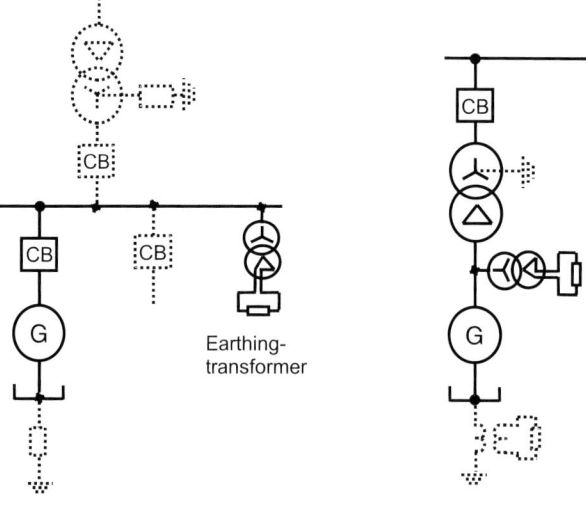

Earthing-transformer

Direct bus connection

Unit connection

Figure 7.2
Generator connections

arise. The objective is to keep the earth fault current below 15 A to avoid melting of the iron and consequently severe damage to the stator core material. The differential protection does not respond to this small current. A sensitive stator earth fault protection must be provided for this purpose.

In the case of the *busbar connection*, the earth current depends on the treatment of the system neutral.

In the event of high impedance neutral earthing (isolated or resonant grounded) which is common with larger generating units, only small earth currents flow, so that a special earth fault protection must be provided. In small systems, the earth current may have to be increased via an earthing transformer with burden resistance to a value of 5 to 10 A to facilitate secure fault detection by the earth current relay.

The differential protection would only operate during a double earth fault when one fault is located in the generator and the other in the system.

With low impedance earthed neutral of the generator or connected network, the fault current during single phase earth faults is larger and can be detected by the differential protection. This neutral earthing method is applied in countries with an Anglo-Saxon protection history. The current magnitude is generally limited by means of a resistance in the neutral to earth connection, so that in the event of an earth fault at the generator terminal, a current in the range between 200 A and 1.5 times generator rated current will flow, depending on the applied earthing method. If the earth fault is not located at the terminals, but on the winding between the terminals and neutral point, the earth current is reduced accordingly:

$$I_E = \frac{l\,[\%]}{100} \cdot \frac{U_n / \sqrt{3}}{R_E} \qquad l = \text{distance of earth fault from the winding neutral}$$

In the event of an earth fault that is located 20% from the generator neutral point, the arising earth current will only be 20% of the earth current flowing during a fault at the generator terminals. If the earth current is restricted to rated generator current, only 20% I_n would flow in this case. This implies that a differential protection with a setting of 20% I_n would cover 80% of the winding from the terminals towards the neutral point. On larger generators, the earth current must be more severely restricted in comparison with the generator rated current. Consequently the normal phase differential protection is not sensitive enough for earth faults. In these cases an additional earth current differential protection is implemented. It may be applied with a lower setpoint if the phase CTs are matched, or if special core balance CTs are used.

Per-phase differential protection

This is the customary short circuit protection on larger generators. A prerequisite for its application is that the three phase conductors on the neutral side are segregated.

The CT sets on the terminal and neutral side almost always have the same transformation ratio so that no ratio matching (by computation or interposing CTs) is necessary.

In general, a stabilised (biased) differential protection is applied. In some countries, a high impedance differential protection may also be used.

Stabilised differential protection

The circuit diagram is shown in Figure 7.3.

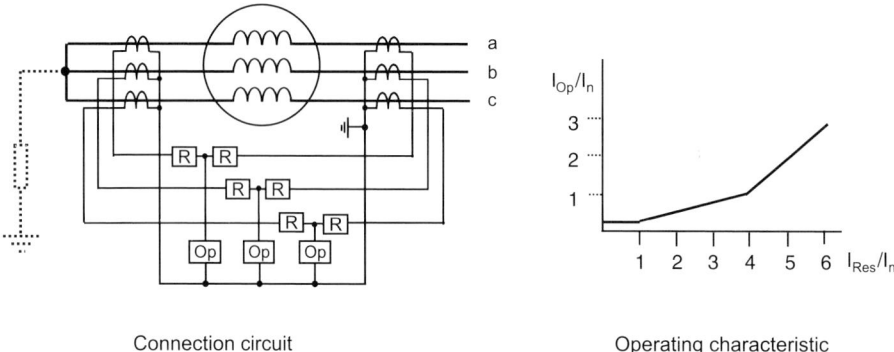

Connection circuit   Operating characteristic

Figure 7.3 Generator differential protection

The squares marked with R designate the points at which restraining current is measured (previously restraining windings), and the squares designated with Op indicate the points at which operating (differential) current is measured (previously differential windings). With conventional technology, the CTs on the terminal side and the neutral side were galvanically connected via the measuring circuit connection. Therefore only the neutral point of the terminal side CTs could be earthed on the secondary side to prevent multiple earthing and parallel paths via earth. With numerical protection, the CT neutrals on both sides must be earthed, as no galvanic connection between the two sides exists.

The stabilised differential characteristic may be very flat for small currents up to approximately rated current (e.g. only 5% slope) as CTs of the same construction will be well matched and additional errors due to interposing CTs are not present. Consequently very good sensitivity is obtained over the entire load range with the customary setting of 10-20% I_n.

One advantage of the stabilised differential protection (low impedance protection) is that other protection devices may be connected in series with it. This was important with conventional individual relay technology. With numerical technology the differential protection is one of the numerous integrated functions of a device (relay 7UM62).

For choosing the CT dimension it must be noted, that the DC time constant of the generator is very long.

Table 7.1 Typical data of a generator in unit connection

Generator rating in MVA	10 to 500 MVA
Sub-transient generator reactance X_d''	10 to 25%
DC time constant of the generator (T_G)	70 to 600ms
Short circuit voltage of the unit transformer	10 to 20%
DC time constant of the unit (T_U)	50 to 300 ms

If the CT dimension must be such that saturation does not take place in the event of a fault at the generator terminals (system fault for busbar connection or fault on low voltage side of unit transformer), then an extremely large over-dimensioning factor is required.

For unit connection without generator circuit breaker, the CT dimension could be reduced so that the generator differential protection maintains stability in the event of faults in the connected power system. In this case, the DC time constant of the unit (series connection of generator and unit transformer) would apply.

Even for this case, very large over-dimensioning factors and correspondingly large CTs would be required for indefinite saturation free transformation.

More moderate CT dimensions are obtained when the saturation free transformation time after fault inception is restricted to the clearing time of external grid faults (e.g. 150 ms).

Acceptable CT costs can only be obtained if a certain degree of saturation is permissible. By applying equivalent burdens to the CTs on both sides, the stability during CT saturation is improved, but this does not guarantee that non-selective tripping does not take place, as differential currents resulting from tolerance of the magnetising curves cannot be eliminated completely.

The integrated saturation detector in numerical protection provides a method for absolutely secure stabilising. The generator provides relatively small short circuit currents, but the duration of the offset is fairly long (long DC time constant). The flux in the CTs therefore rises slowly and the saturation only appears after some time (often only after several cycles). The saturation detector can therefore detect the external fault securely, and initiate the trip blocking time (refer to section 4.2.4).

The blocking time should at least allow for clearance of the external fault, therefore it should be approx. 150 ms, if all close-in short circuits are cleared without time delay. A setting of approx. 500 ms is required if fault clearance by time delayed back-up protection must be considered. However, the blocking is removed in any case if the operating current assumes sinusoidal shape for one or two periods to ensure tripping of a possible consecutive internal fault.

As concerns stability with high through flowing currents, the integrated saturation detector of numerical protection facilitates reduced CT dimensions. If however a low pick-up setting (e.g. 10% I_n) and a low restraint (e.g.12% slope) are used in the range of small currents ($\leq 2 \cdot I_n$), the CTs must be well designed and balanced. For example, a CT over-dimensioning factor $K_{TD} \geq 5$ should be chosen with the relay 7UM62.

Example 7-1: CT dimension for generator differential protection

Given: Generator with unit transformer and no generator circuit breaker

Generator 100 MVA, 6 kV

Sub-transient short circuit reactance: $X_d'' = 10\%$,

DC resistance $r_G = 0.13\%$ ($T_G = 245$ ms)

Unit transformer 6/230 kV, 120 MVA

Short circuit voltage: $u_{X-T} = 15\%$

Ohmic short circuit voltage $u_{R-T} = 0.3\%$ ($T_T = 159$ ms)

Secondary CT burden (relay + connecting cable): $R_B < 1$ VA

Task: Calculate dimension of generator CTs so that stability in the event of external faults is ensured. It may be assumed that close-in system faults are isolated within 150 ms.

Solution: The CT ratio is matched to the generator rated current

$$I_n = \frac{S_n}{U_n \cdot \sqrt{3}} = \frac{100{,}000 \text{ kVA}}{6 \text{ kV} \cdot \sqrt{3}} = 9623 \text{ A}$$

Select CT ratio: 10,000/1 A.

Initially check if stability can be obtained for faults close to the generator: The maximum short circuit current for an external fault close to the generator terminals is:

$$I_{F-G}'' = 1.1 \cdot \frac{100}{X_d'' [\%]} \cdot I_{n-G} = 1.1 \cdot \frac{100}{10} \cdot 9623 \text{ A} = 10.6 \cdot I_{n-CT}$$

The over dimensioning factor required for indefinite saturation free transformation is:

$$K_{TF}' = 1 + \omega \cdot T_G = 1 + 314 \cdot 0.245 = 78$$

The required operational overcurrent factor calculated for the CT with this value is:

$$\text{ALF}' = \frac{I_{F-G}''}{I_{n-CT}} \cdot K_{TF}' = 10.6 \cdot 78 = 827$$

We select a CT of type TPX with $P_n = 30$ VA rating. The internal burden P_i according to the manufacturer's data is assumed to be 20% (6 VA).

The rated accuracy limit factor for the CT is then calculated to be:

$$\text{ALF} = \frac{P_B + P_i}{P_n + P_i} \cdot \text{ALF}' = \frac{1 + 6}{30 + 6} \cdot 827 = 161$$

The CT with 30 VA, ALF > 161 is much too large for practical application. The alternative method with limited saturation free transformation of 150 ms is therefore checked:

$$K_{TF}'^* = 1 + \omega \cdot T_G \cdot \left(1 - e^{-\frac{t_M}{T_G}}\right) = 1 + 314 \cdot 0.245 \cdot \left(1 - e^{-\frac{150}{245}}\right) = 36$$

$$ALF'^* = 10.6 \cdot 36 = 382 \quad \text{and} \quad ALF^* = \frac{1+6}{30+6} \cdot 382 = 74$$

The corresponding CT of 30 VA, $n > 74$ still appears to be excessively large.

It is now also checked to what degree the CT dimension is reduced when stability is only demanded for system faults on the HV side, while non-selective tripping for low voltage side faults on the unit transformer is tolerated (the unit must be switched off in any event, the fault indication is however not selective and determining the cause of the fault is more difficult).

The DC time constant of the unit is calculated as follows:

Impedances of the generator:

$$Z_{n\text{-}G} = \frac{U_n \,[\text{kV}]^2}{S_{n\text{-}G}\,[\text{MVA}]} = \frac{6^2}{100} = 0.360 \ \Omega = 360 \ \text{m}\Omega$$

$$X_d'' = \frac{10\%}{100} \cdot 360 \ \text{m}\Omega = 36 \ \text{m}\Omega \quad \text{and}$$

$$R_G = \frac{0.13\%}{100} \cdot 360 \ \text{m}\Omega = 0.47 \ \text{m}\Omega$$

Impedances of the transformer:

$$Z_{n\text{-}T} = \frac{U_n \,[\text{kV}]^2}{S_{n\text{-}T}\,[\text{MVA}]} = \frac{6^2}{120} = 0.300 \ \Omega = 300 \ \text{m}\Omega$$

$$X_T = \frac{15\%}{100} \cdot 300 = 45 \ \text{m}\Omega \quad \text{and}$$

$$R_T = \frac{0.3\%}{100} \cdot 300 = 0.9 \ \text{m}\Omega$$

The DC time constant for the unit (series connection of generator and unit transformer) is calculated as follows:

$$T_U = \frac{L_G + L_T}{R_G + R_T} = \frac{1}{\omega} \cdot \frac{X_d'' + X_T}{R_G + R_T} = \frac{1}{314} \cdot \frac{36 + 45}{0.47 + 0.9} = 188 \ \text{ms}$$

The fault current in the event of close-in short-circuits on the HV side of the unit transformer is:

$$I_{F\text{-close-in}}'' = \frac{1.1 \cdot 6 / \sqrt{3} \ \text{kV}}{0.036 \ \Omega + 0.045 \ \Omega} = 47 \ \text{kA} = 4.7 \cdot I_{n\text{-}CT}$$

Therefore:

$$K'^*_{TF} = 1 + \omega \cdot T_U \cdot \left(1 - e^{-\frac{t_M}{T_U}}\right) = 1 + 314 \cdot 0.188 \cdot \left(1 - e^{-\frac{150}{188}}\right) = 34$$

$$ALF'^* = 4.7 \cdot 34 = 160 \quad \text{and} \quad ALF^* = \frac{1+6}{30+6} \cdot 160 = 31$$

For the case of restricted stability (150 ms fault clearance time for network faults), a CT with 30 VA, $n > 30$ would be sufficiently dimensioned. Even this would be a relatively large CT.

In the following it is shown that with numerical protection, unlimited stability (also for faults close to the generator) can be achieved at reasonable expense:

The 7UM62 relay is chosen which, due to its integrated saturation detector, only requires an over-dimensioning factor of $K'_{TD} = 5$.

Related to the generator terminal fault, this results in:

$$ALF'^{**} = 10.6 \cdot 5 = 53 \quad \text{and} \quad ALF^{**} = \frac{1+6}{30+6} \cdot 53 = 10.3$$

In this example we maintain the 30 VA CT and specify: CT TPX, 10,000/1, 30 VA, 5P10

In actual fact a CT with a much smaller VA should be selected, better adapted to the low burden of digital protection.

We choose 5 VA and assume again 20% internal burden (1 VA).

The required rated ALF** then is $(1+1)/(5+1) \cdot 53 = 17.7$.

A current transformer 5 VA, 5P20 would be adequate.

It may now be determined when the CT of the selected dimension goes into saturation:

Close-in fault to generator:

The resulting over-dimensioning factor is:

$$K''_{TF \text{ res.}} = \frac{ALF}{ALF^{**}} \cdot K'_{TF} = \frac{20}{17.7} \cdot 5 = 5.6$$

Equation (5-21) in section 5.7 is used to calculate the time until saturation ensues. With $t_M = t_S$ and by solving for t_S the following results:

$$t_{S\text{-Generator-fault}} = T_G \cdot \ln\left(\frac{\omega \cdot T_G}{\omega \cdot T_G - K''_{TF \text{ res.}} + 1}\right) =$$

$$= 0.245 \cdot \ln\frac{314 \cdot 0.245}{314 \cdot 0.245 - 5.6 + 1} = 0.0151 \text{ s} = 15.1 \text{ ms}$$

(The applied equation does not provide accurate results for short saturation-free times, as the AC flux shortly after fault inception still makes up a significant part of the total flux. The accurate calculation with equation (5-22), section 5.7 however shows that the result in this example is quite accurate)

Close-in system fault:

The over-dimensioning factor is increased in this case corresponding to the lower short-circuit current:

$$K'_{\text{TF Net-fault}} = K'_{\text{TF Generator-fault}} \cdot \frac{I''_{F \text{ Gen.}}}{I''_{F \text{ Net}}} = 5.6 \cdot \frac{10.6 \text{ kA}}{4.7 \text{ kA}} = 12.6$$

The following is obtained:

$$t_{\text{S-Net-fault}} = 0.188 \cdot \ln\left(\frac{314 \cdot 0.188}{314 \cdot 0.188 - 12.6 + 1}\right) = 0.041 \text{ s} = 41 \text{ ms}$$

CT remanence and proximity effects

The generator differential protection reacts sensitively to CT errors in the range of low through fault currents. The reason is the generally low set pick-up value (10 to 15% I_n) and the flat increase of the percentage restraint (first slope 10 to 15%).

Practical experience has shown that saturation due to core remanence or mutual magnetic influence from neighbouring high current conductors (proximity effect) are often the reason for false trippings with small fault currents (< two times rated current).

It is therefore recommended to use CTs with 1% accuracy and anti-remanence air gaps (Class TPY to IEC 60044-6 or Class 5PR to IEC 61869-2[1]).

High ratio CTs should be provided with compensation windings to minimize the proximity effect. [5-17, 5-20]

In Germany, linear core CTs (Class TPZ) are usually applied on large generators. Their dimension is chosen such, that saturation free transformation of the maximum short-circuit current is obtained. In this case the CT saturation is not relevant, but higher costs must be accepted.[5-4]

Generator winding connected in delta

In this case, the secondary sides of the CTs allocated in the generator winding triangle must be connected in delta to compensate for the delta-star configuration of the primary circuit. (Figure 7.4)

The CTs in the winding triangle must have the same ratio as the CTs at the generator terminals (although the current in the delta connected windings is lower by the factor $\sqrt{3}$). Numerical relays would however also allow the adaptation to different ratios.

When using a transformer differential relay, the CT secondaries could also be connected in star and the then necessary star-delta transformation could be provided numerically by the integrated vector group adaptation.

---

[1] IEC61869-2: Instrument transformers, Part2: Current transformers (draft) shall replace 60044-1 and 60044-6.

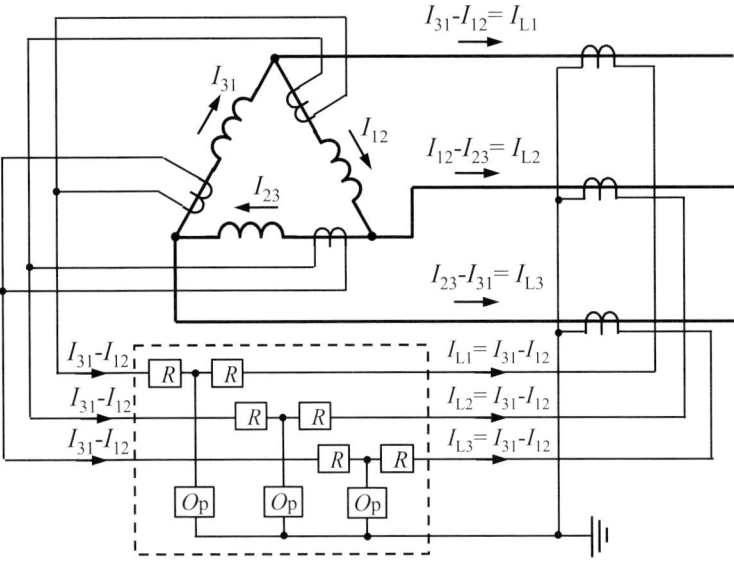

Figure 7.4 Differential protection of delta-connected generator winding

High impedance differential protection

This is an option to cope with the problem of CT saturation.

The principle was described in section 3.4. Figure 7.5 shows the connection diagram.

Dedicated CT cores of the type PX in accordance with IEC 60044-1 (Formerly Class X in accordance with British Standard BS 3938), with the same transformation ratio must be provided. On large generators, this results in no additional cost, as separate cores are generally applied in any event for reasons of redundancy.

The protection system is usually configured for a pick-up sensitivity of 10% I_N.

A check must then be done for the largest through-fault current that can arise, to ascertain whether a voltage limitation (varistor) must be connected in parallel to the relay (refer to section 3.4).

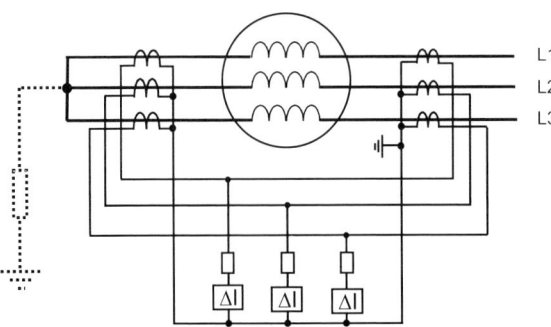

Figure 7.5
High impedance differential
protection per phase

Transverse differential protection

On generators having parallel windings (hydro-electric machines), a transverse differential protection may be applied. For this purpose the parallel windings must have separated terminals. The additional cost for the current transformers is how-ever only justifiable on large generators. The correct current distribution on the parallel paths is monitored. The advantage of this protection is that short circuited turn faults are also detected.

In the event of equal distribution of the phase currents on both windings, the sec-ondary current only circulates through the stabilising windings, similar to the lon-gitudinal differential protection. In the event of an internal fault including short circuited turns, a current circulating through the parallel windings will result, causing a tripping current in the differential path of the relay.

A normal generator differential protection may be applied for this purpose by con-necting it as shown in Figure 7.6.

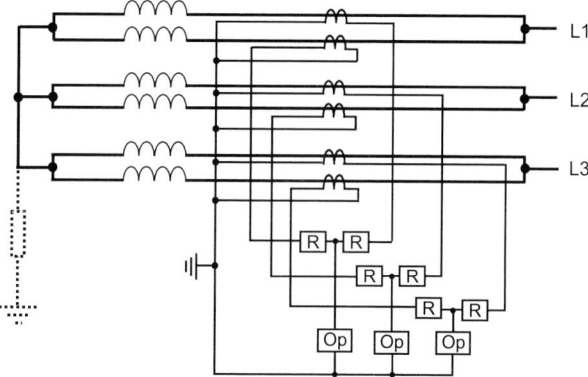

Figure 7.6 Transverse differential protection

To obtain a high sensitivity, the circulating current must be minimised by the man-ufacturer in the range from no load up to rated load of the machine with appropri-ate configuration of the windings. [7-4] Circulating currents smaller $< 2\% \, I_n$ can be achieved in any event. On machines without damping windings, equalising cur-rents arise during external short circuits. By effective matching of the windings, these should be kept to below $0.5\% \, I_n$. If sufficient information regarding the stated aspects is available from the manufacturer, a setting of approximately 2.5% I_n should be possible. (This corresponds to a relay setting of 5%, as the CTs in the parallel windings are usually configured for half the generator rated current).

The application of high impedance protection as transverse differential protection (short circuited turns protection) is also possible. [7-5]

It remains to be said that the transverse differential protection can also be applied in plants with double generators (e.g. Ljungström-machines).

Differential protection with cable mounted window-type CT

On small machines that are connected via cable, a secure differential protection with high sensitivity can be implemented with this technique.

A pre-requisite for this protection is that the three phases on the star point side are available separately, and are routed back through the cable mounted CTs (Figure 7.7). During fault free operation of the machine, the currents neutralise each other in the CT so that no current flows via the differential relay. The current comparison is highly accurate and not subject to saturation problems due to magnetic self-balancing. No stabilising is required; therefore simple over-current relays may be applied. The pick-up threshold may be set to between 2 to 5% of the rated machine current.

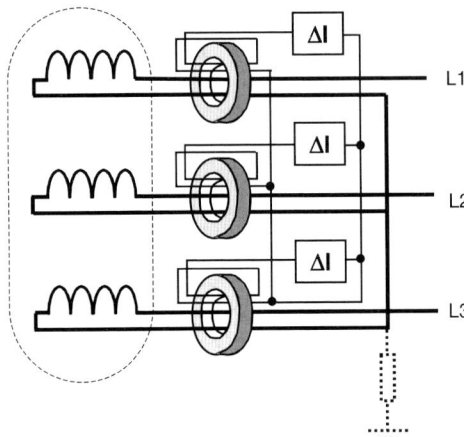

Figure 7.7
Self-balancing differential protection
with cable mounted CTs

Differential earth – fault protection for generators

With low impedance neutral earthing of the generator, earth-faults must be cleared rapidly. An earth current differential protection (restricted earth fault protection) that compares the star-point current with the summated current from the generator terminals is ideally suited for this purpose.

This type of protection is used on smaller generators where the phases on the star-point side are not available separately, i.e. a phase segregated differential protection cannot be applied. On larger machines it is applied in addition to the phase differential protection to achieve greater pick-up sensitivity during earth faults. This applies in particular to winding earth-faults close to the star-point which result in reduced earth-fault currents.

With sensitive setting it is advisable to make the tripping dependent on the presence of displacement voltage (indicator for earth-fault). This prevents mal-operation of the protection if, during external phase short-circuits with large current flow, unequal saturation of the CTs at the generator *terminals* cause an erroneous summated current.

Ideal conditions for the earth-fault differential protection exist when the summated current is provided by a core balance CT with small transformation ratio. In this case, the pick-up sensitivity may be as low as 10 A (primary).

Example 7-2: Setting the phase and earth differential protection

Given: Generator 5 MVA, 4 kV

$X_d'' = 10\%$, DC time constant $T_G = 50$ ms

Star-point earthed via resistance (5.75 Ohm, 400 A)

CT 800/5 A, 5P20, 15 VA, $R_i = 0.16$ Ohm, i.e. $P_i = 4$ VA

Task: The phase and earth differential protection must be set.

The earth-fault protection of the generator winding must cover 90%.

Solution: Initially it is checked to see whether CT saturation is possible during external faults:

$$I_{n\text{-}G} = \frac{5000 \text{ kVA}}{4 \text{ kV} \cdot \sqrt{3}} = 722 \text{ A}$$

$$\frac{I_F''}{I_{n\text{-}CT}} = \frac{1.1 \cdot 10 \cdot 722}{800} = 9.93$$

The required operational accuracy limit factor of the CT for indefinite saturation-free transformation:

$$\text{ALF}' = \frac{I_F''}{I_{n\text{-}CT}} \cdot (1 + \omega \cdot T_G) = 9.93 \cdot (1 + 314 \cdot 0.050) = 156$$

Actual operational accuracy limit factor of the CT:

$$\text{ALF}' = \frac{P_n + P_i}{P_B + P_i} \cdot \text{ALF} = \frac{15 + 4}{1 + 4} \cdot 20 = 76$$

($P_B = 1.0$ VA for the relay + connecting cables estimated)

The actual over-dimensioning factor is:

$$K_{TF}' = \frac{\text{ALF}'}{I_F'' / I_{n\text{-}CT}} = \frac{76}{9.93} = 7.7$$

We use again equation (5-21) of section 5.7 and to determine the expected time to saturation during close-in external faults:

$$t_S = T_G \cdot \ln\left(\frac{\omega \cdot T_G}{\omega \cdot T_G - K_{TF}' + 1}\right) = 0.050 \cdot \ln\frac{314 \cdot 0.050}{314 \cdot 0.050 - 7.7 + 1} =$$

$$= 0.028 \text{ s} = 28 \text{ ms}$$

The numerical protection 7UM62 is selected; it requires an over-dimensioning factor of $K_{TD}' > 5$, and therefore complies with the requirements specified above. A pick-up threshold of 15% $I_n = 120$ A is set for the phase-segregated differential protection.

As the earth current during an earth fault at the generator terminals is 400 A, the phase differential protection only covers from $1 - 120/400 = 70\%$ of the winding.

The earth differential protection of the 7UM62 has a setting range from 0.01 to 1.00 times I_n. For a protection coverage of 90%, the pick-up threshold setting must be $400/10 = 40$ A, i.e. $40/800 = 0.05$ A.

With this very sensitive pick-up threshold, together with the expected CT saturation during external phase faults, the U_0-release ($U_E = 3 \cdot U_0 > 10\% U_{ph\text{-}E}$) must be selected as an additional criterion, so that tripping is only activated in the event of earth faults. For further security against unnecessary operation during through-flowing short-circuit currents, the earth differential protection should be blocked when the overcurrent protection picks up. The user defined logic functions in the 7UM62 are programmed via the CFC supplement to provide a transient blocking function.

It immediately blocks tripping of the earth differential protection when the over-current protection ($I>$) picks up, and resets after the short circuit is removed (reset of $I>$) with a delay on reset (transient blocking time) of approximately 0.5 seconds. This delay on reset is necessary, as the phase overcurrent protection with a higher threshold setting re-sets faster than the sensitive earth fault protection.

The application of core balance CTs mounted around the cable should be considered for applications with such a high sensitivity.

High-impedance restricted earth fault protection (circulating current differential protection)

The connection is the same as for normal earth current differential protection Figure 7.8). Special CTs of the class PX according to IEC 60044-1 (formerly Class X according to BS 3938) are also required in this case similar to the requirement for phase segregated high-impedance protection. This kind of protection is used in countries with Anglo-Saxon protection history on small machines, where phase-segregated access to the winding on the star-point side is not available. A pick-up sensitivity of between 5 and 10% of nominal generator current can be achieved with high security in the event of CT saturation.

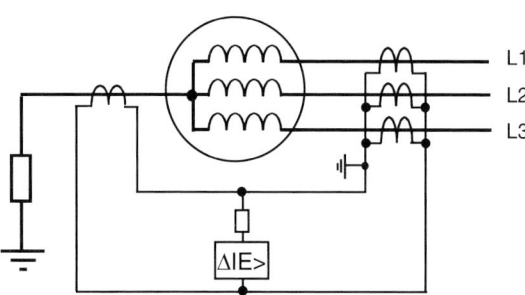

Figure 7.8
HI-restricted earth-fault
protection

With high pick-up sensitivity the CTs must however have a small secondary winding resistance (large conductor cross-section) and a small magnetising current (large core cross-section). This is illustrated in the following example.

Example 7-3: Configuration of a HI-restricted earth-fault protection

Given: Generator 5 MVA, 4 kV, $X_d'' = 10\%$

Star-point earthed wire resistance (5.75 Ohm, 400 A)

CT 800/1, Class PX according to IEC 60044-1:

Secondary winding resistance: $R_{CT2} = 0.7$ Ohm,

Magnetising curve as shown in Figure 7.9

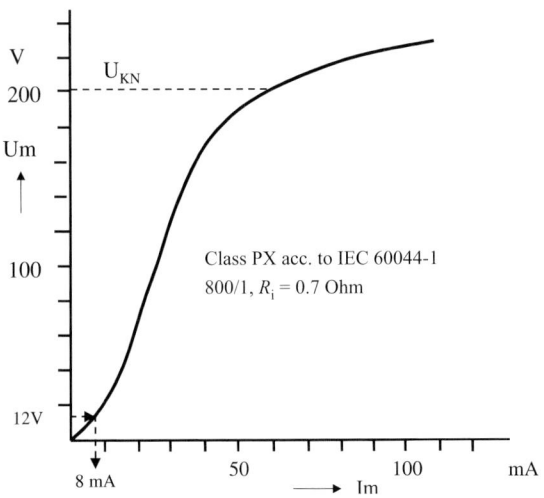

Figure 7.9 Magnetizing curve in example 7-3

Task: Dimension of the CT and protection configuration for 90% earth fault protection coverage of the generator winding.

Solution: The rated current of the generator is:

$$I_{n\text{-}G} = \frac{5000 \text{ kVA}}{4 \text{ kV} \cdot \sqrt{3}} = 722 \text{ A}$$

The HI-protection is connected as shown in section 3.4:

1. Check of stability during external faults:

CT saturation can only occur during phase-phase faults, as the earth current is limited to 400 A by the resistance in the star point.

During all external short-circuits the three phase currents must add up to zero due to the parallel connection (Holmgreen connection), so that no current flows towards the relay or star point CT.

In the event of saturation of the CT in one phase, the other phase connected CTs would feed onto the secondary resistance of this CT. The CT secondary connecting cable burden up to the relay may therefore be neglected in this context. The voltage arising across the secondary resistance of the saturated CT will appear across the high-impedance relay.

For the 3-phase fault (worst case) the following applies:

$$I''_{F\,max.} = 1.1 \cdot \frac{100}{X''_d[\%]} \cdot I_{n\text{-}G} = 1.1 \cdot \frac{100}{10} \cdot 722 = 7942$$

The maximum shunt-voltage then is:

$$U_{\Delta FD} = \frac{I''_{F\,max.}}{r_{CT}} \cdot R_{CT2} = \frac{7942\ A}{800/1} \cdot 0.7\ \Omega = 6.95\ V$$

For security reasons the relay pick-up threshold should be at least 20% above this. The next higher setting on the relay 7VH60 is chosen: 12 V.

2. Checking the pick-up sensitivity:

The magnetising current of the CT at the relay setting threshold can be obtained from the magnetising curve: 8 mA.

The relay current at the pick-up threshold (12 V) according to the data sheet is 20 mA.

Converted to a primary current, this pick-up threshold is:

$$I_{min.} = r_{CT} \cdot (n \cdot I_{mR} + I_R) = \frac{800}{1} \cdot (4 \cdot 8 + 20) \cdot 10^{-3} = 41.6\ A$$

Referred to the 400 A earth current during an earth fault at the machine terminals, this is $(41.6/400) \cdot 100 = 10.4\%$, i.e. a protection coverage of 89.6% is achieved.

In principle the knee-point voltage would only have to be 2 times 12 = 24 Volt. To obtain a small magnetising current at the relay pick-up voltage, a CT with $U_{KN} = 200$ Volt is selected (this approximately corresponds in magnitude to 5P20, 10 VA)

3. Check to see if a varistor is required to limit the voltage during internal earth faults

With a setting of 12 V and a current of 20 mA the internal resistance of the relay can be calculated: $12/0.020 = 600$ Ohm.

During an internal fault (earth fault), a theoretical voltage across the relay can be calculated if CT saturation is not considered:

$$U_F = \frac{I_{E\text{-}max.}}{r_{CT}} \cdot R_{i\text{-}Relay} = \frac{400}{800/1} \cdot 600\ \Omega = 300\ V$$

If CT saturation is considered, the following is obtained:

$$U_{peak} = 2 \cdot \sqrt{2 \cdot U_{KN} \cdot U_F} = 2 \cdot \sqrt{2 \cdot 200 \cdot 300} = 693\ V$$

This value is far below the peak of the 2 kV rated insulation voltage (2.8 kV) and therefore no varistor is required.

7.2 Motor Differential Protection

The same protection principles as described for the generator protection also apply here.

Synchronous motors are directly comparable with synchronous generators in the event of a short circuit.

Asynchronous motors have a somewhat different response. During system faults, they feed a short circuit current back into the system; this current however rapidly decays to zero.

For stability in the event of fault current flowing through the machine, and the required CT dimension, the starting process is decisive Figure 7.10.

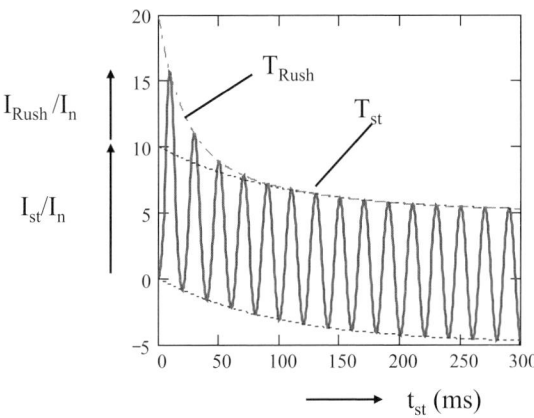

Figure 7.10
Starting current of asynchronous motors

The starting current I_{St} is in the range between 4 to 7 times I_n with a DC time constant T_{St} of approx. 40 ms (<1 MW) to 70 ms (> 1 MW).

An in-rush current with the same order of magnitude is super-imposed; it however decays in two to three cycles corresponding to a time constant T_{Rush} of approx. 20 ms.

The total in-rush starting current therefore is approx. 2 to 2.5 times I_{St} equal to 8 to 20 times I_N.

Example 7-4: Configuring the CTs for motor differential protection

Given: Asynchronous motor: $P_n = 2$ MVA, $U_n = 5$ kV
              $I_{St} = 5 \cdot I_n$, $T_{St} = 50$ ms

Task: Calculate the dimension of the CT for the differential protection

Solution:

$$I_n = \frac{P_n}{\sqrt{3} \cdot U_n} = \frac{2000 \text{ kVA}}{\sqrt{3} \cdot 5 \text{ kV}} = 231 \text{ A}$$

CTs with ratio 250/1 A are selected.

The total in-rush starting current is estimated to be
$I_{tot} = 2.5 \cdot I_{St} = 2.5 \cdot 5 = 12.5 \cdot I_n$.

This corresponds to $12.5 \cdot (231/250) = 11.5$ times CT nominal current.

$K_{TF} = 1 + \omega \cdot T_{St} = 1 + 314 \cdot 0.05 = 15.7$

$$\text{ALF}' = K_{TF} \cdot \frac{I_{tot}}{I_{n\text{-}CT}} = 15.7 \cdot 11.5 = 181$$

The selected differential protection 7UM62 allows a certain degree of saturation so that an over-dimensioning factor of $K_{TD} \geq 5$ is sufficient. (The integrated switch-over to higher stablity during starting is not considered in this example.)

Resulting from this: $\text{ALF}' = 5 \cdot 11.5 = 57.5$

The following CT type is chosen: $P_n = 5$ VA nominal rating (internal burden according to manufacturer: $P_i < 1$ VA)

The connected burden (CT connecting cable plus relay) is estimated to be $P_B < 1$ VA.

Therefore the required rated accuracy limit factor can be calculated as follows:

$$\text{ALF} = \frac{P_B + P_i}{P_n + P_i} \cdot \text{ALF}' = \frac{1+1}{5+1} \cdot 57.5 = 19.2$$

A standard CT with ALF > 20 is selected and the following is specified in the order:

CT 250/1A, 5P20, 5 VA, $P_i < 1$ VA ($R_i < 1$ Ohm)

The recommendation to use CTs with anti-remanence air-gaps (Class PR or TPY) for differential protection is also valid here.

8 Transformer Differential Protection

Transformers are important system components available in many different constructions. The range of HV transformers reaches from small distribution transformers (from 100 kVA) up to large transformers having several hundred MVA. Apart from the large number of simple two and three-winding transformers, a range of complex constructions in the form of multi-winding and regulating transformers also exist.

Differential protection provides fast and selective short-circuit protection on its own, or as a supplement to Buchholz (gas pressure) protection.

It is usually applied on transformers above approx. 1 MVA. On larger units above approx. 5 MVA it is standard.

The transformer differential protection contains a number of supplementary functions (adaptation to transformation ratio and vector group, stabilisation against in-rush and over-excitation) and therefore requires some fundamental consideration for the configuration and setting calculation.

8.1 Basic physics

To better understand the protection response during short-circuits and switching operations, the physical principles of the transformer are initially covered in detail. [8-1]

Equivalent circuit of a transformer

The primary and secondary winding are linked via a magnetic core by means of the main flux Φ Figure 8.1. To obtain the flux, the magnetising current (excitation current) I_m according to the magnetising curve is required. In the electrical equivalent

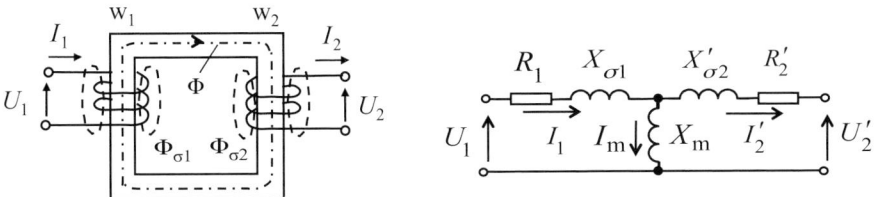

Equivalent electromagnetic circuit   Equivalent electric circuit

Figure 8.1 Equivalent circuit of a transformer

circuit, this excitation requirement corresponds to the main reactance X_m. The leakage flux $\Phi_{\sigma 1}$ and $\Phi_{\sigma 2}$ are only linked to their respective own windings and make up the leakage reactances $X_{\sigma 1}$ and $X'_{\sigma 2}$. R_1 and R'_2 are the respective winding resistances. All currents and impedances are referred to the primary side.

$X_m = U/I_m$ corresponds to the slope of the magnetising curve. During load and particularly in the event of short-circuits, the operating point is below the knee-point in the steep portion of the curve. The magnetising current at nominal voltage only amounts to approx. $0.2\% I_n$, i.e. in the non-saturated segment of the curve X_m is approx. 500 times larger than the nominal impedance of the transformer and approx. 5000 times greater than the leakage reactances. During load and short-circuit conditions, a simplified equivalent circuit may therefore be used for the calculations (Figure 8.2).

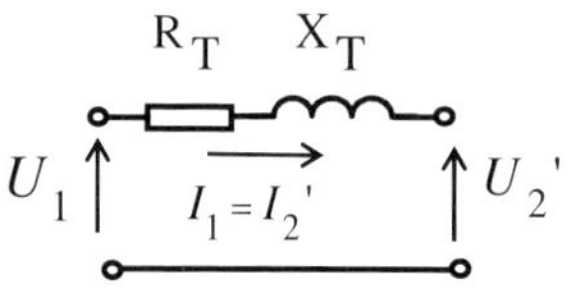

$$I_1 \cdot w_1 = I_2 \cdot w_2$$
$$X_T = X_{\sigma 1} + X'_{\sigma 2}$$
$$R_T = R_1 + R'_2$$

Figure 8.2
Simplified transformer equivalent circuit

The series reactance X_T corresponds to the short-circuit voltage in %, relative to the nominal impedance of the transformer:

$$X_T = \frac{u_{X\text{-}T}\,[\%]}{100} \cdot X_{Tn} \tag{8-1}$$

$$X_{Tn} = \frac{U_n}{\sqrt{3} \cdot I_n} = \frac{U_n^2}{P_n} \tag{8-2}$$

The series resistance corresponds to the ohmic short-circuit voltage in %, and is also based on the nominal impedance. For calculation of the short-circuit current, the resistance may be neglected; it must only be considered when calculating the DC time constant.

Table 8.1 lists typical transformer data. [8-2]

Table 8.1 Typical transformer data

Rating MVA	Transformation ratio (kV/kV)	Short-circuit voltage $u_{X\text{-}T}$ in %	Open circuit current % I_N
600	400/230	19	0.25
300	230/110	24	0.1
40	110/10	17	0.1
16	30/10	8.0	0.2
6.3	30/10	7.5	0.2
0.63	10/0.4	4.0	0.15

In-rush [8-3 to 8-7]

When energising a transformer, one-sided over-excitation results, due to remanance causing large magnetising current flow (in-rush current).

The flux does not return to zero when the transformer is switched off, but remains at the remanance point Φ_{Rem}, which may be above 80% of the nominal induction. When the transformer is re-energised, the flux increase starts at this point. Depending on the energising instant on the sinusoidal voltage (point on wave), an off-set course of the flux can result. For the large flux values in the saturation range, a very large magnetising current is required, and cyclic current peaks will result. The curve form corresponds with the sinusoidal half-waves of a simple half-wave rectified AC current that decays with a very large time constant (Figure 8.3).

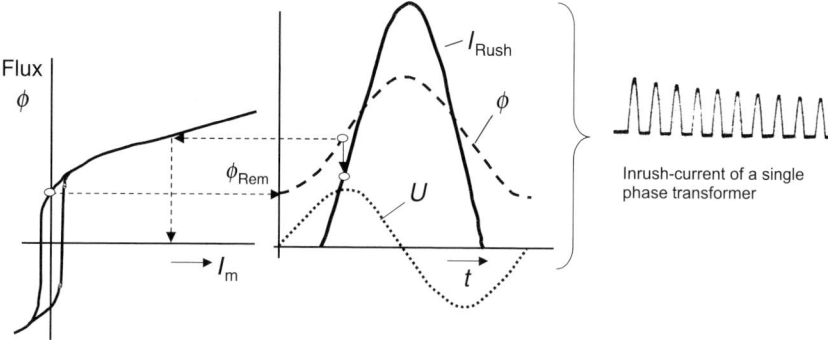

Figure 8.3 Origin of in-rush current

The rush current is particularly large when cores of cold rolled steel with a nominal induction (1.6 to 1.8 Tesla) are operated close to the saturation induction (approx. 2 Tesla).

On a three-phase transformer, a three-phase rush current will result, which depends on the vector group and the method of star-point earthing on the trans-

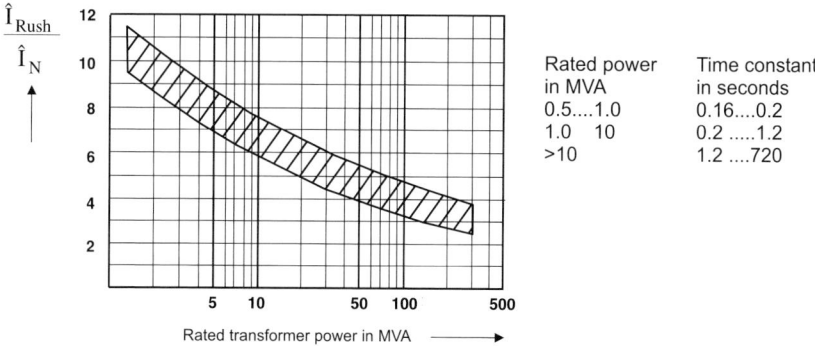

Rated power in MVA	Time constant in seconds
0.5....1.0	0.16....0.2
1.0 10	0.2 .....1.2
>10	1.2 720

Figure 8.4 Typical rush current of a star delta transformer

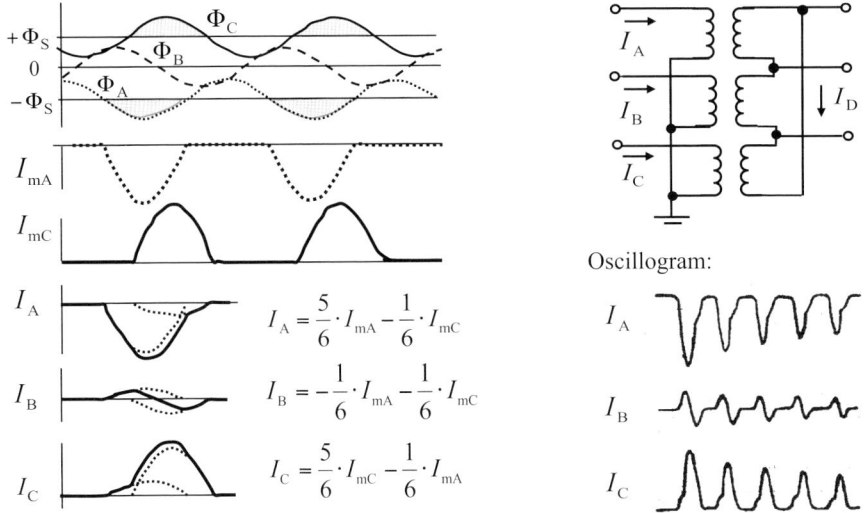

The equations shown in the figure:

$$I_A = \frac{5}{6} \cdot I_{mA} - \frac{1}{6} \cdot I_{mC}$$

$$I_B = -\frac{1}{6} \cdot I_{mA} - \frac{1}{6} \cdot I_{mC}$$

$$I_C = \frac{5}{6} \cdot I_{mC} - \frac{1}{6} \cdot I_{mA}$$

Oscillogram:

Figure 8.5 Typcal in-rush current data

former. [8-3 and 8-7] In general, two phases will saturate and draw large magnetising currents. On star delta transformers, these currents are coupled to the non-saturated phase via the delta winding. This causes the typical rush currents as shown in Figure 8.5.

The rush currents in the three phases can be calculated from the required magnetisation (I_{mA} and I_{mC}) of the two saturated core-limbs A and C with the given equations. The current on phase B thereby corresponds to the current in the delta winding I_D. Please refer to the literature for the theoretical analysis [8-7]. The shown oscillogram of an in-rush occurrence confirms the calculated curves.

Amplitude and time constant depend on the transformer size. (Figure 8.4)

It must be noted that a similar rush current also arises when a close-in external short–circuit is switched off and the transformer is re-magnetised by the recovery of the voltage. It however is substantially smaller than the in-rush following energising of a switched-off transformer.

Large rush currents can also occur when asynchronous systems are switched together via a transformer, as the large voltage difference can cause transient saturation of the core. [8-8]

Sympathetic in-rush

When transformers were connected in parallel, it was observed that the differential protection of the transformer that was in service issued a trip. The reason for this is sympathetic in-rush current, which results from the rush current of the transformer that is being energised (Figure 8.6).

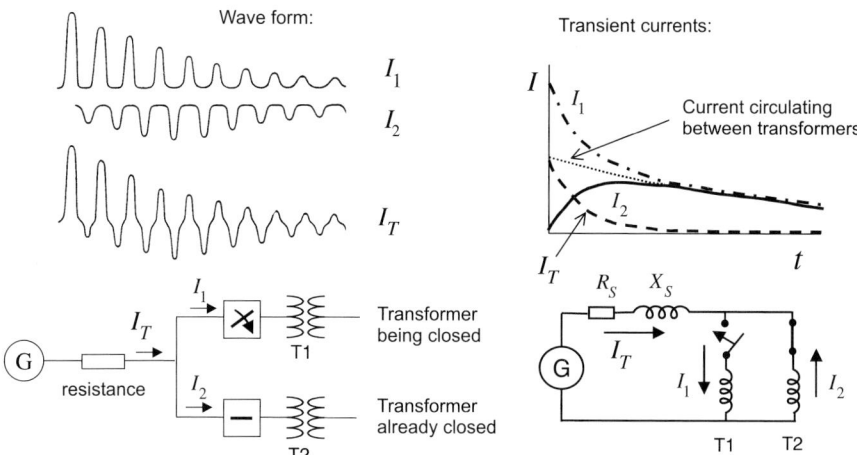

Figure 8.6 Sympathetic inrush current

The voltage drop resulting from the initial rush current across the source resistance of the in-feed affects the second transformer in parallel and causes the sympathetic in-rush current (I_2). The current from the system (I_T) decays rapidly; however a current still circulates between the two transformers due to the small damping (large time constant $\tau = X/R$ of the windings). [8-9 and 8-10]

In-rush blocking

The in-rush current flows into the protected object from a single side and appears as an internal fault. The transformer differential protection must therefore be stabilised against this phenomenon. The large amount of second harmonic in the rush current was already used with conventional protection for this purpose. The second harmonic is filtered out of the differential current (operating current) by means of a filter, and is then used as additional restraint current in the measuring bridge. When it was above approximately 15% in relation to the 50(60) Hz fundamental, a very large additional restraint was introduced to prevent tripping. Other manufacturers compared the 100(120) and 50(60) Hz components directly with a separate bridge circuit, which then directly blocked the protection, as it is now done in the software of numerical protection.

The 100(120) Hz component in the rush current depends on the base width of the sinusoidal caps (Figure 8.7).

It decreases as the base width B increases.

Investigations have shown that a base width greater than 240° hardly ever arises in practice [8-7 and 8-12], which implies a minimum second harmonic component of 17.5%. A setting of 15% therefore makes sense for in-rush blocking.

The 3rd harmonic may not be used for in-rush blocking, as it is strongly represented in the short-circuit current when CT saturation takes place.

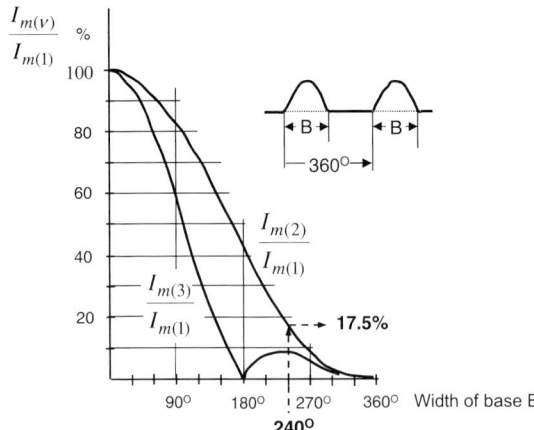

Figure 8.7
Harmonic content
of in-rush current

A more sensitive setting than 15% of the second harmonic should normally not be applied as the off-set short-circuit current will also have a second harmonic component in case of CT saturation.

In rare cases, for example with weak in-feed, a soft energisation with a very small second harmonic component may occur. Under these conditions, a reduced setting of e.g. 12% may be considered. Preference should however be given to the cross-blocking function.

Cross-blocking

This function, which was already applied in conventional relays, is now available in all numerical relays and may be activated, if desired. It takes into account that the second harmonic component in the individual phases is different and may not be sufficient, in the phase with the smallest component, to activate the blocking.

The measuring system in all phases is therefore blocked when a single phase detects the rush blocking condition.

Transformer over-fluxing

If the transformer is operated with excessively high voltage, then the required magnetisation is also increased. The magnetising current rises sharply when the operating point on the magnetising curve is close to the point of saturation. The wave form becomes more and more distorted with increasing odd harmonic content. (Figure 8.8)

The increased magnetising current appears as a tripping current in the differential protection with large over-voltage; this can cause tripping, depending on the configuration of the transformer.

Over-voltages can occur in the system, due to the distribution of reactive power flow in the event of tap changer or AGC problems, or following load shedding. This is particularly true for geographically large systems with long lines. One critical case is the switching off of a power station under full load conditions which results

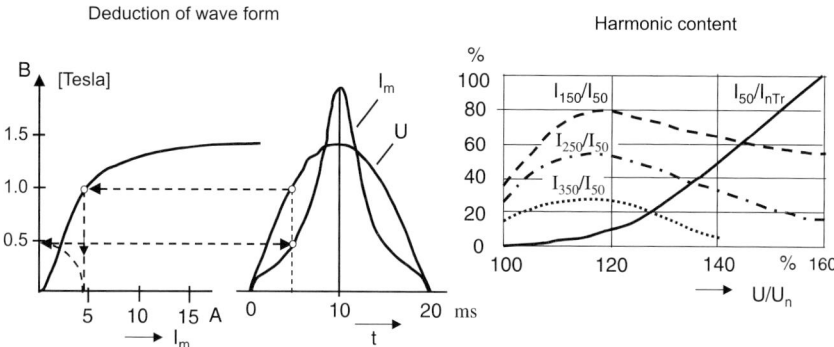

Figure 8.8 Magnetising current in the event of transformer over-fluxing

in a severe over-voltage condition at the unit transformer as a result of the large excitation of the generator.

The transformer can tolerate the over-excitation, which causes heating, for a given time without sustaining damage. During this time the system regulation must ensure that the voltage returns to the permissible range.

Only if this does not occur, must the transformer be isolated by a special over-excitation protection having a U/f-dependent time delay. Tripping by the differential protection with a fast measurement due to these conditions must be avoided at all cost.

Modern numerical relays therefore provide an integrated blocking of the trip in the event of over-excitation (over-fluxing). It is based on the large 5th harmonic component in the tripping current which clearly indicates over-fluxing. Tripping is blocked when the ratio I_{150Hz}/I_{50Hz} exceeds a set value. For this setting it must be noted that the 5th harmonic component again decreases if the over-voltage is very large (Figure 8.8). A typical setting is 30% [8-12]

If the over-voltage is very large, blocking is no longer sensible as the transformer is at risk. The blocking can therefore again be re-set when the 5th harmonic component is above a set ratio of the 50 Hz component, which increases as the over-voltage increases.[8-6]

8.2 Numerical measured value processing

The conventional connection of the measured value inputs to the differential protection and the transformer ratio and vector group adaption where covered in section 5.8.

The numerical protection follows the same basic principle and implements this by means of numerical computation. External interposing transformers are therefore no longer required. The required filtering and measuring functions are implemented in a more flexible manner, providing increased accuracy.

Numerical measured value adaptation

Prior to the current comparison, two adaptations are required:

On the earthed star-point winding the zero-sequence component must be elimi-nated or compensated for (refer section 5.8)

On delta windings, the vector group must be compensated for (see section 5.8)

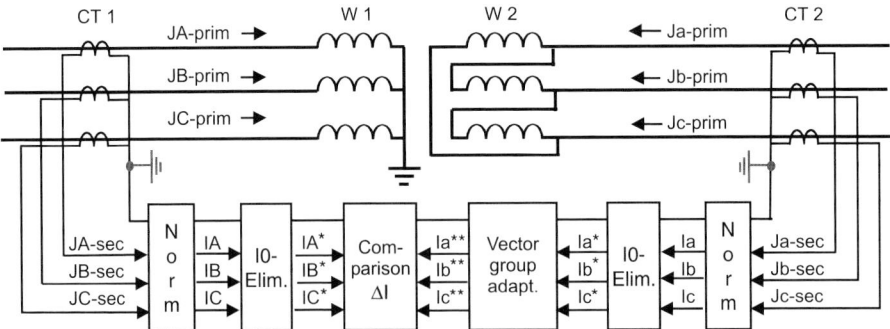

Figure 8.9 Numerical current adaptation for the comparison

As an example, the Siemens device 7UT6 is used here. The computation rules will be based on the designations according to Figure 8.9.

Initially the secondary CT currents on both sides are brought to a common base (normalised) in relation to the transformer rated power S_n (on multi-winding transformers, the winding with the larger rating is used as reference):

$$I_{\text{n-Transf.-W1}} = \frac{S_n}{\sqrt{3} \cdot U_{n\text{-}1}} \quad \text{and} \quad I_{\text{N-Transf.-W2}} = \frac{S_n}{\sqrt{3} \cdot U_{n\text{-}2}} \tag{8-3}$$

$$\begin{bmatrix} I_A \\ I_B \\ I_C \end{bmatrix} = \frac{I_{\text{n-prim.-CT1}}}{I_{\text{n-Transf.-W1}}} \cdot \begin{bmatrix} J_{\text{A-sec.}} \\ J_{\text{B-sec.}} \\ J_{\text{C-sec.}} \end{bmatrix} = k_{\text{CT-1}} \cdot \begin{bmatrix} J_{\text{A-sec.}} \\ J_{\text{B-sec.}} \\ J_{\text{C-sec.}} \end{bmatrix} \tag{8-4}$$

and $$\begin{bmatrix} I_a \\ I_b \\ I_c \end{bmatrix} = \frac{I_{\text{n-prim.-CT2}}}{I_{\text{n-Transf.-W2}}} \cdot \begin{bmatrix} J_{\text{a-sec.}} \\ J_{\text{b-sec.}} \\ J_{\text{c-sec.}} \end{bmatrix} = k_{\text{CT-2}} \cdot \begin{bmatrix} J_{\text{a-sec.}} \\ J_{\text{b-sec.}} \\ J_{\text{c-sec.}} \end{bmatrix} \tag{8-5}$$

Adaptation of the secondary CT nominal current from 1 or 5A is done by means of input transformers with link-selectable ratio adaption in the relay. In the 7UT6 relay, each input can be set to 1 or 5 A by means of jumpers. In special cases, the 5 A relay input may also be applied to advantage with a secondary CT current of 1 A (refer example 8-2).

The correction of the ratio deviation of the primary CT rated current to the corresponding transformer nominal current is then done by the software.

Subsequently the zero sequence current component is eliminated. On the star-point side this is absolutely essential if the winding star-point is earthed. If the star-point is not earthed, or the winding is delta connected, this is not necessary.

The zero sequence current is: $I_0 = \frac{1}{3} \cdot (I_A + I_B + I_C)$

The elimination in the phase currents is calculated with the following equations:

$$
\begin{aligned}
I_A^* &= I_A - I_0 \\
I_B^* &= I_B - I_0 \quad \text{or in matrix-form} \\
I_C^* &= I_C - I_0
\end{aligned}
\qquad
\begin{bmatrix} I_A^* \\ I_B^* \\ I_C^* \end{bmatrix}
= \frac{1}{3} \cdot
\begin{bmatrix} 2 & -1 & -1 \\ -1 & 2 & -1 \\ -1 & -1 & 2 \end{bmatrix}
\cdot
\begin{bmatrix} I_A \\ I_B \\ I_C \end{bmatrix}
$$

Thereafter, the vector group can be considered (on three winding transformers there are two). The high voltage winding is always used as a reference according to the vector group designation i.e. for example the star connected winding of a Yd5 transformer.

The three phase system of the low voltage winding(s) lags by the vector group number in each phase.

The adaptation can be determined directly from the connection of the winding. Alternatively, symmetrical components may be applied.

The following general equation applies, whereby k is the vector group number,

$$
\begin{bmatrix} I_a^{**} \\ I_b^{**} \\ I_c^{**} \end{bmatrix}
= \frac{2}{3} \cdot
\begin{bmatrix}
\cos[k \cdot 30°] & \cos[(k+4) \cdot 30°] & \cos[(k-4) \cdot 30°] \\
\cos[(k-4) \cdot 30°] & \cos[k \cdot 30°] & \cos[(k+4) \cdot 30°] \\
\cos[(n+4) \cdot 30°] & \cos[(k-4) \cdot 30°] & \cos[k \cdot 30°]
\end{bmatrix}
\cdot
\begin{bmatrix} I_a^* \\ I_b^* \\ I_c^* \end{bmatrix}
\qquad (8\text{-}6)
$$

For a Yd5-transformer with $k = 5$ the following result is obtained:

$$
\begin{bmatrix} I_a^{**} \\ I_b^{**} \\ I_c^{**} \end{bmatrix}
= \frac{1}{\sqrt{3}} \cdot
\begin{bmatrix} -1 & 0 & 1 \\ 1 & -1 & 0 \\ 0 & 1 & -1 \end{bmatrix}
\cdot
\begin{bmatrix} I_a^* \\ I_b^* \\ I_c^* \end{bmatrix}.
$$

The measured values for comparison in the differential protection therefore arise as follows:

$$
\begin{bmatrix} I_{\Delta\text{-}A} \\ I_{\Delta\text{-}B} \\ I_{\Delta\text{-}C} \end{bmatrix}
=
\begin{bmatrix} I_A^* \\ I_B^* \\ I_C^* \end{bmatrix}
+
\begin{bmatrix} I_a^{**} \\ I_b^{**} \\ I_c^{**} \end{bmatrix}
$$

Thereby it must be noted that currents flowing into the protected object are considered with a positive sense.

Example 8-1: Measured value adaption in numerical protection

Given:　　　Transformer Yd5 in accordance with Figure 8.10.

Wanted:　　Sequence of computation in a numerical differential protection for the illustrated external fault. Proof that the protection remains stable.

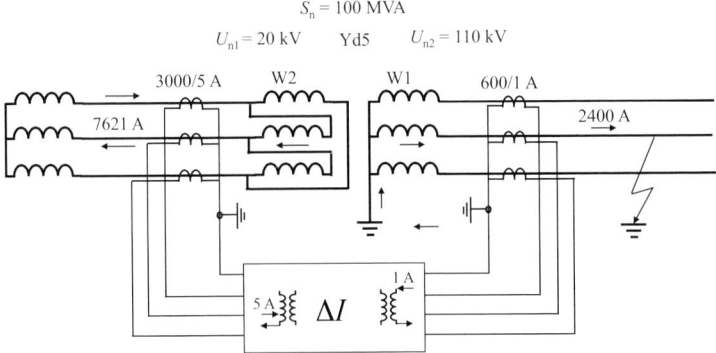

Figure 8.10 Circuit and data for example 8-1

Solution:　　The single phase fault current of 2400 A on the 110 kV side appears on the 20 kV side as a two phase current:

$$I_{F-2} = \frac{U_{1n}}{U_{2n}} \cdot \frac{1}{\sqrt{3}} \cdot I_{F-1} = \frac{110}{20} \cdot \frac{1}{\sqrt{3}} \cdot 2400 = 7621 \text{ A}$$

The high voltage winding is selected as reference winding W1.

Adaptation on the 110kV-side (winding W1):

$$I_{n\text{-Transf.-W1}} = \frac{100 \text{ MVA}}{\sqrt{3} \cdot 110 \text{ kV}} = 525 \text{ A}$$

The input transformer of the protection is set to 1 A nominal current.

Referred to the full-load current, the following results are obtained:

$$J_{A, B, C\text{-sec.}} = \frac{1}{600} \cdot 2400 = 4.0 \text{ A} \quad \text{and} \quad I_{Norm} = \frac{600}{525} \cdot 4 = 4.57 \text{ A}$$

The I_0-elimination is activated:

$$\begin{bmatrix} I_a^* \\ I_b^* \\ I_c^* \end{bmatrix} = \frac{1}{3} \cdot \begin{bmatrix} 2 & -1 & -1 \\ -1 & 2 & -1 \\ -1 & -1 & 2 \end{bmatrix} \cdot \begin{bmatrix} 0 \\ -4.57 \\ 0 \end{bmatrix}, \quad \text{therefore:} \quad \begin{aligned} I_a^* &= 4{,}57/3 \\ I_b^* &= -2 \cdot 4{,}57/3 \\ I_c^* &= 4{,}57/3 \end{aligned}$$

Adaptation on the 20 kV-side (winding W2):

$$I_{n\text{-Transf.-W2}} = \frac{100 \text{ MVA}}{\sqrt{3} \cdot 20 \text{ kV}} = 2887 \text{ A}$$

The input transformer of the protection for winding 2 is set to 5 A.

Referred to the full-load current, the following results are obtained:

$$J_{a, b, c\text{-sec.}} = \frac{1}{3000} \cdot 13200 /\sqrt{3} = 4.4 /\sqrt{3} \text{ A} \quad \text{and}$$

$$I_{\text{Norm}} = \frac{3000}{2887} \cdot 4.4 / \sqrt{3} = 4.57 / \sqrt{3} \text{ A}$$

The I_0-elimination is not required on this side, and therefore not activated.

The vector group adaptation for Yd5 is as follows:

$$\begin{bmatrix} I_a^{**} \\ I_b^{**} \\ I_c^{**} \end{bmatrix} = \frac{1}{\sqrt{3}} \cdot \begin{bmatrix} -1 & 0 & 1 \\ 1 & -1 & 0 \\ 0 & 1 & -1 \end{bmatrix} \cdot \begin{bmatrix} 4.57 / \sqrt{3} \\ -4.57 / \sqrt{3} \\ 0 \end{bmatrix}, \quad \text{therefore:} \quad \begin{matrix} I_a^{**} = -4.57/3 \\ I_b^{**} = 2 \cdot 4.57/3 \\ I_c^{**} = -4.57/3 \end{matrix}$$

The resultant tripping current (differential current) therefore is:

$$I_{\Delta\text{-A}} = I_A^* + I_a^{**} = 4.57 / \sqrt{3} - 4.57 / \sqrt{3} = 0$$

$$I_{\Delta\text{-B}} = I_B^* + I_b^{**} = -2 \cdot 4.57 / \sqrt{3} + 2 \cdot 4.57 / \sqrt{3} = 0$$

$$I_{\Delta\text{-C}} = I_C^* + I_c^{**} = 4.57 / \sqrt{3} - 4.57 / \sqrt{3} = 0$$

The protection therefore remains stable.

An external interposing CT may in some cases be required. A typical example would be on a three winding transformer when one winding is only rated for a very small load. This is illustrated with the following practical example.

Example 8-2: Application example for external interposing transformer

Given: Three winding transformer according to Figure 8.11.

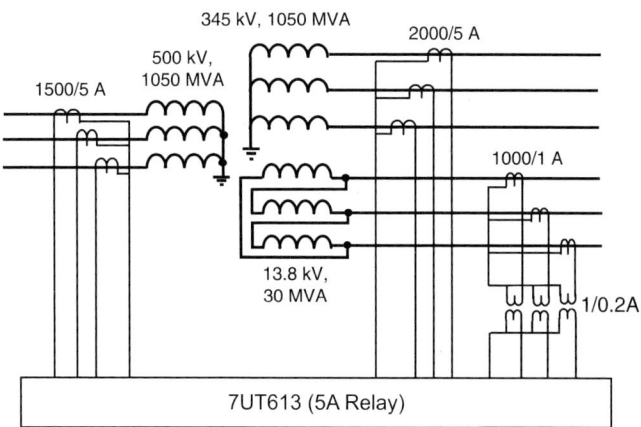

Figure 8.11 Connection diagram and data for example 8-2

Wanted: Configuration and setting of the measured value adaption

Solution: The delta winding has an extremely small rating in comparison with the other windings. The primary current of the CTs 1000/1 is matched to this small rating (30 MVA at 13.8 kV corresponds to a rated current of 1255 A).

The rated current on the high voltage winding (500 kV) is:

$$I_{n\text{-Transf.-W1}} = \frac{1050 \text{ MVA}}{500 \cdot \sqrt{3}} = 1213 \text{ A}$$

The 1500/5 A CT is matched to this. ($k_{CT\text{-}1} = \dfrac{1500}{1213} = 1.27$)

The rated current of the 345 kV winding is:

$$I_{n\text{-Transf.-W2}} = \frac{1050 \text{ MVA}}{345 \cdot \sqrt{3}} = 1757 \text{ A}$$

The 2000/5 A CT approximately corresponds to this current.

($k_{CT\text{-}2} = \dfrac{2000}{1797} = 1.11$)

The rated current of the tertiary winding is:

$$I_{n\text{-Transf.-W3}} = \frac{30 \text{ MVA}}{13.8 \cdot \sqrt{3}} = 1255 \text{ A}$$

The CT 1000/1 is provided for this current. It is roughly matched to the output rating (30 MVA) and provides the correct transformation ratio for an overcurrent back-up protection in this location. It however has an extreme mismatch to the rating of the main winding (1050 MVA), to which the differential protection must be adapted.

An adaption to 1050 MVA would correspond to a current of I_{W3} = 1213·500/13.8 = 43,949 A on the 13.8 kV side. An adaptation by the factor $k_{CT\text{-}3}$ = 1000/43,949 = 1/44 would therefore be required.

With the selection of 1 A as CT secondary current, with a 5 A relay, a part of the adaption (factor 5) is already obtained for the mismatch of the input CT. A mismatch of 5/44 = 1/8.8 still remains.

With the relay 7UT6 it is recommended, due to internal processing accuracy, to limit the ratio for the internal adaptation to $1/4 \leq k_{CT} \leq 4$. An external interposing CT 5/1 (1 to 0.2 A) is therefore applied, so that the adaptation factor is further reduced to 5/8.8 = 1/1.76.

The current transformation chain consisting of CT + interposing CT and 5/1 false adaptation of the relay input CTs results in a total ratio of (1000/1)·5/1)·(5/1) = 25,000. The relay setting point "Primary rated current of the CT" must therefore be set to this effective CT ratio of 25,000 A for the tertiary winding.

When defining the dimension of the interposing CT, it must be noted that the current level is very small during short-circuits at the tertiary side of the transformer. Referred to 1000/1 A, not much more than ten times rated current is to be expected. The interposing CT can therefore be dimensioned for 1/0.2 A. With the multi-tap type 4AM5170 (see section 5.8, Figure 5.35) the correct choice would be 8 to 40 turns.

The above example illustrates the combined effect of CT, external interposing transformer, input transformer and numeric calculation for the adaptation of the measured values. The interposing CT and the measuring input can be included in equation (8-4) to obtain the following result for the internal current that is processed in the relay (winding 1 as example):

$$\begin{bmatrix} I_A \\ I_B \\ I_C \end{bmatrix} = \frac{I_{\text{n-Interpr.CT-relay side}}}{I_{\text{n-Interpr.CT-c.t. side}}} \cdot \frac{I_{\text{n-sec.-CT-1}}}{I_{\text{n-relay-MI-1}}} \cdot \frac{I_{\text{n-prim.-CT1}}}{I_{\text{n-Transf.-W1}}} \cdot \begin{bmatrix} J_{\text{A-sec.}} \\ J_{\text{B-sec.}} \\ J_{\text{C-sec.}} \end{bmatrix} \qquad (8\text{-}7)$$

$I_{\text{n-relay-MI-1}}$ is the nominal current of the measuring input 1 of the relay. Generally, the nominal current of each measuring input is the same as the relay nominal current. If the secondary nominal currents of the CTs on the transformer terminals are different, the measuring inputs must however be matched by jumper selection.

In the above example, the mismatch is used to advantage to provide the equivalent of a 5/1 interposing CT. The following results:

$$\begin{bmatrix} I_A \\ I_B \\ I_C \end{bmatrix} = \frac{0.2\ \text{A}}{1\ \text{A}} \cdot \frac{1\ \text{A}}{5\ \text{A}} \cdot \frac{25000\ \text{A}}{44000\ \text{A}} \cdot \begin{bmatrix} J_{\text{A-sec.}} \\ J_{\text{B-sec.}} \\ J_{\text{C-sec.}} \end{bmatrix} = \frac{1000}{43949} \cdot \begin{bmatrix} J_{\text{A-sec.}} \\ J_{\text{B-sec.}} \\ J_{\text{C-sec.}} \end{bmatrix}$$

That means an adaptation to the rating 1050 MVA.

Measuring algorithm

The adapted measured values are subjected to numeric evaluation according to the differential protection measuring principle.

The pick-up characteristic of relay 7UT6 has three sections which is typical for numerical protection. (Figure 8.12)

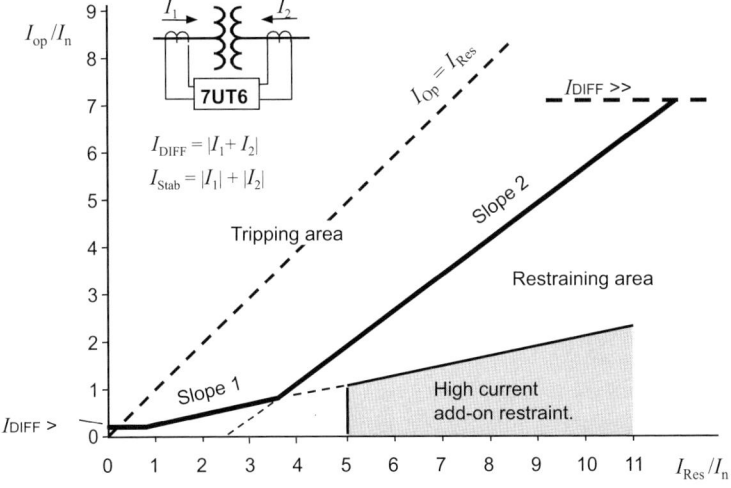

Figure 8.12 Pick-up characteristic of the transformer differential protection (Relay 7UT6)

Compared to generator protection, the basic pick-up threshold $I_{DIFF}>$, and the slope of the first branch must have less sensitive settings as the magnetising currents of the transformer, and ratio errors due to tap changers result in false differential currents. Adaptation of the transformation ratio in accordance with the tap changer tap position would in theory be possible, is however not implemented in practice due to the increased complexity.

A common setting value is 20% I_n for transformers without tap changer and up to 30% on transformers with tap changers having ±22% tap changer range.

Setting the slope of the first branch to 25% and the second to 50% is appropriate for normal applications. Stability in the event of large differential currents that arise in the event of CT saturation is ensured by the integrated saturation detector with temporary blocking of the trip output

The saturation detector is activated when the restraint current enters the area of add-on restraint, shown in Figure 8.12. This occurs in case of through flowing currents during the first milli-seconds after fault inception, i.e. in the initial saturation-free time of the CTs. (Refer to section 4.2.4, Figure 4.17)

The pick-up value can be set to allow adaptation to the expected transient performance of the CTs. ($I_{Res}/I_n = 5$ in the shown diagram, corresponding to a through flowing current of 2.5 times rated current.)

False tripping due to in-rush currents or due to over-excitation is prevented by the harmonic restraint functions described above. During internal faults with severe CT saturation, harmonics may also be present, which can cause delay of the trip output. In this case; the high set stage $I_{DIFF}>>$ will however respond.

High differential current stage

The short-circuit current flowing through a transformer is limited by the short-circuit reactance to $I_n \cdot (100/u_T \%)$. In the event of external faults, the differential current can therefore also not be greater than this, even under the most severe non-symmetrical saturation of the CTs.

If the tripping currents are greater than this maximum current, then tripping without any stabilising may be carried out without delay.

The transformer differential protection 7UT6 has a corresponding non-stabilised high-current stage. It ensures fast tripping also in those cases when during internal faults with large DC off-set short-circuit current and severe CT saturation, a temporary pick-up of the rush-blocking takes place. As extreme saturation is not to be expected with the limited short-circuit currents flowing through the transformer, $I_{DIFF}>>$ may also be set to a value below the current corresponding to the short-circuit reactance.

The high-current stage responds to the fundamental wave of the short circuit current which means that the DC component and the high order harmonics of the rush currents are eliminated. Including a security margin of 20%, a setting of approximately 60% of $\hat{I}_{Rush}/\sqrt{2}$ may be applied as the maximum value of the fundamental wave in the rush current is only approximately 50% of the rush current peak. The factory pre-setting is $7.5 \cdot I_{\text{n-Transformer}}$. It should be appropriate for most applications.

In case of internal transformer faults near the terminals, high fault currents may lead to fast CT saturation. This is always the case when small transformers are connected to a system with high short-circuit power.

Therefore, a fast momentary value processing very high-set element $I_{\text{DIFF}}\!>\!>\!>$ is additionally provided in the relay 7UT6. It operates when two samples exceed twice the $I_{\text{DIFF}}\!>\!>$ setting value. This ensures ultra high speed tripping before saturation occurs. The measuring principle was discussed in section 4.2.2.

Earth-fault differential protection

During an earth-fault on an earthed transformer winding, short-circuit currents that can cause severe damage will flow. On the in-feed side the corresponding currents may be relatively small if the short-circuit current is only linked by a few turns of the secondary winding.

On solidly earthed winding star-points the ratio is extreme if the fault is only a few turns away from the star-point (Figure 8.13).

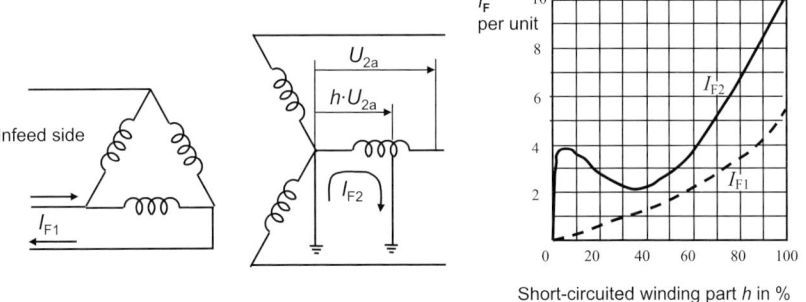

Figure 8.13 Earth-fault on a transformer winding with solid earthing [8-13]

The reactance is inversely proportional to the square of the short-circuited turns, while the inducing voltage decreases linearly. Accordingly the parabolic curve results, whereby the winding resistance has a limiting action close to the star-point.

Figure 8.13 also applies for inter-turn short-circuit whereby the current then does not return via earth but flows via the short-circuit bridge without earth connection. It must be noted that the fault current I_{F} in this case is not detected by a relay connected at the star-point earth.

If the star-point is earthed via an impedance, the conditions shown in Figure 8.14 apply. The secondary current is linearly proportional to the number of short-circuited turns, while the current at the in-feed side is inversely proportional to the square of the number of short-circuited turns.

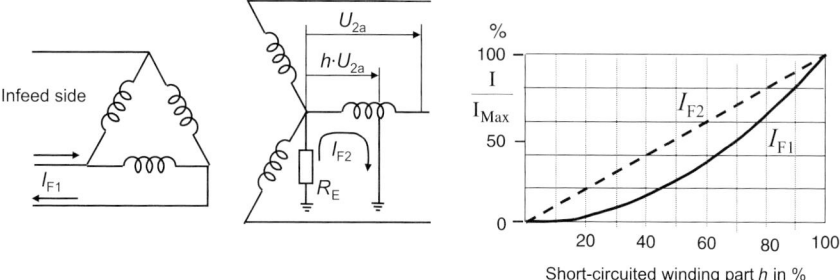

Figure 8.14 Earth-fault on a transformer winding with resistive earthing [8-13]

The fault currents may in this case easily be calculated:

$$I_{F2} = \frac{h \cdot U_{2a}}{R_E} \quad \text{and} \quad I_{F1} = h \cdot \frac{U_{2n}}{U_{1n} \cdot \sqrt{3}} \cdot I_{F2} = h^2 \cdot \frac{1}{\sqrt{3}} \cdot \frac{U_{2n}}{U_{1n}} \cdot \frac{U_{2a}}{R_E} \quad (8\text{-}8)$$

The primary short-circuit current is very small for faults close to the star-point, so that the differential protection will only pick up for earth-faults closer to the transformer terminals.

The following is an example of this:

Example 8-3: Currents during transformer earth-faults

Given: Transformer 20 MVA, U_{1n} = 132 kV, U_{2n} = 13.8 kV

             Circuit according to Figure 8.14.

             CT on the 132 kV side: 100/1 A

             Earth current limited by R_E to 2000 A

             Fault location 20 % from the star-point.

Wanted: Will the differential protection detect the fault (setting 25 % I_N)?

Solution: The primary short-circuit current (= tripping current) is:

$$I_{F1} = \left(\frac{20\%}{100}\right)^2 \cdot \frac{13.8}{132} \cdot \frac{1}{\sqrt{3}} \cdot 2000 \text{ A} = 4.8 \text{ A}$$

With a setting of 25 % I_n, which is equivalent to 25 A, the differential protection would fail to pick up.

The protection coverage during earth-faults on the 13.8 kV winding can be estimated by using formula (8-8):

$$\left(\frac{h[\%]}{100}\right)^2 \cdot \frac{13.8}{132} \cdot \frac{1}{\sqrt{3}} \cdot 2000 \text{ A} = 25$$

Solving this equation, we get h = 45.4 %

The earth fault protection range would therefore only cover 100 – 45.5 = 54.5 % of the 13.8 kV winding counted from the winding terminals down to the winding neutral.

In the example at hand, it is therefore advisable to connect a CT with earth current relay (pick-up threshold 200 A) in the earth connection of the star-point, to increase the range of protection coverage to 90%. This protection however requires a large time delay setting as the earth current relays in the system must trip faster to maintain selectivity.

Below, the earth current differential protection is described. It facilitates non-delayed tripping for this fault condition.

Restricted earth-fault protection

The earth current differential protection (restricted earth-fault protection) is an ideal supplement of the phase fault protection, in particular on transformer windings with star-point earthing via an impedance (earth current limiter). Thereby the pick-up sensitivity during earth-faults is improved. [8-14]

The protection principle makes use of a comparison of the star-point current I_0^* with the summated (residual) current of the feeder I_0^{**} (Figure 8.15).

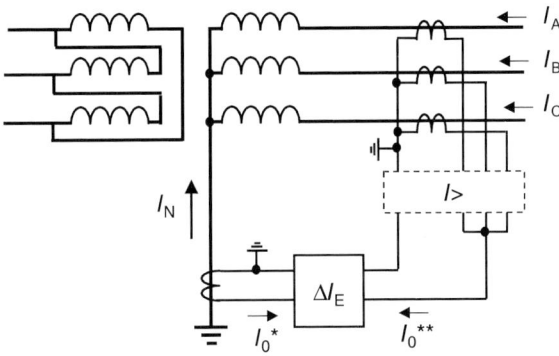

Figure 8.15 Restricted earth-fault protection

To improve the pick-up sensitivity while maintaining good selectivity, the so-called product relays were applied with conventional technology. These polarise the tripping current with the current of the star-point so that the protection obtains its highest sensitivity (maximum torque) when the tripping current (differential current) and the star-point current are in the same direction (have the same polarity).

The same was reached by the rectifier bridge comparator with moving coil relay used for directional discrimination.

The numerical protection 7UT6 uses this amplitude comparator principle with numerical technology. (Figure 8.16)

The following relationships apply:

$$I_0^* = I_N$$

$$I_0^{**} = I_R + I_S + I_T = 3 \cdot I_0$$

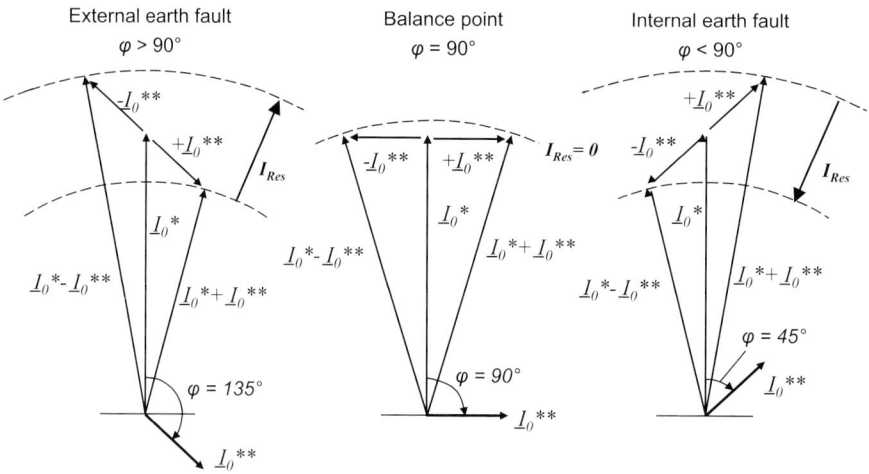

Figure 8.16 Restraint current of REF protection in 7UT613

Two angle ranges must be distinguished for the tripping criterion:

The *basic tripping range* is defined by the approximate phase coincidence of $\underline{I}_0^*$ and $\underline{I}_0^{**}$: $-90° \le \varphi(\underline{I}_0^*/\underline{I}_0^{**}) \le +90°$. In this range, no stabilising applies. The tripping current corresponds to the star-point current $\underline{I}_0^*$ so that the protection has a constantly low pick-up threshold: $I_{Op} = I_0^* > I_{set}$.

The *extended tripping range* allows angle differences $\varphi > 90°$, which, during internal faults, may occur in the presence of extreme CT saturation. This range however is dynamically limited by an additional restraint:

$$I_{Res} = \left| \underline{I}_0^* - \underline{I}_0^{**} \right| - \left| \underline{I}_0^* + \underline{I}_0^{**} \right| \tag{8-9}$$

It becomes effective when I_{Res} is positive, that means only in the angle range $+90° \le \varphi(\underline{I}_0^*/\underline{I}_0^{**}) \le +270°$.

The tripping condition in this extended range is as follows:

$$\left| I_0^* \right| > I_{set} + k_0 \cdot I_{Res} \tag{8-10}$$

Therefore:

$$\left| I_0^* \right| > I_{set} + k_0 \cdot \left(\left| \underline{I}_0^* - \underline{I}_0^{**} \right| - \left| \underline{I}_0^* + \underline{I}_0^{**} \right| \right) \tag{8-11}$$

During *external earth-faults*, $\underline{I}_0^*$ and $\underline{I}_0^{**}$ are in the same direction (have opposite signs), so that only restraint current flows and no differential current results. I_{Res} is positive and therefore effective.

During *internal earth-faults* $\underline{I}_0^*$ and $\underline{I}_0^{**}$ flow into the protected object and therefore have the same sign (opposite direction). Only if the difference angle φ exceeds 90° the restraint becomes increasingly effective.

The following equation can be derived for the operating characteristic:

$$\frac{I_0^*}{I_{set}} = \frac{1}{1 + \sqrt{2} \cdot k_0 \cdot \left(\sqrt{1 + \dfrac{I_0^{**}}{I_0^*} \cdot \cos \varphi} - \sqrt{1 - \dfrac{I_0^{**}}{I_0^*} \cdot \cos \varphi} \right)} \qquad (8\text{-}12)$$

The angle φ corresponds to the phase shift between $\underline{I}_0^{**}$ and $\underline{I}_0^*$.

In Figure 8.17 the pick-up characteristic is shown in polar coordinates for $\underline{I}_0^{**}/\underline{I}_0^* = 1$. The amplitude scale in this diagram is referred to the set threshold value ($\underline{I}_0^*/I_{set}$).

The pick-up current dependence on the phase angle between $\underline{I}_0^*$ and $\underline{I}_0^{**}$ is clearly visible.

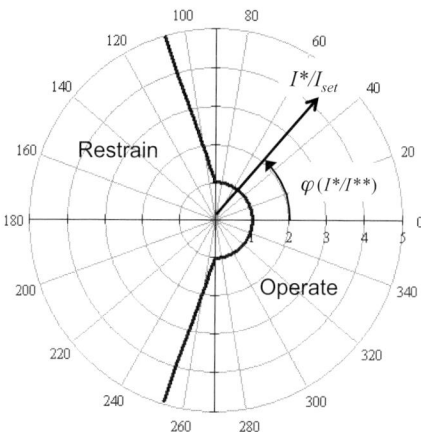

Figure 8.17
Polar characteristic of the earth-current differential protection (device 7UT613)

The right hand side angle range ($\varphi = \pm 90$) of the diagram designates the area of internal faults. Here, the protection has high sensitivity. The "ideal" internal fault is located at $\varphi(\underline{I}_0^* / \underline{I}_0^{**}) = 0°$.

The extended tripping range extends from 90° ($k_0 \to \infty$) up to an angle resulting from the set biasing factor k_0 (here 110° for $k_0 = 2$). The larger the k_0-factor setting is, the more the tripping range will be restricted. (The relay 7UT6 uses a fixed setting $k_0 = 4$, corresponding to 100°)

The „ideal" external fault appears on the left hand side at 180°.

In the angle range $\varphi = \pm 90$ the pick-up threshold is small. When the angle 90° is exceeded, the pick-up threshold increases sharply depending on the k_0 setting.

In this manner, high sensitivity for internal faults and high stability during external faults is obtained.

8.3 High impedance differential protection

The HI protection on transformers is applied in two versions – as restricted earth-fault protection and as differential protection on auto-transformers.

HI restricted earth-fault protection

The normal restricted earth-fault protection was described in the previous section. The HI version is often applied in Anglo-Saxon protection practice because of its simple construction and high stability during external faults with CT saturation. Separate CT cores of type Class X (PX to IEC 60044-1) with equal transformation ratio are however required. (Refer to section 3.4)

It is usually applied on earthed Y-connected windings. The three phase CTs and the star-point CT are connected in parallel to the relay (+ series resistance).

On a delta winding, in an earthed system, the HI protection may also be applied as shown on the left hand side in Figure 8.18. The sum of the phase currents must always be 0 so that no large voltage appears across the differential branch. In the event of an earth-fault in the delta winding, the current sum is no longer 0, and the HI protection will pick up.

The lay-out is done the same as on the star connected winding, only in this case the 4th CT is omitted.

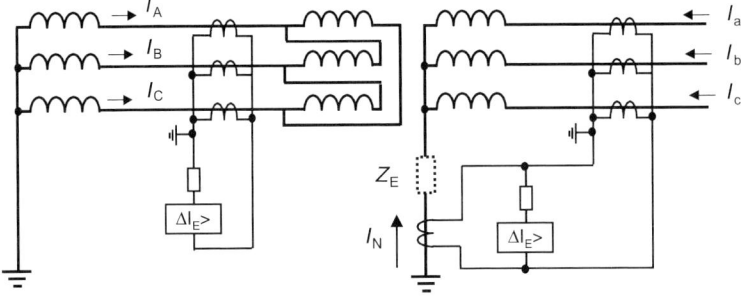

Figure 8.18 Application of the HI restricted earth-fault protection

HI-protection on auto transformers

On auto transformers, the HI protection can be applied across the galvanically connected windings (Figure 8.19). For this purpose, the winding ends at the star-point side must be accessible. This provides an alternative to the normal differential protection. It must however be noted that the delta connected winding is not covered by this protection. On large transformer banks, both protection principles are sometimes applied in parallel.

If the winding ends at the star-point side are not individually accessible, then a HI restricted earth-fault protection can be applied to improve the pick-up sensitivity during earth-faults, as was already demonstrated for machines and normal transformers (Figure 8.20).

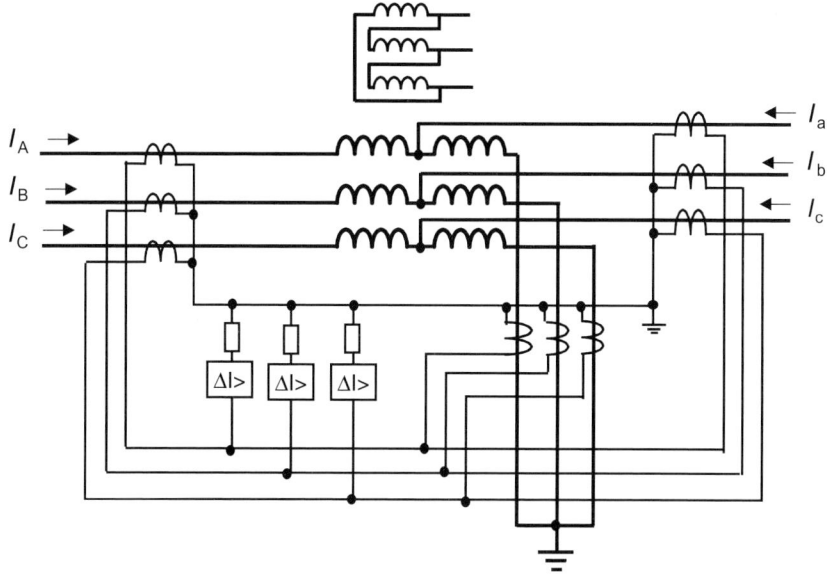

Figure 8.19 HI protection for an auto transformer

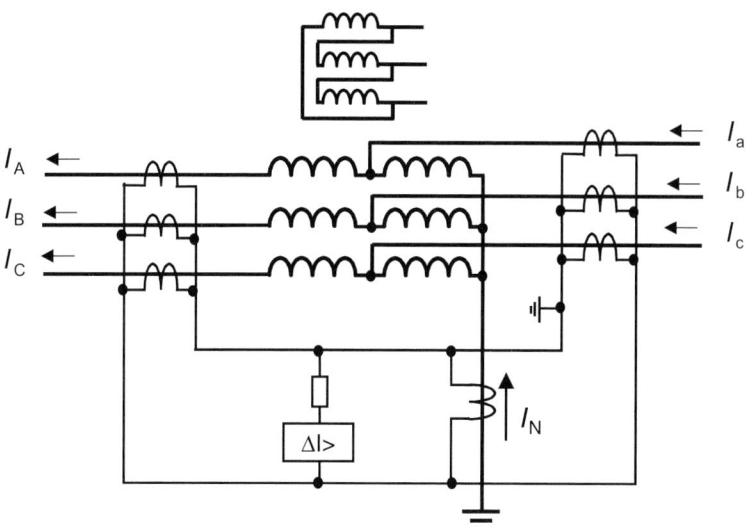

Figure 8.20 HI-earth differential protection on auto transformers

Setting the high impedance protection on transformers

In principle, the rules stated in section 3.4 apply here. Detailed rules of application for the various forms of high impedance protection can be found in the British ESI (Electricity Supply Industry) Standard 48-3: "Instantaneous High-Impedance Differential Protection".

The following recommendations are given there:

Pick-up threshold:

Solid earthed system: 10 to 60% of the rated winding current

System with earth current limiting: 10 to 25% of the terminal short-circuit current

The stability must be calculated with the maximum fault current flowing through the protected winding (if no detailed data is available: $16 \times I_{n\text{-}T}$).

8.4 Relays for transformer differential protection

The standard relays are suitable for two or three winding transformers (7UT612/613). [8-15, 8-16]

For application on 1-1/2-circuit breaker plants (refer to the examples hereafter) a version suitable for up to 5 current inputs is also available (7UT635).

Apart from the differential protection, the relays also contain further protection functions such as overload protection, overcurrent protection, restricted earth-fault protection, breaker failure protection and optional tank protection. The application is shown in the following examples.

The tank protection is a peculiarity in the French protection technology and is mostly applied instead of the differential protection. For this purpose, the transformer is mounted with isolation to earth and an overcurrent relay is connected in the tank to earth connection (Figure 8.21). The relay responds to the earth current that flows when there is a fault between the winding and the tank.

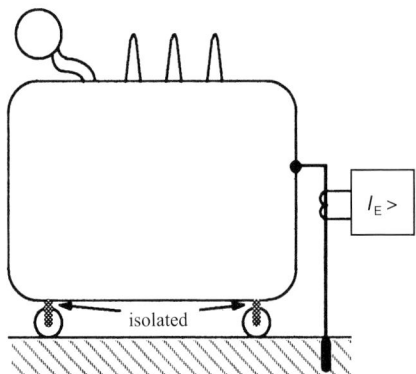

Figure 8.21
Tank protection

On the 7UT6 relay, a sensitive current input (0.01 to $1.0 \times I_n$) is provided for this function.

Figure 8.22 shows the various connection alternatives of the relay 7UT613.

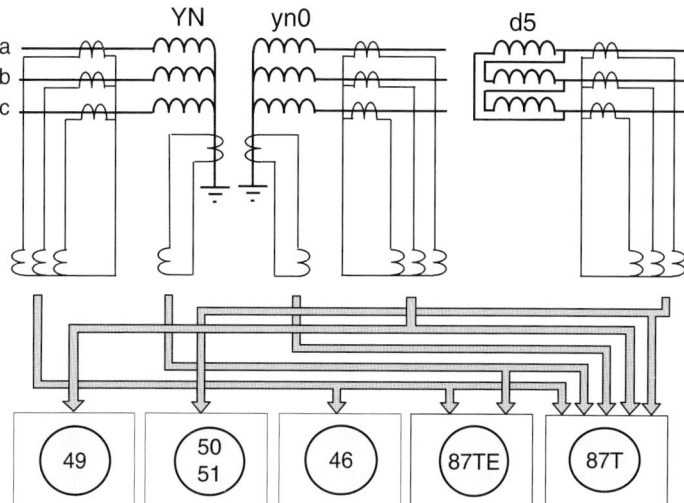

Figure 8.22 Relay 7UT613, protection functions and connection

The two inputs for the earth currents flowing in the star-points may be used to increase the pick-up sensitivity during earth-faults; either as I_0-correction of the differential protection 87T or as separate restricted earth-fault protection 87TE. One input may alternatively also be applied for tank protection.

The allocation of protection functions to the windings can be determined via setting parameters. If the relay 7UT613 is applied to a two winding transformer, the third phase measuring input together with the earth current input may for example be applied as restricted earth-fault protection of a separate earthing transformer (refer to relay manual).

Relays for high impedance protection

Due to its principle, the HI protection cannot be implemented in its entirety with digital measures. It is only possible for the measuring element. In the numerical relays 7UT613 and 7UM62 a measuring input and sensitive current relay function for the restricted earth-fault protection is available.

It is however common practice to still apply simple analog static relays.

Some relays have an integrated series resistance and are calibrated as voltage relays while others are simple current relays with separate series resistor.

The relay 7VH60 (single phase) available from Siemens is a current relay with a pick-up threshold of 20 mA and integrated series resistance. The pick-up threshold of the voltage can be set between 6 and 60 or 24 and 240 V. The additional varistor required for limiting the voltage must be connected externally. The relay documentation contains detailed configuration notes.

8.5 Application examples for transformer protection

The following examples show the application of numerical relays for transformer protection.

The individual functions are designated, using ANSI codes (American National Standard C37.2: "Electrical Power System Device Function Numbers", refer to the list in the addendum).

The integrated supplementary functions in each relay may be applied advantageously. It must however be noted that back-up protection functions for redundancy must be provided in a separate hardware (a further relay). Therefore, the overcurrent protection included in the differential protection 7UT6 can only be implemented as back-up protection for external faults in the connected system. The back-up protection for the transformer itself must be provided by a separate overcurrent relay (e.g. 7SJ600). The Buchholz (gas pressure) protection providing fast short-circuit protection is provided together with the transformer.

Two-winding transformer protection

Apart from the Buchholz protection (Bu), the differential protection 7UT612 is provided as the second fast short-circuit protection. (Figure 8.23)

The integrated time graded overcurrent protection (51) provides back-up protection for faults in the supplied system. A separate overcurrent protection on the low voltage winding side is therefore not necessary. The relay 7SJ600 provides the

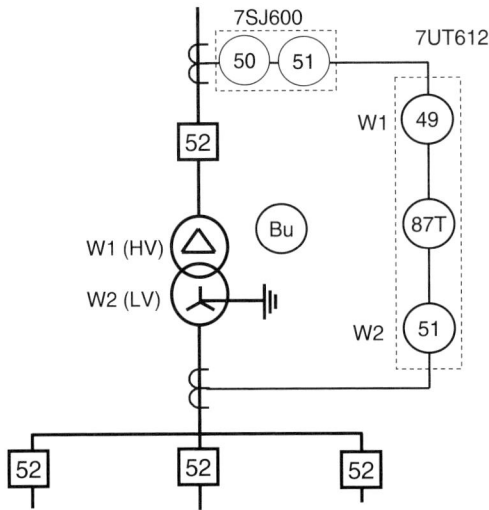

Figure 8.23
Two-winding transformer
protection

back-up protection for circuits in the transformer, and additional back-up protection for low voltage side faults. The high-set tripping stage $I\!>\!>$ (50) must be set above the through-flowing short-circuit current so that it does not pick up in the case of faults at the low voltage winding side. The delayed tripping $I\!>$,t (51) must be time graded to not trip faster than the overcurrent protection in the 7UT612.

Three-winding transformer protection

The recommendations made for the two-winding transformer in principle also apply here. The relay for three-winding transformers 7UT613 must however be used in this case. The negative sequnce O/C-function (46) is here used as sensitive back-up protection against unsymmetrical faults. (Figure 8.24) In the event of very small rating of the tertiary winding and corresponding CT transformation ratio it must be checked whether an external interposing CT is necessary (refer example 8-2).

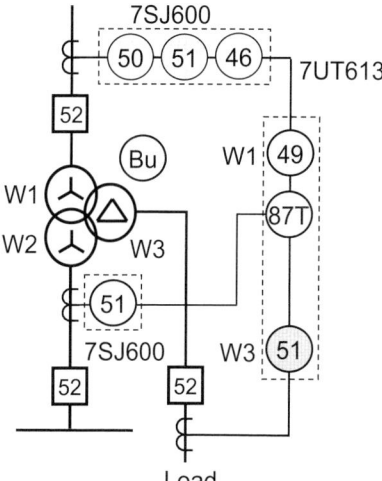

Figure 8.24
Three-winding transformer protection

Transformer protection with restricted earth-fault protection

The restricted earth-fault protection 87TE (REF for short), can always be applied at all windings with earthed neutral. It should in particular be applied where the winding neutral is earthed via a resistance or reactance, i.e. when the earth-fault current is restricted (Figure 8.25). A CT must however be available in the winding neutral to earth connection.

The REF provides fast and selective tripping in the event of an earth-fault on the winding W2. This was referred to in section 8.2, Figure 8.14. A pick-up sensitivity ≤10% of the earth-fault current at the transformer terminals (90% protection coverage) should be strived for. The supplementary function 87TE is integrated in the device 7UT613

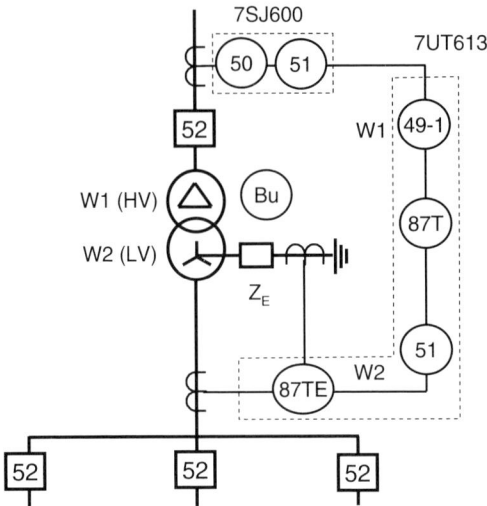

Figure 8.25
Transformer protection
with restricted earth-fault
protection

Transformer protection for 1-½ circuit breaker arrangements

For 1-½ CB applications, the transformer differential protection must be connected via separate stabilising current inputs to two sets of CTs. If this is the case at the high as well as low voltage side, then a total of 5 stabilising current inputs are required for a three-winding transformer (Figure 8.26).

A parallel connection of the CTs, as is shown with the distance protection, is not permissible with the differential protection, as fault currents flowing through the two CTs connected in parallel would cancel each other out, so that no stabilising

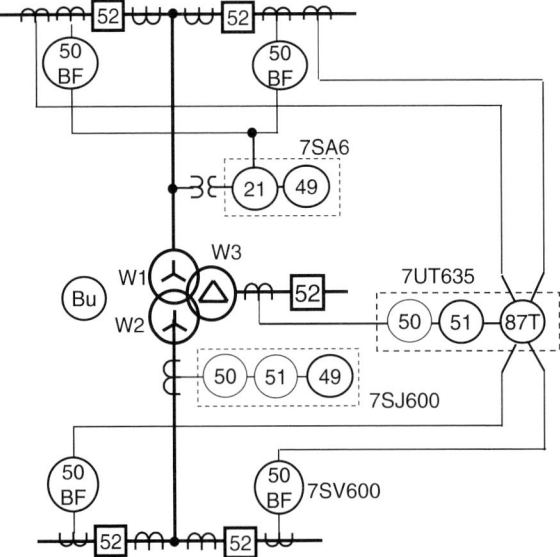

Figure 8.26 Transformer protection for 1-½-CB arrangements

current is generated. With the usual setting of 20-30% I_n the not stabilised differential protection would then already tend to mal-operate for very small deviations in the CT response (1% CT non-symmetry at $20 \cdot I_n$ through fault current would already result in 20% I_n differential current) . A less sensitive setting of the pick-up threshold is not acceptable due to the loss of sensitivity for internal faults.

On the high voltage side, a distance protection (21) is assumed, for providing back-up protection. On the low voltage side this is provided by an overcurrent protection (50/51). The overcurrent protection is connected to the bushing CTs of the transformer. The breaker-failure protection on busbars with 1-$^1/_2$ CB configuration is provided by a dedicated relay for each CB as shown in the Figure. The winding currents are significant for the overload protection (49). For the high voltage winding the integrated function in the distance protection can therefore be used, while the corresponding function in the overcurrent protection can be applied for the low voltage.

Auto-transformer protection

The protection configuration (Figure 8.27) is in principle the same as that of a normal separate-winding transformer.

A differential protection 87T suitable for a three-winding transformer (7UT613) must be selected so that the delta-connected tertiary winding can be included in the zone of protection.

To increase the earth-fault sensitivity, it is Anglo-Saxon protection practice to often apply an additional restrictive earth-fault protection 87TE. The single-phase HI relay (7VH600) is shunt-connected to the CTs of the transformer terminals and

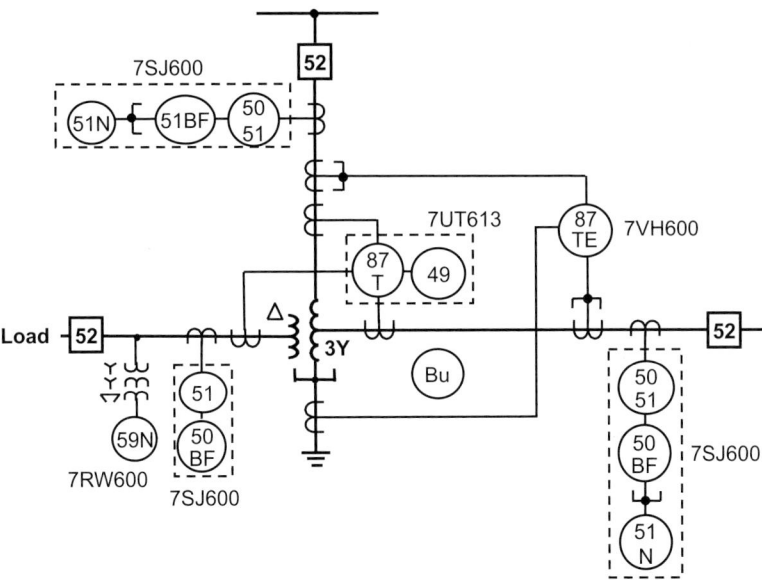

Figure 8.27 Auto-transformer protection

star-point. These 7 CTs must however have the same class according to Class PX of IEC 60044-1 and are dedicated to this protection function only. A pick-up sensitivity of $10\% I_n$ is generally strived for. Alternatively, an overcurrent relay $I_E>,t$ (51N) may also be applied in the transformer star-point. This would however have to be time graded with the system earth-fault relays to ensure selective tripping.

The delta connected winding which is often only used for auxiliary supply is provided with a separate overcurrent relay (51) to cover external phase faults. The voltage relay $U_E>$ (59N) connected to the open (broken) delta winding of the voltage transformers measures the displacement voltage $3 \cdot U_0$, which indicates an earth-fault in the tertiary winding or in the connected distribution system.

At the high and low voltage side of the auto-transformer, an overcurrent relay 7SJ600 is connected with its high set tripping stage $I>>$ (50) and time delayed stage for phase faults $I>,t$ (51) and earth-faults $I_E>,t$ (51N).

At each terminal the breaker failure protection (50BF) is activated in the corresponding relay.

As was shown above in the example 8-2 for the three-winding transformer, it must again be checked here whether an interposing CT is required on the delta winding for the differential protection 87T.

Protection for a large transformer bank

For very large transformer banks, for example a system coupler, a redundant protection with high speed tripping should be provided.

In central Europe the differential protection is usually duplicated and connected to separate CT cores. (87TP and 87TS)

Anglo-Saxon protection usually applies a normal low impedance differential protection (87TP) connected to the transformer bushing CTs together with a HI differential protection 87TS connected to CTs at the switch gear (Figure 8.28). In particular when connecting in $1\frac{1}{2}$ CB applications, implementing 87TS as HI protection has the advantage that the numerous CTs can simply be connected in parallel. With normal low impedance protection, the connection to CTs of $1\frac{1}{2}$ CB switchgear requires a four-terminal differential protection version (compare previous examples).

The following should be further considered for planning of the auto-transformer protection:

Relay 87TS is not influenced by the tap changer (changing transformer ratio), however it does not protect the tertiary winding.

A distance protection (21) is applied to both the high and low voltage side. The fast tripping stage is set to reach about 80% into the transformer thereby providing fast protection in this range. It is also possible to apply a zone that reaches through the transformer in combination with a directional comparison logic to obtain 100% protection coverage. In the latter case, the redundant differential protection may be omitted.

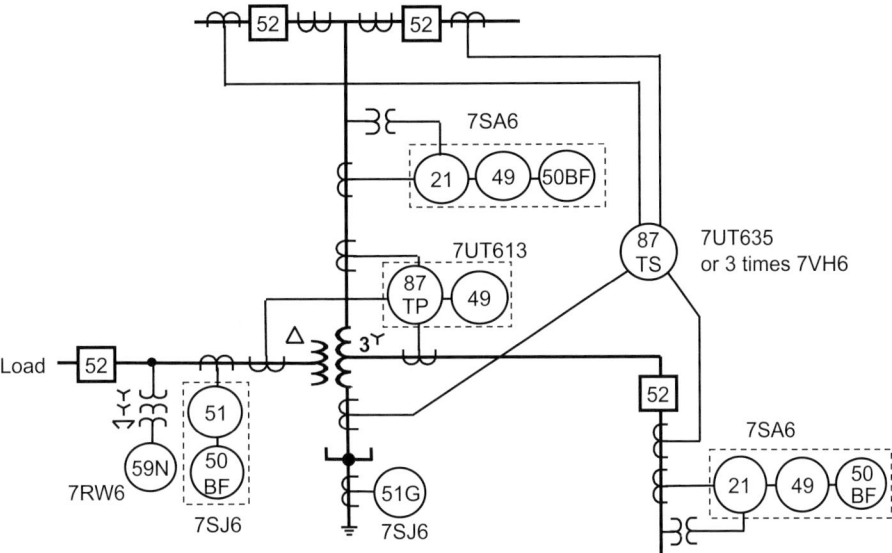

Figure 8.28 Protection of a large transformer bank

For the remaining protection functions the comments made in the previous example apply also here.

Protection of regulating transformers

The regulating transformer adds a longitudinal or quadrature variable voltage to the feeder voltage. With the longitudinal voltage, the voltage magnitude and reactive power flow is controlled while the quadrature voltage regulates the active (real) power flow. A combined regulating transformer contains both longitudinal and quadrilateral regulation.

The complete regulating unit consists of a phase regulator and an exciting transformer which can be housed in a single tank or in separate tanks. Various designs and winding arrangements occur in practice. [8-17 to 8-19]

Depending on the construction, the excitation transformer can be connected in star or delta configuration.

The protection configuration for these complex transformers demands a detailed analysis and model simulation of the load and fault current distribution. Figure 8.29 shows an example which was described in detail in several publications. [8-20, 8-21]

The primary differential protection 87TP is the protection for the auto transformer which also covers the primary winding of the exciting transformer. The numerical relay 7UT613 is well suited for this purpose. Alternatively, a high impedance protection could also be applied for this purpose, as shown in Figure 8.28.

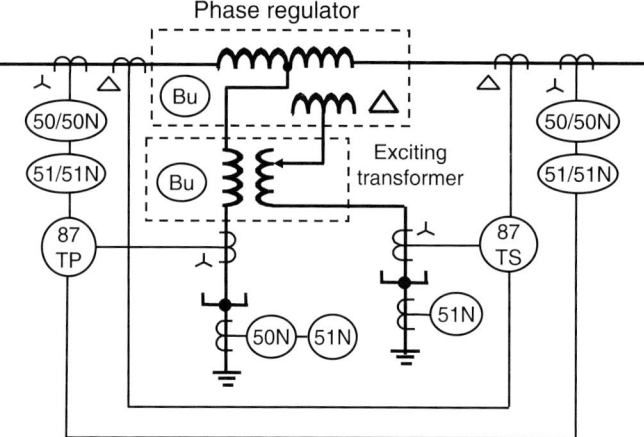

Figure 8.29 Protection of a phase shifting regulating transformer

The phase regulating transformer connection makes a star delta conversion necessary for the secondary differential protection 87TS (illustrated by a triangle in the figure). The 7UT613 is also suitable for this purpose.

It must however be checked if saturation of the regulator series winding is possible during external faults considering the lower voltage rating (40 to 50% phase to neutral rated voltage). The saturation would unbalance the ampere turns ratio of the phase angle regulator and may consequently cause false operation of the differential protection 87TS. Desensitising or blocking the 87S protection in case of overfluxing would then be necessary, for example by a V/Hz monitoring function. [8-21, 8-25]

The analysis of current distribution and measured value adaptation goes beyond the scope of this book. It is covered in detail in the referred publications and may be used as an illustrative example. [8-20, 8-21, 8-25, 8-26]

Additional current relays for earth-faults and back-up protection must be applied as shown in Figure 8.29. Alternatively distance protection may also be applied as back-up protection.

For further application cases of phase regulator protection, refer to [8-23, 8-24].

The modern relaying technology allows not only the numerical vector group correction of star-delta transformers in steps of 30° but also a finer angle adaptation necessary for phase shifting transformers. With the knowledge of the tap position of the phase regulator, the relay can calculate the current shift between high and low voltage side and automatically adapt the currents for comparison in the overall differential protection. [8-27]

Protection of generator transformer units

The unit protection encompasses a number of protection components.[7-1] Only the differential protection is elucidated on here.

The common configurations are shown in Figure 8.30 (without unit auxiliaries).

The necessary ratio and vector group correction is represented in the diagram by interposing CTs. In the case of numerical relays this adaptation is done by numerical computation in the relay. This was explained in detail in section 8.2.

On small generators, the differential protection covers both the generator and transformer (overall differential protection).

On larger units, a dedicated differential protection is provided for the generator, on the one hand for the higher sensitivity (10-15% instead of 25-30% I_N), and on the other hand for the selective indication of generator faults.

The variants b) and c) are available for this purpose. The German Power System Relaying Committee recommends variant b).

On a unit with generator breaker, the variant d) may be applied, because the unit transformer which also provides the auxiliary supply may be started and operated separately.

The complete arrangement of the differential protection on a large unit is shown in Figure 8.31. The alternatives for the CT connections are shown with dashed lines. If the differential protection of the unit transformer is connected to the CT core in front (at the generator side) of the auxiliary supply transformer (standard connection according to the German Power System Relaying Committee), then the CT primary current should be matched to the generator CTs. In any event it should not be smaller by more than factor 4, as an external interposing CT would be required in this case (refer to example 8-2).

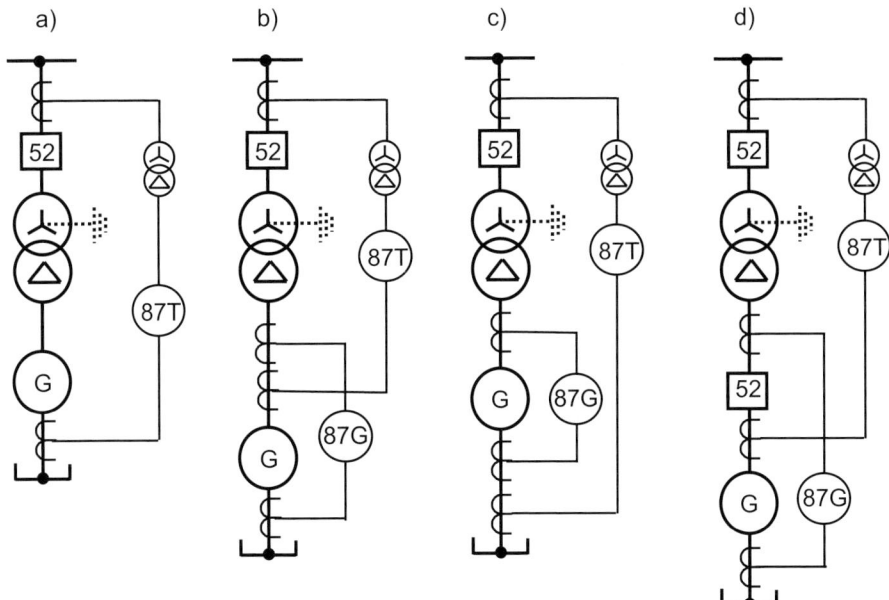

Figure 8.30 Various differential protection schemes for generator units

Based on the rated current of the auxiliary transformer, the CT must be dimensioned for a large overcurrent factor if the protection of the unit transformer must remain stable in the event of a short circuit at the auxiliary transformer terminals on the generator side. The current flowing from the unit transformer is $6.7 \times I_{\text{n-Transf.}}$, at $u_T = 15\%$, and from the generator it is $10 \times I_{\text{n-Gen.}}$, at $X_d'' = 10\%$; together therefore approximately $17 \times I_{\text{n-Gen.Unit}}$ if the rating of the generator and unit transformer are assumed to be approximately equal. If the rating of the auxiliary transformer is approximately 10% of the unit rating, the resulting short-circuit current is therefore $170 \times I_{\text{n-Aux.-Transf.}}$, and furthermore this current has an extremely long DC time constant of up to several 100 ms (refer to Table 7.1 of section 7.1). The ideal but expensive solution at this point is a linear TPZ core with the same data as the set of CTs at the generator terminals. The alternative connection to a set of CTs at the auxiliaries' side (dashed line alternative Figure 8.31) avoids this large over dimensioning, however at the cost of selectivity.

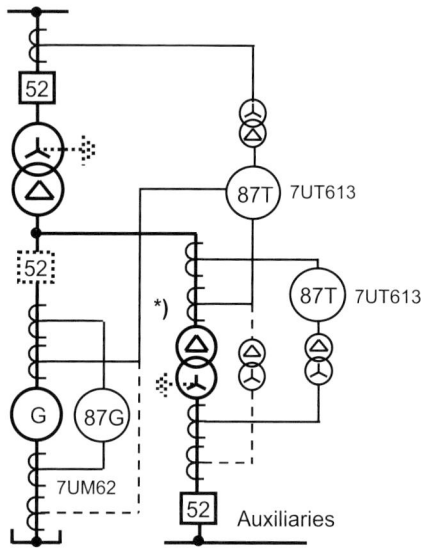

*) same ratio as generator CTs

Figure 8.31
Differential protection
for a generator unit

The CT cores that are connected to the differential protection of the auxiliary transformer on the generator side must be dimensioned such that the protection trips without delay during internal faults. This remains a very stringent requirement even if, in the case of numerical relays (7UT612), a saturation free conversion time of only less than a quarter cycle is required (refer to the following example). Stability for fault currents flowing through the transformer, i.e. for faults in the auxiliaries, on the other hand is less critical due to the low current magnitude.

Example 8-4: Dimensioning of the CTs at the auxiliary transformer

Given: Generator unit 230 MVA (f_n = 50 Hz)

Generator: 200 MVA, 10.5 kV, X''_d = 15%, R_G = 0.63·10^{-3} Ohm

Unit transformer: 230 MVA, 110/10.5 kV, $u_{X\text{-}T}$ = 13.2%, $u_{R\text{-}T}$ = 0.14%

Auxiliary transformer: 25 MVA, 10.5/5 kV, $u_{X\text{-}Aux\text{-}T}$ = 14%,

$u_{R\text{-}Aux\text{-}T}$ = 0.64%

Task: Calculate the dimensions for the CTs of the differential protection across the auxiliary transformer

Solution: The following criteria apply to the differential protection 7UT612 with regard to the CT dimensions:

– To ensure fast tripping during internal faults: time to saturation ≥ 4 ms

For stability during an external fault: over-dimensioning factor K_{TD} ≥ 4

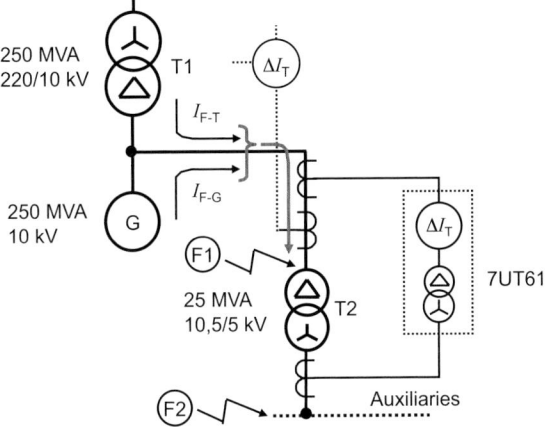

Figure 8.32 Circuit diagram for example 8-4

For internal faults the critical case is a short-circuit at the terminals on the generator side (F1):

Rated current of the generator:

$$I_{n\text{-}G} = \frac{S_n\,[\text{MVA}]\cdot 10^3}{U_n\,[\text{kV}]\cdot\sqrt{3}} = \frac{200\cdot 10^3}{10.5\cdot\sqrt{3}} = 11\ \text{kA}$$

Short-circuit current from the generator:

$$I_{F\text{-}G} = \frac{1.1\cdot I_{n\text{-}G}}{X''_d\,[\%]} = \frac{1.1\cdot 11}{0.15} = 81\ \text{kA}$$

$$X_{n\text{-}G} = \frac{U^2_{n\text{-}G}\,[\text{kV}^2]}{S_{n\text{-}G}\,[\text{MVA}]} = \frac{10.5^2}{200} = 0.551\ \Omega$$

$$X_G = \frac{X''_d\,[\%]}{100}\cdot X_{n\text{-}G} = \frac{15}{100}\cdot 0.551 = 0.083\ \Omega$$

DC time constant:

$$T_G = \frac{L_G}{R_G} = \frac{X_G}{\omega \cdot R_G} = \frac{0.083}{314 \cdot 0.63 \cdot 10^{-3}} = 0.42 \text{ s}$$

Rated current of the unit transformer:

$$I_{n\text{-}T} = \frac{S_{n\text{-}T} \, [\text{MVA}] \cdot 10^3}{U_n \, [\text{kV}] \cdot \sqrt{3}} = \frac{230 \cdot 10^3}{10.5 \cdot \sqrt{3}} = 12.6 \text{ kA}$$

Short-circuit current from the unit transformer:

$$I_{F\text{-}T} = \frac{1.1 \cdot I_{n\text{-}T}}{u_{X\text{-}T}[\%]} = \frac{1.1 \cdot 12.6}{0.132} = 105 \text{ kA}$$

$$X_{n\text{-}T} = \frac{U_{n\text{-}T}^2 \, [\text{kV}^2]}{S_{n\text{-}T} \, [\text{MVA}]} = \frac{10.5^2}{230} = 0.48 \ \Omega$$

$$X_T = \frac{u_{X\text{-}T}[\%]}{100} \cdot X_{n\text{-}T} = \frac{13.2}{100} \cdot 0.48 = 0.063 \ \Omega$$

$$R_T \approx \frac{u_{R\text{-}T}[\%]}{100} \cdot X_{n\text{-}T} = \frac{0.14}{100} \cdot 0.48 = 0.67 \cdot 10^{-3} \ \Omega$$

DC time constant:

$$T_T = \frac{L_T}{R_T} = \frac{X_T}{\omega \cdot R_T} = \frac{0.063}{314 \cdot 0.67 \cdot 10^{-3}} = 0.30 \text{ s}$$

The total current is:

$$I_F = I_{F\text{-}G} + I_{F\text{-}T} = 81 + 105 = 186 \text{ kA}$$

The corresponding equivalent time constant of the total current is calculated with the equation (5-26) in section 5.7:

$$T = \frac{I_{F\text{-}G} \cdot T_G + I_{F\text{-}T} \cdot T_T}{I_{F\text{-}G} + I_{F\text{-}T}} = \frac{81 \cdot 0.42 + 105 \cdot 0.30}{81 + 105} = 0.35 \text{ s}$$

The transient factor for 4 ms saturation free conversion is obtained with the equations 5-24 and 5-25, in section 5.7. The value is extracted from the corresponding diagram in Figure 5.16: $K_{TF}'' \approx 0.75$

The rated current of the auxiliary transformer is:

$$I_{n\text{-}Aux\text{-}T} = \frac{S_{n\text{-}Aux\text{-}T} \, [\text{MVA}] \cdot 10^3}{U_n \, [\text{kV}] \cdot \sqrt{3}} = \frac{25 \cdot 10^3}{10.5 \cdot \sqrt{3}} = 1380 \text{ A}$$

Current transformers of the type 5P with a ratio 2000/1 and 10 VA rated burden are chosen. According to the manufacturer's data, the internal burden is approximately 20% of the rated burden. The connection cables plus relay have a burden of < 1 VA.

The required operational accuracy limit factor for internal faults is:

$$\text{ALF}' = K_{TF}'' \cdot \frac{I_F}{I_{n\text{-}CT}} = 0.75 \cdot \frac{186 \text{ kA}}{2 \text{ kA}} = 70$$

To check the stability during through fault currents, the fault F2 must be analysed:

The short-circuit current is estimated by neglecting the source impedances:

$$I_{\text{F-Aux-T}} = \frac{1.1 \cdot I_{\text{n-Aux-T}}}{u_{\text{X-Aux-T}}[\%]} = \frac{1.1 \cdot 1.38}{0.14} = 11 \text{ kA}$$

The resulting operational accuracy limit factor taking into consideration the required over-dimensioning factor $K_{\text{TD}} = 4$ is as follows:

$$\text{ALF}' = K_{\text{TD}} \cdot \frac{I_{\text{F-Aux-T}}}{I_{\text{n-CT}}} = 4 \cdot \frac{11}{2} = 22$$

Therefore the criterion for internal faults must be considered.

And finally the required accuracy limit factor is obtained:

$$\text{ALF} = \frac{P_i + P_{\text{connected}}}{P_i + P_{\text{rated}}} \cdot \text{ALF}' = \frac{2+1}{2+10} \cdot 70 = 17.5$$

A value $n = 20$ is chosen and specified:

CT 2000/1, 5P20, 10 VA, $P_i < 2$ VA

9 Line Differential Protection

The principle of line differential protection and the properties of the communication channels were explained in chapters 3 and 6. Here, the devices and their applications will be discussed. The Siemens SIPROTEC product range will be used for this purpose, describing the state of the art. For the products of other manufacturers, reference can be made to the literature (Books [A15 to A17] and technical papers [9-7 to 9-10]).

A number of variants adapted to the various types of signal transmission (Table 9.1) are available in the SIPROTEC range (numerical devices).

Table 9.1 Variants for digital feeder differential protection

Relay (operating principle)	Communication channel	Maximum distance	Max. number of line terminals	Communication
7SD600 (differential, composite current)	Pilot wire, two cores, Loop resistance max. 1200 Ω	ca.12 km	2 ends	Voltage comparison, 50/60 Hz, $U_Q < 200$ V, I-core < 4mA / I_n
7SD80 (phase comparison)	Dedicated OF Pilot wire	ca. 40 km ca. 20 km	2 ends	Serial digital OF: up to 512 Kbit/s PW: 64, 128 Kbit/s
7SD61, 7SD84 (differential)	Dedicated OF Pilot wire with O/E signal converter Data networks	ca. 100 km ca. 15 km > 100 km	2 ends	Serial digital, HDLC Protocol, 64 to 512 Kbit/s Data Interfaces: IEEE C37.94, X.21 or G.703 by external converter
7SD52 7SD86/87 (differential)	as 7SD61/84	as 7SD61	6 ends	as 7SD61/84
7SD51* (phase comparison)	Dedicated OF, Data networks	ca. 20 km > 100 km	2 ends	Serial digital, IEC 60870-5 19.2 Kbit/s X.21 or G.703 by external converter

* first relay with digital line end to end communication, now phased out

Devices for pilot wire connections were already implemented with conventional technology. [9-1 and 9-2] Corresponding successor devices in digital technology were primarily implemented for protection refurbishment in urban and industrial networks as the installed copper cables are usually kept in service due to financial constraints [9-3].

Common practice at present is to apply devices with serial data transmission via FO connections or digital communication networks. The differential protection in this form can also be implemented on feeders above 100 km length. At the EHV level it is often applied as second main protection next to the distance relay. On feeders with more than two terminals it is the ideal solution for fast tripping on 100% of the protected object.

Since some time, relays are also offered with digital communication via pilot wires up to a distance of about 20 km.

9.1 Three core pilot wire (triplet) differential protection

This is a classic current differential protection which however applies two measuring circuits in series, one at each line terminal. Figure 9.1 The three-phase CT current is summated at each line end by means of summation CTs as described in section 3.3.2. The summated current that flows via the three pilot wire cores equals 100 mA during symmetrical three-phase nominal current.

The restraint current $I_{Res} = |\underline{I}_1| + |\underline{I}_2|$ is computed numerically. The operating current $I_{Op} = |\underline{I}_1 + \underline{I}_2|$ is available twice, once directly via the measuring input at the differential core and once via the numerical computation. This may be implemented for plausibility check as well as for monitoring.

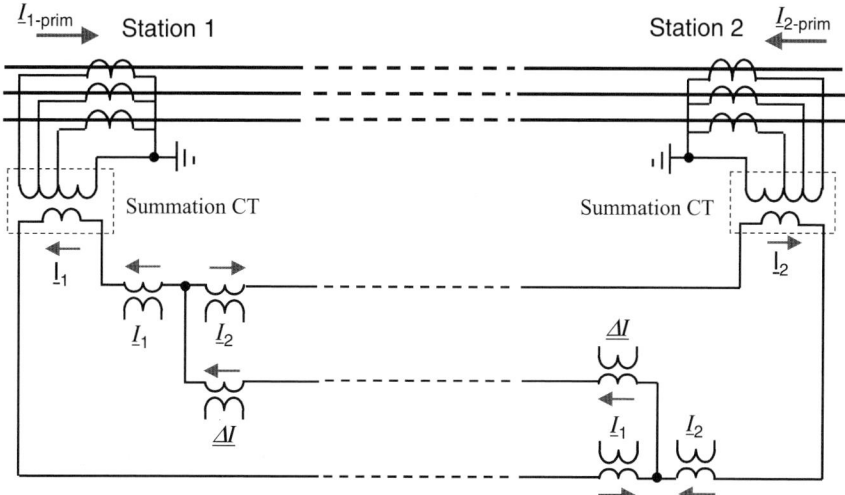

Figure 9.1 Circuit diagram of three pilot wire core differential protection

The three core pilot wire differential protection has been used mainly in Germany as conventional electro-mechanical (7SD95) and static types (7SD10/74).

Also a numerical version (7SD503) compatible with the conventional relays was manufactured for some years but was later phased out when digital communication via pilot wires became available with special modems. (See section 9.3 below.)

Due to the high pilot wire current level, these three core differential relays were relatively insensitive to electromagnetic influences compared to the classical pilot wire relays.

For short distances, even normal control cables could be used instead of special pilot wire connections. For feeder lengths above approx. 1 km special twisted and high voltage insulated protection cables were however recommended. (See section 6.1.1)

The maximum permissible core resistance here was 200 Ohm. For the standard protection cables with 1.4 mm core diameter (11.9 Ω/km) this corresponds to a distance of 16.8 km.

It must be noted that the burden of these on large core length can become very high. In the case of single phase faults in phase L1 (highest pick-up sensitivity however with almost 3 times higher core current) the basic burden is 3.1 VA. A burden of 9.0 VA per 100 Ω core resistance had to be be added! It is therefore not practical to apply this kind of protection above 10 km feeder length unless pilot wire cables with larger cross section are available.

The tripping characteristic of the relay 7SD503 consisted of three stages, similar to the transformer differential protection. The basic pick-up threshold $I_{DIFF}>$ was however set above nominal current so that pilot wire faults do not result in incorrect tripping.

In addition to this, the relay had a local current threshold $I>$, which must be exceeded to release the trip signal. This was applied when the protection must only trip at the feeder end at which short-circuit power is available, i.e. at the terminal where current flows.

9.2 Two core (twisted pair) pilot wire differential protection

This version of the feeder differential protection is intended for application with twisted pair pilots (telephone cores). With the device 7SD600 the voltage comparison principle is implemented, whereby in the case of load or external faults, no current flows via the pilot wire cores. During internal faults, the tripping current flows via the pilot wire loop. In conventional devices, this was a constant value of 4 mA per I_n with in-feed at both ends. With the numerical devices, the core current is a maximum of about 8 mA/I_n and decays to about 4 mA/I_n as the pilot cable length increases.

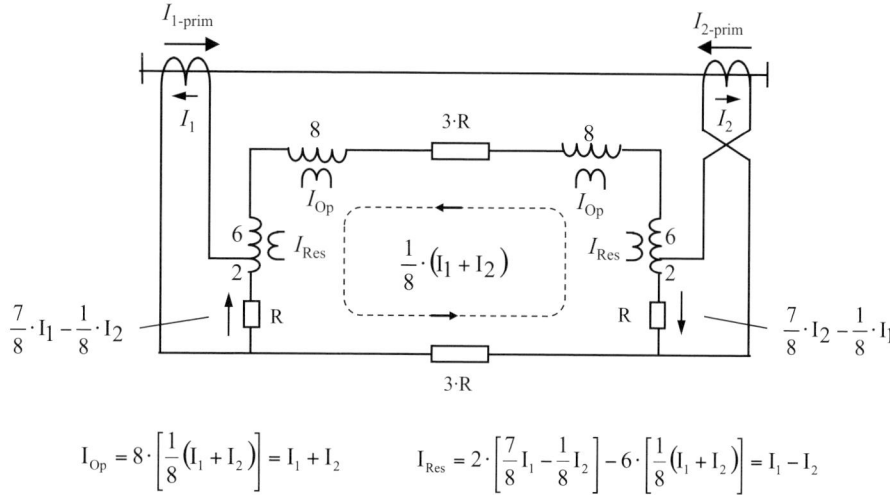

$$I_{Op} = 8 \cdot \left[\frac{1}{8}(I_1 + I_2) \right] = I_1 + I_2 \qquad I_{Res} = 2 \cdot \left[\frac{7}{8}I_1 - \frac{1}{8}I_2 \right] - 6 \cdot \left[\frac{1}{8}(I_1 + I_2) \right] = I_1 - I_2$$

Figure 9.2 Conventional twisted pair pilot wire differential protection

The operating mode of the voltage comparison protection is best explained by means of the conventional protection (7SD92) as shown in Figure 9.2.

For the sake of simplicity, the summation CT s are not shown. It is assumed that the CT connected to the feeder directly transforms the current to the secondary summation CT current (16 mA for the relay 7SD92). The ratio of core resistance to shunt resistance has a fixed value 3. This was achieved by applying a corresponding setting of the shunt resistance in the relay.

To calculate the current distribution in the resistance circuit the superposition principle is applied. Therefore the current distribution from the left-hand and right-hand CT is considered separately and then the two are summated. The current transformer acts as a current source so that an infinitely large internal resistance may be assumed. Consequently the connection to the CT may be considered as an open circuit if no primary current is flowing.

At each end only 1/8th of the current emanating from the CT (16 mA) flows into the pilot wire cores (2 mA). 7/8th flow via the shunt resistance.

Summing up the currents from both ends, results in the shown current distribution.

The restraining current I_{Res} and operating current I_{Op} are obtained via small interposing transformers with the shown turns ratio.

The equations in Figure 9.2 show the result. Thereby it must again be noted that the current flowing into the feeder is considered as being positive. Therefore the sign of the right side current reverses when current flows through the feeder (external fault), so that the current difference is given by I_{Op} and the current sum is given by I_{Res}.

Numerical pilot wire (twisted pair) differential protection 7SD600

With numerical technology, no fixed ratio of core to shunt resistance is required, so that the resistance adjustment in the device is not necessary.

A fixed shunt resistance of $R_b = 220$ Ohm is provided. The loop resistance of the pilot wire cable R_x is entered as a setting parameter. The current distribution is calculated numerically. In this context it must be noted that at each device a series resistance of $R_a = 220$ Ohm is connected between the relay and the pilot wires, so that the total loop resistance is $2 \cdot R_a + R_x$.

Note:

When barrier transformers are applied, an additional resistance of 60 Ohms per transformer (7XR95) must be added to the loop resistance R_x.

For the pilot wire monitoring, if this is included in the device, a resistance of 100 Ohms must be added as well.

The corresponding equivalent circuit for relay 7SD600 is shown in Figure 9.3.

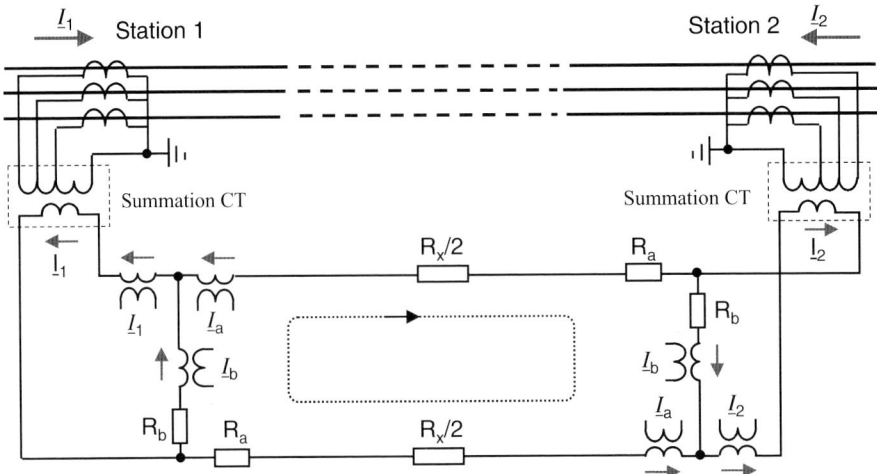

Figure 9.3
Equivalent circuit of the numerical pilot wire (twisted pair) differential protection 7SD600

For the loop shown with dashed line, the following equation is obtained:

$$(\underline{I}_1 - \underline{I}_a) \cdot R_b - \underline{I}_a \cdot (R_x/2 + R_a) + (\underline{I}_2 - \underline{I}_a) \cdot R_b - \underline{I}_a \cdot (R_x/2 + R_a) = 0 \quad (9\text{-}1)$$

The following is introduced:

$$x = \frac{R_x/2 + R_a}{R} \quad (9\text{-}2)$$

The result is:

$$(\underline{I}_1 + \underline{I}_2) = 2 \cdot (1 + x) \cdot \underline{I}_a \qquad (9\text{-}3)$$

On the left-hand side, the currents $\underline{I}_1$ und $\underline{I}_a$ are known by direct measurement. The unknown $\underline{I}_2$ can be calculated with the loop equation (9-1):

$$\underline{I}_2 = (\underline{I}_1 + \underline{I}_2) - \underline{I}_1 = 2 \cdot (1 + x) \cdot \underline{I}_a - \underline{I}_1 \qquad (9\text{-}4)$$

The left relay therefore has all the values it requires for the calculation of the operating and restraint current:

$$I_{\text{OP-1}} = |\underline{I}_1 + \underline{I}_2| = |2 \cdot (1 + x) \cdot \underline{I}_a| \qquad (9\text{-}5)$$

$$I_{\text{Res-1}} = |\underline{I}_1| + |\underline{I}_2| = |\underline{I}_1| + |2 \cdot (1 + x) \cdot \underline{I}_a - \underline{I}_1| \qquad (9\text{-}6)$$

In station 2 the valuess $\underline{I}_2$ and $\underline{I}_a$ are measured, while $\underline{I}_1$ must be calculated. The following analogous equations apply:

$$I_{\text{Op-2}} = |\underline{I}_2 + \underline{I}_1| = |2 \cdot (1 + x) \cdot \underline{I}_a| \qquad (9\text{-}7)$$

$$I_{\text{Res-2}} = |\underline{I}_2| + |\underline{I}_1| = |\underline{I}_2| + |2 \cdot (1 + x) \cdot \underline{I}_a - \underline{I}_2| \qquad (9\text{-}8)$$

The familiar tripping characteristic with dual slope is also applied here. (Figure 3.14 in section 3.2.4) The basic pick-up threshold $I_{\text{DIFF}}>$ is set above load current.

The above equations may also be used to calculate the current and voltage conditions on the pilot wire cores. From equation (9-5) the current in the pilot wire cable is obtained:

Current in the pilot wire cores:

$$\underline{I}_a = \frac{\underline{I}_1 + \underline{I}_2}{2 \cdot (1 + x)} \qquad (9\text{-}9)$$

Current in the shunt resistance in station 1:

$$\underline{I}_{b1} = \underline{I}_1 - \underline{I}_a \qquad (9\text{-}10)$$

Current in the shunt resistance in station 2:

$$\underline{I}_{b2} = \underline{I}_2 - \underline{I}_a \qquad (9\text{-}11)$$

The shunt voltages across the pilot wire loop in station 1 and 2 therefore are:

$$\underline{U}_{x1} = R_b \cdot \underline{I}_b - R_a \cdot \underline{I}_a \quad \text{and} \quad \underline{U}_{x2} = R_b \cdot \underline{I}_b - R_a \cdot \underline{I}_a \qquad (9\text{-}12) \text{ and } (9\text{-}13)$$

Example 9-1: Current and voltage on the pilot wires during an internal short-circuit L2-L3 with in-feed from both ends

Given: $I_{F1} = 10 \cdot I_n$ and $I_{F2} = 5 \cdot I_n$

Parameters of the pilot wire cable: 15 km, $R'_x = 73.2$ Ohm/km

Solution: The loop resistance is: $R_x = 15 \cdot 73.2 = 1098$ Ohm

Using equation (9-2) the following is calculated

$x = (1098/2 + 220)/220 = 3.5$

With three-phase nominal current, a summation CT current of 20 mA flows.

In the case of two phase nominal current L2-L3 the current is larger by a factor of 1.15 (refer section 3.3.2, Table 3.1)

The summation CT currents in station 1 and 2 therefore are:

$I_1 = 10 \cdot 20 \cdot 1.15 = 230$ mA and $I_2 = 5 \cdot 20 \cdot 1.15 = 115$ mA

With equation (9-9) the current in the pilot wires is obtained:

$$I_a = \frac{230 \text{ mA} + 115 \text{ mA}}{2 \cdot (1 + 3.5)} = 38.37 \text{ mA}$$

The current in the shunt path at station 1 is:
$I_{b1} = 230 - 38.37 = 191.63$ mA

The shunt voltage across the pilot wires at station 1 is:
$U_{x1} = 191.63 \text{ mA} \cdot 220 \ \Omega - 38.37 \text{ mA} \cdot 220 \ \Omega = 33.72$ V

In station 2 the results are as follows:
$I_{b2} = 115 - 38.37 = 76.63$ mA

$U_{x2} = 76.63 \text{ mA} \cdot 220 \ \Omega - 38.37 \text{ mA} \cdot 220 \ \Omega = 8.39$ V

The voltage distribution is graphically shown in Figure 9.4.

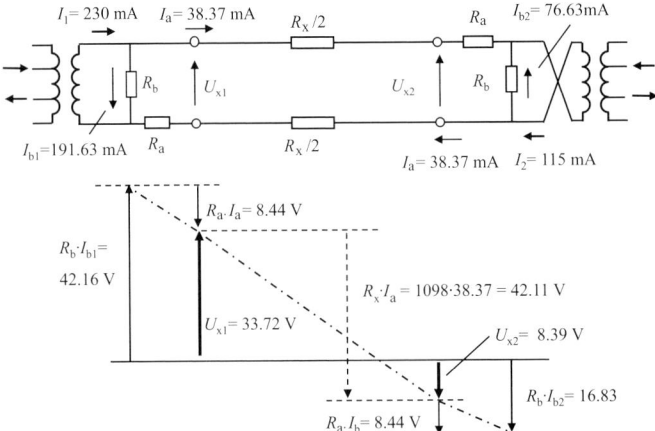

Figure 9.4 Two core pilot wire differential protection, example 9-1, resulting current and voltage distribution

The protection is designed so that for currents up to approximately 20 times I_n the shunt voltage will not rise above approx. 300 V. Above this level, the voltage across the summation CT is limited by means of a varistor.

In the event of an external fault, no current flows via the pilot wire cores (pilot wire capacitance neglected). Consequently the entire summation current flows via R_b and the shunt voltage is $U_b = I_b \cdot R_b$. In the case of three-phase nominal current flowing through the feeder, the following is obtained: $U_b = 20\text{mA} \cdot 220\,\Omega = 4.4$ V.

For a three-phase external fault current of $I_F = 20$ times I_n flowing through the feeder, the shunt voltage would therefore be 88 V. On the other hand, the current during a single phase fault in L1 would result in a summation current that is larger by the factor 2.89, and the shunt voltage would consequently be 254 V.

If very high earth-fault currents of more than 20 times rated current are expected, it may be decided to change the taps at the summation CT from sensitive earth-current connection to three-phase connection to reduce the composite current level (change from equation (3-17) to (3-18) in section 3.3.2). Herewith the pilot core to core voltage would be reduced correspondingly.

Permissible interference voltage

The numerical relay 7SD600 has an external summation CT with 2 kV insulation against the pilot wires. The maximum permissible longitudinal voltage (common mode voltage of both pilot wire terminals to ground) is 60% of the nominal isolation i.e. 1.2 kV. If larger longitudinal voltages can occur, a summation transformer or additional barrier transformer with greater insulation must be provided (refer section 6.1.1)

The induced longitudinal voltage (between both cores and ground) also causes a shunt voltage (between the two pilot wires) due to the non-symmetry of the pilot wire cores (refer section 6.1.1). A circulating current through the pilot wire cores and the connected devices will result from this voltage:

$$I_{\text{Interf.}} = \frac{U_{\text{Q-Interf.}}}{2 \cdot R_a + 2 \cdot R_b + R_x} = \frac{U_{\text{Q-Interf.}}}{2 \cdot (1 + x) \cdot R_b} \tag{9-14}$$

This interference current acts as a tripping quantity when it exceeds the pick-up threshold of the protection ($I_{\text{DIFF}>}/I_n$) i.e. the relay will trip when:

$$I_{\text{Interf.}} > I_{\text{a-trip}} = \frac{(I_1 + I_2)_{\text{trip}}}{2 \cdot (1 + x)} = \frac{(I_{\text{DIFF}>}/I_n) \cdot 20 \text{ mA}}{2 \cdot (1 + x)}$$

The permissible interference voltage therefore is:

$$U_{\text{Q-Interf.}} = I_{\text{Q-Interf.}} \cdot 2(1 + x) \cdot R_b < R_b \cdot (I_{\text{DIFF}>}/I_n) \cdot 20 \text{ mA} =$$
$$= (I_{\text{DIFF}>}/I_n) \cdot 220\,\Omega \cdot 20 \text{ mA}$$

As a result the following simple equation is obtained:

$$U_{\text{Q-Interf.}} < (I_{\text{DIFF>}} / I_n) \cdot 4.4 \text{ V} \qquad\qquad (9\text{-}15)$$

The protection pilot wire cables (refer to section 6.1.1, Table 6.1) have a non-symmetry factor of $< 10^{-4}$ at 50 Hz. Therefore a longitudinal voltage of 10 kV would result in a shunt voltage of only 1 V. The interference immunity is therefore relatively large when these special protection cables are applied. For other cable types, it is recommended to estimate the degree of interference according to section 6.1.1 and to check the interference immunity.

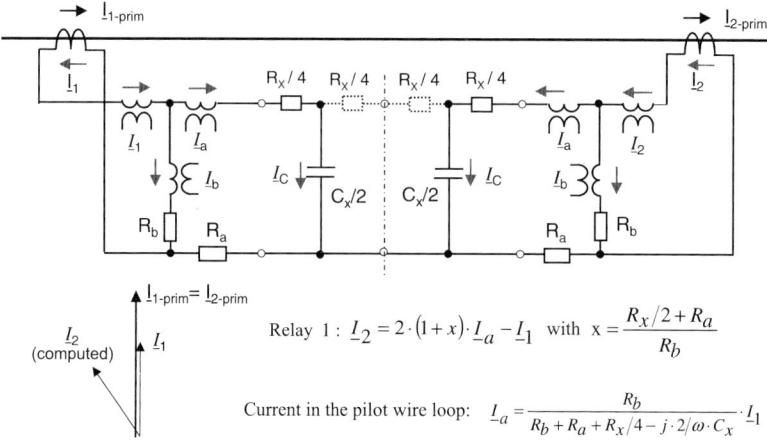

Figure 9.5 Influence of core capacitance on the measurement

Impact of the pilot wire capacitance

On telephone cables, a shunt capacitance of 60 nF/km must be reckoned with. In accordance with the shunt voltage that is applied, a charging current flows in the pilot wires. This was not considered in the above equivalent circuits and therefore causes a measuring error. In the event of external faults, this may result in maloperation (Figure 9.5).

During external faults, the shunt voltage at the two ends has the same magnitude but opposite polarities. Consequently no resistive current flows through the pilot wires. The charging current appears as capacitive current at both ends and causes a phase shift of the calculated current for the remote station.

In Figure 9.5 the pilot wire cable is represented by a 4–terminal model with two T-sections, which is an exact replica at nominal frequency (50/60 Hz). As the circuit is symmetrical and no current flows through the centre point of the pilot cable, the left and right hand sides may be considered separately. The following equation for the current $\underline{I}_a$ which equals the current $\underline{I}_C$ flowing through half the core capacitance can be derived.

$$\underline{I}_a = \underline{I}_C = \frac{R_b}{R_b + R_a + R_x/4 - j2/\omega C_x} \cdot \underline{I}_1 \tag{9-16}$$

$\underline{I}_a$ directly corresponds to the tripping current of the relay. The effect on the stability during external faults can therefore be estimated.

For the left hand side the following applies:

$$\underline{I}_{Op} = \underline{I}_1 + \underline{I}_2 = 2 \cdot (1 + x) \cdot \underline{I}_a =$$

$$= 2 \cdot (1 + x) \cdot \frac{R_b}{R_b + R_a + R_x/4 - j2/\omega C_x} \cdot \underline{I}_1 \tag{9-17}$$

The same applies to the right hand side.

Example 9-2: Influence of pilot wire capacitance on the measuring accuracy of relay 7SD600

Given: Pilot wire 20 km, $R'_x = 73.2 \ \Omega/km$, $C'_x = 60 \ nF/km$

Question: Is stable differential protection without compensation of the pilot wire capacitance possible?

Solution: The loop resistance of the pilot wire is $R_x = 20 \cdot 73.2 = 1464 \ \Omega$

The core capacitance is: $C_x = 60nF/km \cdot 20 \ km = 1200 \ nF = 1.2 \cdot 10^{-6} \ F$

With (9-2) the following is obtained:

$$x = \frac{1464/2 + 220}{220} = 4.33$$

The operating current results from (9-17):

$$\underline{I}_{Op} = 2 \cdot (1 + 7.66) \cdot \frac{200}{220 + 220 + 1464/4 - j2/(314 \cdot 1.2 \cdot 10^{-6})} \cdot \underline{I}_1 =$$

$$= (0.066 + j0.432) \cdot \underline{I}_1$$

$$|\underline{I}_{Op}| = 0.437 \cdot I_1$$

The current $\underline{I}_2$ is calculated as follows:

$$\underline{I}_2 = \underline{I}_{Op} - \underline{I}_1 = (0.066 + j0.432) \cdot \underline{I}_1 - \underline{I}_1 = (-0.934 + j0.432) \cdot \underline{I}_1$$

$$|\underline{I}_2| = 1.029 \cdot I_1$$

Therefore

$$I_{Res} = |\underline{I}_1| + |\underline{I}_2| = 2.03 \cdot I_1$$

$$k = \frac{I_{Op}}{I_{Res}} = \frac{0.437 \cdot I_1}{2.03 \cdot I_1} = 0.22$$

Stability during external faults is ensured if the slope of the tripping characteristic is set significantly greater than 0.22.

Stable operation of relay 7SD600 is possible with $k_1 = 0.33$ (see dual slope operating characteristic of Figure 3.14 in chapter 3.2.4).

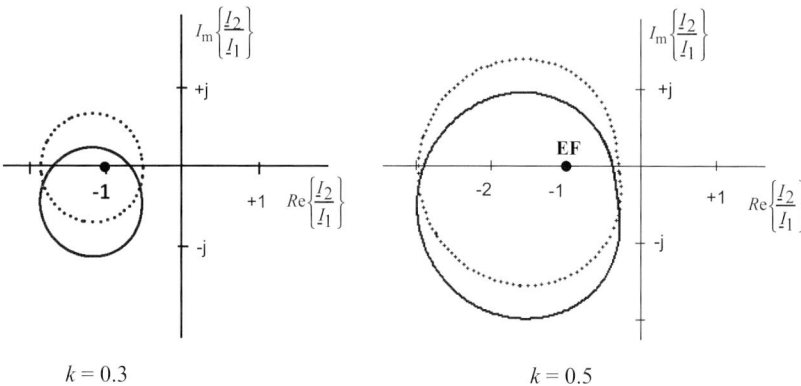

$k = 0.3$ $k = 0.5$

Figure 9.6 Influence and compensation of the pilot wire capacitance shown in the polar phasor diagram

The influence of the pilot wire capacitance is best visualised by the polar diagram (Figure 9.6). The shown diagrams correspond to a biased characteristic $I_{Op} = k \cdot I_{Res}$. (See section 3.2.4) They are valid for the first slope of the dual slope operating characteristic of relay 7SA600.

The area inside the solid circle characteristic depicts the range in which the protection detects external faults and remains stable. Outside the characteristic the relay trips. (It is assumed that the basic pick-up threshold $I_{Op} > I_B$ is true, which means the currents are above nominal current). The operating point EF for through-flowing load and short-circuit currents is located at $\underline{I}_2/\underline{I}_1 = -1$ on the real axis. Without pilot wire capacitance (this applies to short feeders) the pick-up characteristic is also located symmetrically about the real axis (shown in dashed lines) and encloses the external fault point EF with a large security margin. Consequently large measuring errors are tolerated without resulting in mal-operation by the protection.

Under the influence of pilot wire capacitance (in this example 1.2 μF per 20 km pilot wire length) the characteristic drifts downwards into the 4th quadrant. With the small amount of stabilising of $k = 0.3$, the external fault point EF appears closer to the border line of the operating characteristic so that the protection tends to over-function. Non-symmetrical CT saturation could readily result in mal-operation.

With the increased bias of $k = 0.5$, the protection would keep a higher security margin however at the expense of sensitivity.

The compensation for the pilot wire capacitance described below rotates the characteristic back so that it is almost equivalent to its ideal state as shown in Figure 9.6.

In numerical devices this function can be implemented as add-on in the firmware. Relay 7SA502 is an example for this.[1] The capacitive charging current is calculated

---

[1] The relay 7SA502 has in the mean time been phased out because the new relay type 7SD61 also allows digital data communication via pilot wires, thus avoiding the impact of pilot capacitances.

according to equation (9-16) and subtracted from the measured pilot wire current $\underline{I}_a$, which means, it is subtracted from the operating current.

The pilot wire capacitance C_x must be entered as a parameter. It can be derived from the cable data. In the case of relay 7SD502 it could be directly measured by the protection device during commissioning.

For this purpose the pilot wires were open circuited at one end while test current was injected from the other end. From the applied voltage U_b and the charging current flowing into the pilot wires, $I_a = I_C$, the capacitance could then be directly calculated by the relay (Figure 9.7).

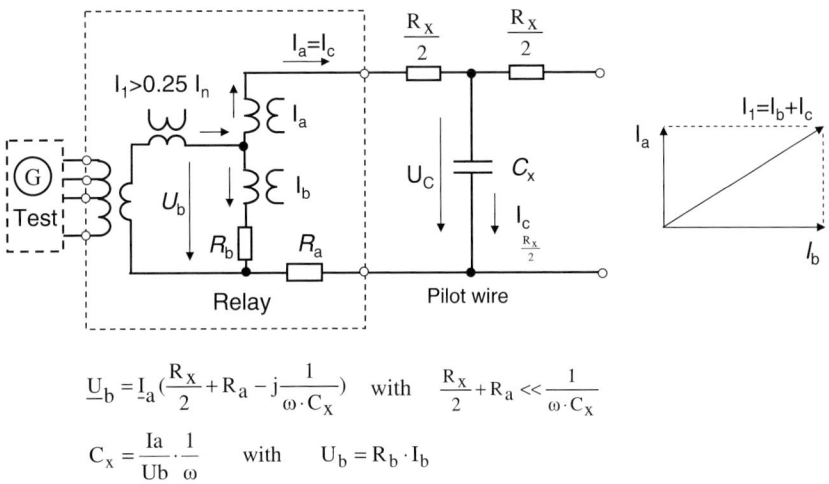

$$\underline{U}_b = \underline{I}_a \left(\frac{R_X}{2} + R_a - j\frac{1}{\omega \cdot C_X} \right) \quad \text{with} \quad \frac{R_X}{2} + R_a \ll \frac{1}{\omega \cdot C_X}$$

$$C_X = \frac{Ia}{Ub} \cdot \frac{1}{\omega} \quad \text{with} \quad U_b = R_b \cdot I_b$$

Figure 9.7 Measuring the pilot wire capacitance

Two core pilot wire differential protection 7SD600

This relay is an economic version of the two core pilot wire differential protection. It is mainly used for cable protection in distribution networks in particular in urban areas where the traditional protection via pilot wires has been common practice. (New installations use optic fiber cables but existing pilot wires will be further used and only replaced step by step in the next years.)

The summation CT of this device is applied externally (Figure 9.8). The varistor limits the pilot core to core voltage to about 300 V.

The maximum pilot cable length (telephone cables) amounts to approx. 15 km (1200 Ω loop resistance and 1 μF pilot capacitance). Compensation of the pilot capacitance is in this case not provided.

The relay is equipped with saturation detector and 2nd harmonic inrush stabilisation.

The tripping time is in the range of 20 to 30 ms.

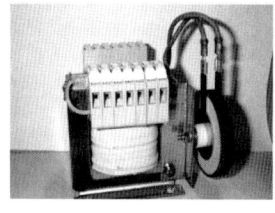

Figure 9.8
Relay 7SD600 with
summation CT

The configuration of the pilot wire connection should be done according to section 6.1.1. The CT dimensioning includes the requirement that the operational accuracy limit factors at both ends of the line maintain symmetry in the range $3/4 \leq \mathrm{ALF}'_1/\mathrm{ALF}'_2 \leq 4/3$. Furthermore, the CTs should not go into saturation when the maximum fault current flows through the line whereby the current is considered as pure AC current for this condition (over-dimensioning factor $K_{\mathrm{TD}} \geq 1$).

To ensure fast tripping in the event of internal faults with very large short-circuit current, the saturation free time should be at least 5 ms.

The pilot wire resistance can be measured by the relay itself in a closed pilot wire loop with single ended injection of test current. (Figure 9.9) The relay calculates the resistance according to the given equation. The indicated value must then be set.

The pilot wires are continuously monitored by a superimposed 2 kHz loop current (2 mA). By amplitude modulation, this monitoring current can also be used for transfer trip and alarming functions.

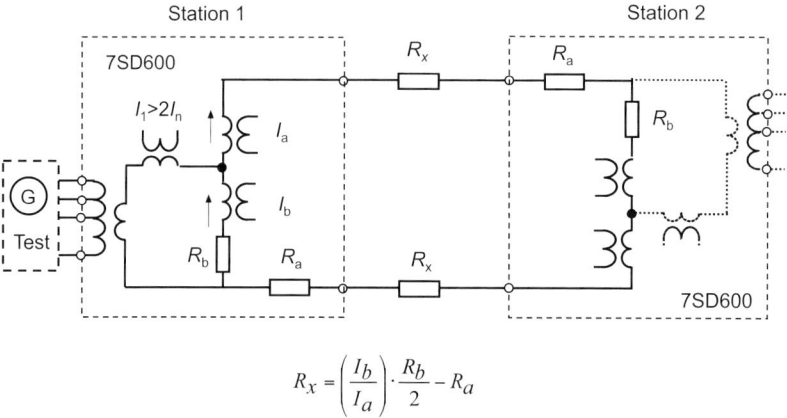

$$R_x = \left(\frac{I_b}{I_a}\right) \cdot \frac{R_b}{2} - R_a$$

Figure 9.9 Measuring the pilot wire resistances

9.3 Line differential protection with digital communication

The progress in optic fiber technology and data communication has led to a tremendous performance upgrade and proliferation of line differential protection. The modern relays can be used with dedicated fibers or data networks, but also with digital microwave. Even the use with pilot wires is possible by advanced signal modulation techniques.

In the following the relay design and application aspects are discussed.

9.3.1 Devices and system configuration

Device variants for feeders with two ends 7SD61/7SD84[1] and for feeder configurations with up to 6 ends (7SD52/86/87)[2] are available. [9-4, 9-5 and 11-1]

These relays utilise the combination of micro-processor technology and modern communication in an optimal manner:

- Numerical differential protection with phasors (phase selective measurement)
- Fast synchronous data communication 64 up to 512 Kbit/s
- Operation via dedicated FO channels with the following options:

Integrated FO module	Connector	FO fiber	Optical wavelength	Optic budget	Maximum distance ca.
FO5	ST	Multi-mode 62,5/125 μm	820 nm	8 dB	1.5 km
FO6	ST	Multi-mode 62,5/125 μm	820 nm	16 dB	3.5 km
FO18	LC	Mono-mode 9/125 μm	1300 nm	16 dB	60 km
FO19	LC	Mono-mode 9/125 μm	1550 nm	29 dB	100 km

- Operation via digital communication networks (X.21 or G703.1) with integrated channel delay compensation, optionally with GPS synchronisation
- Operation via ISDN networks or wire-bound links with special signal converters.
- Internet compatible local or remote operation with a standard web browser

The multi-end relay variants 7SD86/87/7SD52 can be configured for ring and chain topology communication (Figure 9.10).

---

[1] The relay type 7SD84 belongs to the Siemens SIPROTEC5 series. The corresponding type of the SIPROTEC4 series is 7SD61.

[2] The relay variants 7SD86/87 belong to the Siemens SIPROTEC5 series. The corresponding type of the SIPROTEC4 series is 7SD52

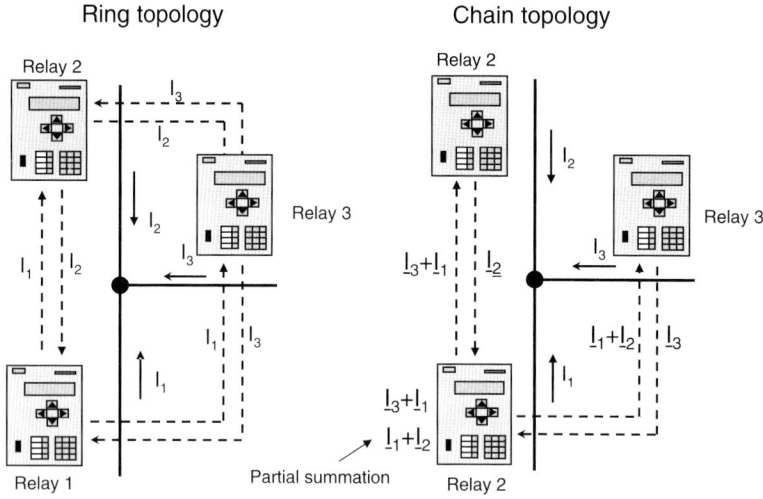

Figure 9.10 Alternative topologies with relays 7SD52/86/87

The ring topology has the basic advantage that a redundant connection exists. This means that in the event of the loss of a data connection an automatic change-over (within 20 ms) to the chain topology can be done so that the protection functionality is fully maintained.

In the chain topology the currents are partially summated and forwarded as partial sums to the next closest device.

The communication within a protection system may also be routed via various types of data communication channels. Figure 9.11 depicts an application example with two direct FO connections and one data link via a communication network.

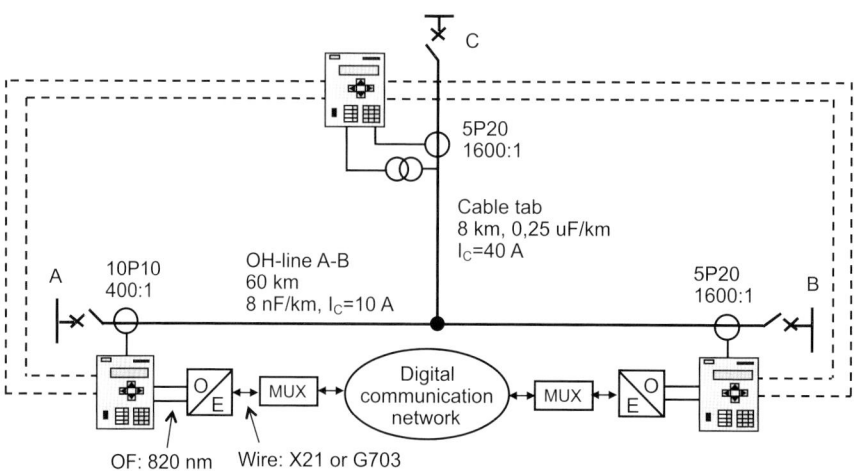

Figure 9.11 Application example with mixed communication channels

Transformers may also be included in the protected object (feeder with up to 6 terminals). The ratio and vector group adaptation as well as the in-rush stabilising are integrated in the devices 7SD52/86/87.

9.3.2 Measuring technique

The basics of differential protection with phasors and the synchronisation via data communication channels were covered in section 4.2.

The devices 7SD61/84 and 7SD52/86/87 use a new biasing method to get high sensitivity for internal faults even in the presence of high through-flowing load currents. The classic percentage biased differential protection with constant biasing factor is no longer applied. Instead only the actual measuring and communication errors of the components in the protection system are summated and applied as restraining signal for comparison with the operating (differential) current (the vectorial sum of the feeder currents). In other words an adaptive stabilising is applied which under normal circumstances (no CT saturation and error free data communication) is small and therefore allows for high pick-up sensitivity. As a result, even high resistance faults can be detected.

In total, the restraint consists of the following components:

– *Set pick-up threshold* $I_{\mathrm{DIFF}}>$ as constant value
– *Synchronisation error* (jitter and non-symmetrical channel delay times) $(\sum e_{\mathrm{Sync}})$
– *CT errors* $(\sum e_{\mathrm{CTn}})$

$$I_{\mathrm{Res}} = I_{\mathrm{Diff}}> + \sum e_{\mathrm{Sync}} + |\underline{I}_1| \cdot e_{\mathrm{CT1}} + |\underline{I}_2| \cdot e_{\mathrm{CT2}} + |\underline{I}_3| \cdot e_{\mathrm{CT1}} + \dots + |\underline{I}_n| \cdot e_{\mathrm{CTn}} \quad (9\text{-}18)$$

The operating current corresponds, as usual, to the vector sum of the currents.

$$I_{\mathrm{Op}} = |\underline{I}_1 + \underline{I}_2 + \underline{I}_3 + \dots + \underline{I}_n| \quad (9\text{-}19)$$

Figure 9.12 shows the resulting characteristic. There may be up to 6 line terminals ($n = 6$).

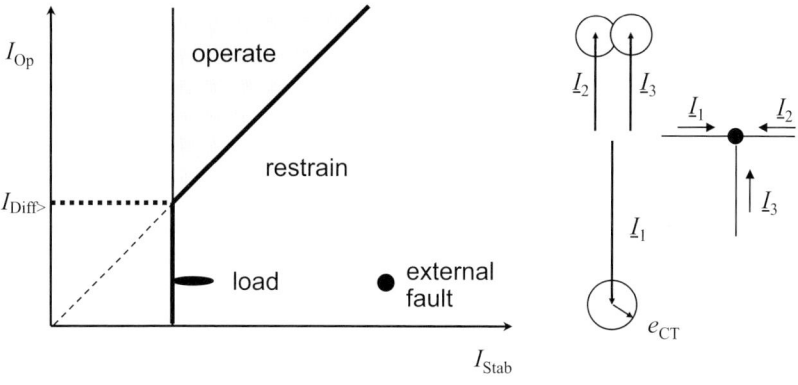

Figure 9.12 Pick-up characteristic of the relays 7SD61/84 and 7SD52/86/87

In the normal case, the synchronisation errors are small (may even be neglected with a dedicated optic fiber connection) and the CT errors cause the main restraint in addition to the basic pick-up setting $I_{\text{Diff}}>$.

The errors determined at one end are transmitted to all the other ends via exchanged communication telegrams, so that each device can calculate the total error condition.

The individual stabilising components are:

Pick-up threshold

In this context, a differentiation between devices with and without charging current compensation must be made.

In devices without compensation the setting of $I_{\text{Diff}}>$ is primarily determined by the charging current of the feeder which appears as erroneous differential current and acts in a tripping sense. (Figure 9.13)

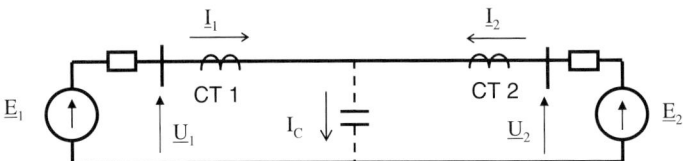

Figure 9.13 Charging current of a feeder

The charging current is proportional to the voltage applied to the feeder:

$$I_C[\text{A}] = \frac{U_n[\text{V}]}{\sqrt{3}} \cdot \omega\left[\frac{1}{\text{s}}\right] \cdot C_b'\left[\frac{\text{F}}{\text{km}}\right] \cdot s[\text{km}] =$$
$$= 3.63 \cdot 10^{-6} \cdot U_n[\text{kV}] \cdot f_n[\text{Hz}] \cdot C_b'[\text{nF}] \cdot s[\text{km}] \qquad (9\text{-}20)$$

with I_C Charging current of the feeder
  U_n Nominal voltage phase of the feeder (Ph-Ph)
  f_n System frequency
  C_b' Capacitance of the feeder per km (effective capacitance)

$I_{\text{Diff}}>$ should be set ≥ 2.5 times I_C, however not less than 0.15 times I_n.

Example 9-3: Charging current of a cable

Given: 110 kV cable, length = 16 km, $C_b' = 310$ nF/km (50 Hz-system)

Wanted: Charging current

Solution: $I_C = 3.36 \cdot 10^{-6} \cdot 110 \cdot 50 \cdot 310 \cdot 16 = 99$ A

In devices with charging current compensation, the charging current may be ignored, and a setting of approximately 15-20% of nominal current (the maximum load current of the feeder) may be applied. This allows for high sensitivity pick-up threshold also on long cable or overhead line feeders.

Synchronisation errors

The relay continuously measures the communication channel delay time and applies a corresponding angle correction to the measured current phasors (refer to section 4.2.3). The applied synchronising method is however based on equal channel delay in the transmit and receive path via the communication channel. If the channel delays differ, then half this difference appears as angle error in the measurement ($\Delta\varphi[°] = 0.5 \cdot \Delta t\,[\mathrm{ms}]/18°$ for $f_N = 50$ Hz). A channel delay difference of 1 ms therefore causes a phase angle deviation of $0.5 \cdot 18 = 9°$. This corresponds to a differential current of $\Delta I = 2 \cdot \sin(9°/2) = 0.16$, in other words 16%. The maximum expected channel delay difference can be set on the relay via parameter. The relay then correspondingly increases the restraint.

In practice propagation time differences of only about 0.2 ms are accepted and compensated in this way. In case of larger time differences GPS synchronisation is generally applied.

CT errors

In the load and short-circuit operating range the CT error can be approximated with a corresponding slope as shown in Figure 9.14. The CT error definitions according to IEC 60044-1 are used for this purpose. The error in the load range (normal operation) is e_L. With normal burden connected to the CT, this factor applies up to the value $(\mathrm{ALF}'/\mathrm{ALF}_n) \cdot I_n$. Here ALF' is the operational accuracy limit factor and ALF_n the nominal accuracy limit factor (refer to section 5.2).

During short-circuits, the total error (5% for 5P and 10% for 10P CTs) is reached with $\mathrm{ALF}' \cdot I_n$. The error e_F in the short-circuit range applies in this case.

Both error values must be applied as setting parameters.

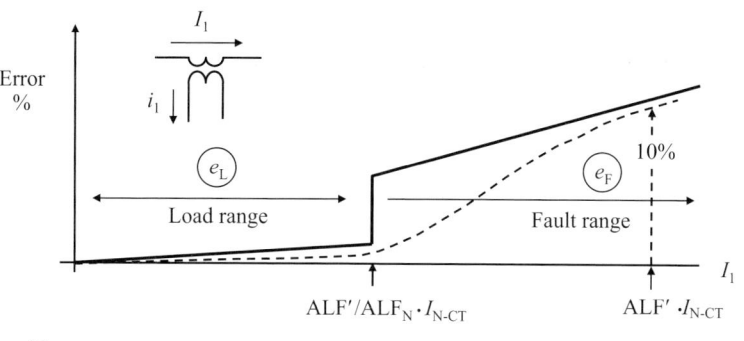

Figure 9.14 Approximation of the CT error in 7SD61/84 and 7SD52/86/87

Example 9-4: Stabilising with differential protection 7SD52/86/87

Given: Feeder configuration and CT data according to Figure 9.15

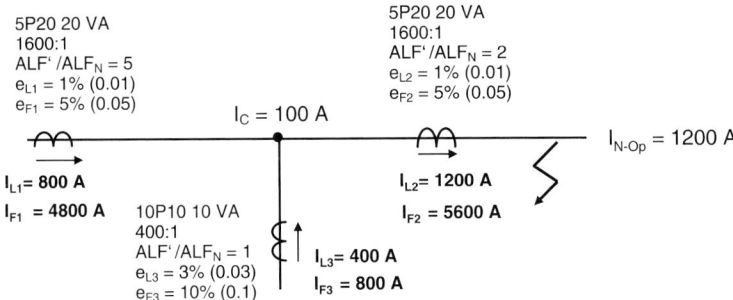

Figure 9.15 Circuit diagram and data for example 9-4

Wanted: Stabilising and tripping current

Solution: For the load range, the following is obtained with the data in Figure 9.15:

$$I_{Res} = 2.5 \cdot I_C + |I_1| \cdot e_{L1} + |I_2| \cdot e_{L2} + |I_3| \cdot e_{L3}$$

$$I_{Res} = 2.5 \cdot 100 \text{ A} + 800 \text{ A} \cdot 0.01 + 1200 \text{ A} \cdot 0.01 + 400 \text{ A} \cdot 0.03 = 282 \text{ A}$$

$$I_{Op} = |I_1 + I_2 + I_3| = I_C = 100 \text{ A}$$

In a similar fashion, the external short-circuit provides the following results:

$$I_{Res} = 2.5 \cdot 40 \text{ A} + 4800 \text{ A} \cdot 0.01 + 5600 \text{ A} \cdot 0.05 + 800 \text{ A} \cdot 0.1 = 508 \text{ A}$$

Here it is considered that I_c drops to $I_c = 40$ A as the voltage collapses due to the close-in fault.

For the computation of tripping currents it was assumed that the CTs operate without errors. In reality, the results will be influenced by the actual errors that arise. A security margin of 50% for the CT error setting is recommended.

In the event of CT saturation, the CT errors are significantly larger. The measuring algorithm detects this by means of harmonic analysis of the distorted currents. The harmonic content of each CT is converted into an additional saturation factor f_S. Restraint is automatically increased thereby:

$$I_{Res} = I_{Diff} > + \sum e_{Sync} + \sum |I_n| \cdot f_{sn} \cdot e_{CTn} \qquad (9\text{-}21)$$

Fast charge comparison

The Fourier-transformation for computation of the phasors requires a data window of one cycle length. The data communication and processing takes further time so that a trip command by the differential protection with phasor computation requires approximately 1+1/2 cycles (30 ms in case of $f_N = 50$ Hz). To achieve faster tripping times, the 7SD61/84 and 7SD52/86/87 relays execute a charge comparison in parallel, thereby obtaining tripping times below one cycle with communication channel speed of 512 kBit/s. For this function the charge (current-time integral) over a quarter cycle (5 ms) is calculated in each half wave at each line end and transferred to the other line ends. The integration is done by addition of the product of the sampled values and sampling interval. This principle is shown in Figure 9.16.

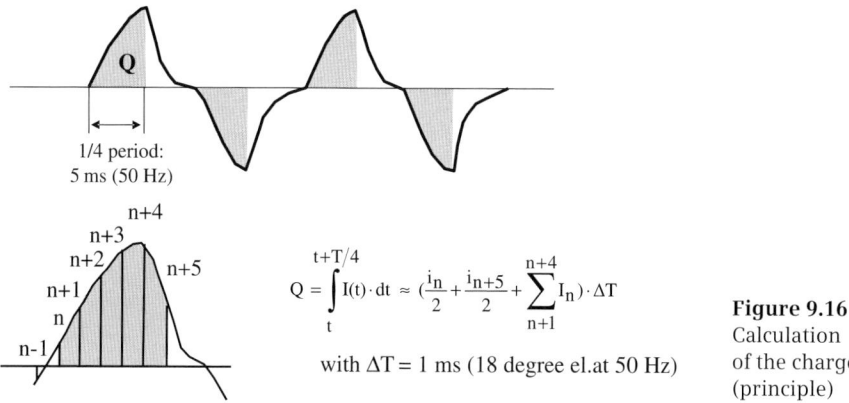

$$Q = \int_{t}^{t+T/4} I(t) \cdot dt \approx (\frac{i_n}{2} + \frac{i_{n+5}}{2} + \sum_{n+1}^{n+4} I_n) \cdot \Delta T$$

with $\Delta T = 1$ ms (18 degree el.at 50 Hz)

Figure 9.16
Calculation
of the charge
(principle)

This comparison only functions correctly if the integration is done at all line ends at the same (synchronised) time interval. This is achieved by calculated interpolation.

The resultant charge values are then compared with the same principle that is applied for the phasor based differential protection.

This technique is only effective for currents above nominal current (released by a high set current detector ($I\gg$) and can reach a result in a very short time due to the short data window. In the case of CT saturation, in other words with very severe distortion of the sign wave, the saturation detector will disable the charge comparison function.

CT requirements for 7SD61/84 and 7SD52/86/87 relays

For fault current flowing through the protected object the CTs may not saturate inside the first 5 ms. This corresponds to an over dimensioning factor $K_{TD} = 1.2$. For internal faults, the largest short-circuit current must still allow for saturation-free transformation over the first 3 ms ($K_{TD} \geq 0.5$), to ensure non-delayed tripping

The relays can be adapted to different transformation ratios of the CTs. The primary nominal current of the CT at the various line ends should however not differ by more than the factor 8.

9.3.3 Communication variants

The protection devices can be connected directly with each other via FO (mono-mode fibers) up to a distance of approx. 100 km.

For communication via data networks, ISDN or telephone cables, signal converters are however required to adapt the communication interface (Figure 9.17).

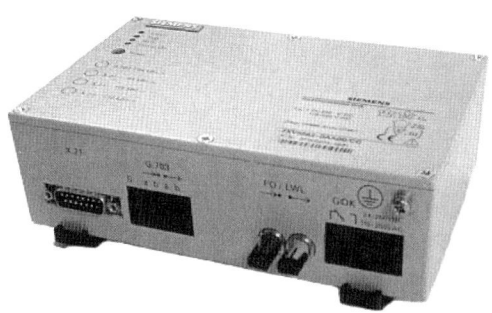

Figure 9.17
Signal converter 7XV56

As the interfaces are usually wire-bound with low signal levels of only a few volt, the signal converters are located in close proximity to the data communication terminal devices and connected via screened cable to these. The connection to the protection device is implemented via FO connection which is largely immune to interference (multimode cable, maximum 1.5 km).

Transmission via digital communication networks

In this case the signal converter adapts the FO signal coming from the protection to a standard interface X21 or G703 provided by the data network. It is installed close to the PCM-multiplexer. The connection is made via X21 with a 15 core data cable having a DSUB connector. For G703 a 5-pole connection with screw terminals is provided.

Since 2002 the IEEE Standard C37.94 for optical interfaces between relays and multiplexers exists. [6-31] It allows a direct OF connection from the relay to the multiplexer of the data network. Modems and the EMC critical wiring are in this case no more necessary.

Transmission via ISDN

In this case the converter contains the modem for transmission in accordance with the ISDN standard. The basic connection S0 (UK0-interface) is used. The protection utilises the two B channels each having 64 Kbit/s. Dialling or switching functions are not supported. A standing point to point connection is a pre-requisite.

Transmission via (twisted pair) pilot wires

For the data transmission via wire-bound connection (twisted pair) the interference voltages acting on the pilot wires must be considered (refer to section 6.1.1).

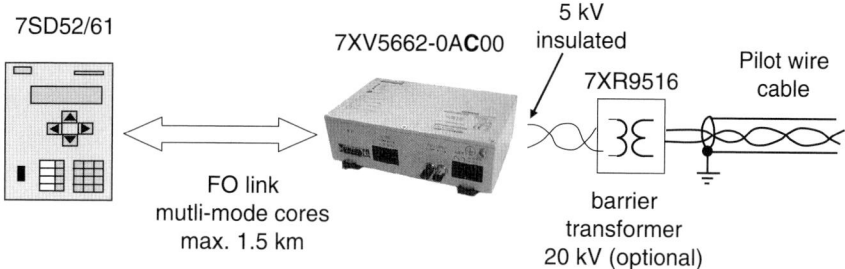

Figure 9.18 Digital communication via twisted pairs

The signal converter is therefore fitted with a 5 kV isolated interface (Figure 9.18). For particularly critical applications, an additional external barrier transformer with 20 kV isolation can be applied.

The transmission with 64 Kbit/s data stream is done fully bi-directional and synchronous with a technique that slightly varies from the ISDN using a 4 level (quad) digital code (2B1Q). The frequency spectrum that needs to be transferred for this purpose has a centre frequency of approx. 80 kHz. The signal level reaches a maximum of 2.5 V. With common telephone cable twisted pairs, a distance of approx. 15 km can be bridged with this technique (maximum loop resistance 1400 Ohm and signal damping < 40 dB at 80 kHz).

9.3.4 Additional functions and application notes

The modern line differential relays provide complete feeder protection with numerous additional functions, including overcurrent protection, overload protection, breaker failure protection, high impedance earth fault protection and auto-reclosure.

The overcurrent protection may be applied as back-up protection for the downstream busbars and the feeders connected to these. The overload protection is applicable for cable feeders.

In addition to the measured values, blocking or release signals, trip commands and control signals may be transferred to the remote line end through the existing data link.

The relays can also be equipped with two interfaces for redundant communication.

"Two-in-One" differential and distance protection

The relays 7SD523 and 7SL86/87 unite full differential and distance protection in one device with all supplements (51BF, 25, 79, 67N, fault locator, etc.)

The *distance protection* is usually applied as back-up protection for the remote busbar and the following lines. In the event of a communication failure it may also assume the task of main protection, even providing 100% protection coverage using a redundant signal communication path.

On special applications (e.g. three terminal lines without feeding condition) it may also be sensible to apply an under-reaching distance zone in parallel with the differential protection for non-delayed tripping independent of the communication.

The *earth fault directional comparison protection* may be applied to increase the pick-up sensitivity in the event of high resistance earth faults.

For important lines, dissimilar protection and communication may be used to set up fully redundant teleprotection systems. For this purpose two combined differential and distance protection relays (7SD523 or 7SL86/87) would be used. The differential protection would be each operated with digital communication through a data network, while the distance protections would each use an existing power line carrier signalling system. Alternatively the distance protection scheme could be operated through a second data link provided by dedicated optic fiber or digital micro-wave.

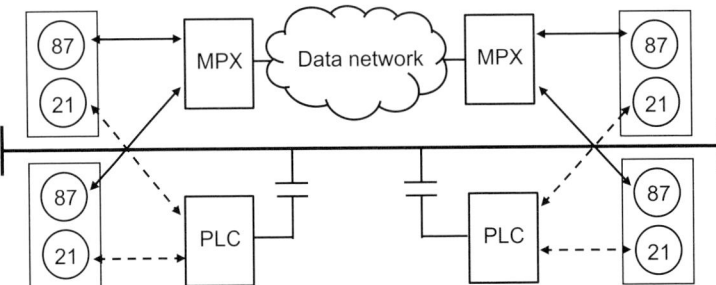

Figure 9.19 Redundant line protection

Two-sided fault locator

Numerical relays typically include a single-sided fault locator using measuring quantities of one line end only.

The differential (and distance) relays of the SIPROTEC 5 series offer as an option a novel two-sided variant.

It evaluates the synchronised voltage and current phasors from both line ends and thereby achieves a greater measuring accuracy.

In addition to the current phasors, needed for the differential protection, voltage phasors are exchanged via the digital communication link.

In principle, the algorithm calculates the line voltage drop from each end. The fault is located where both voltage trajectories meet. (Figure 9.20)

In reality however the measured values are applied to a real line model and the fault location is calculated by iteration using least squares techniques. [9-16, 9-17]

The algorithm is based on positive sequence quantities only. The measurement is therefore independent of the zero-sequence line data and is also not affected by zero-sequence parallel line coupling.

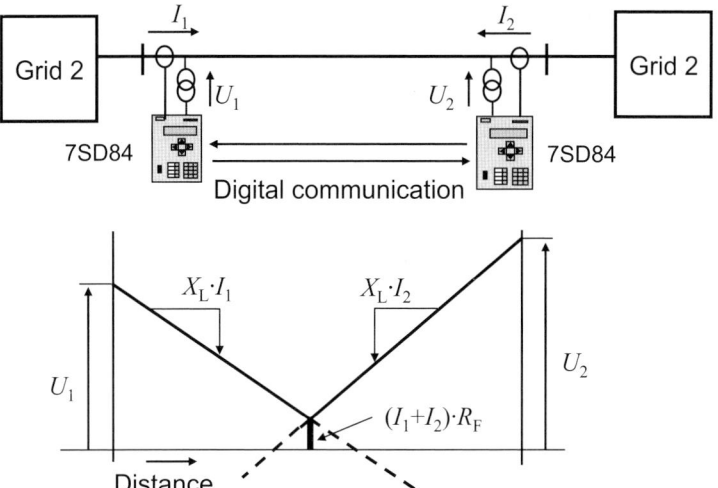

Figure 9.20 Double-sided fault locator

Application in breaker-and-one-half or ring type substations

In these kinds of substations, the feeders are each connected to two circuit breakers and two sets of current transformers.

In the traditional relaying practice, the CT secondary sides are connected in parallel to feed the relay with the summated current, equal to the line current.

This parallel connection may however result in false differential currents during external faults due to different CT magnetising characteristics or unequal core remanence. The most critical case occurs when high fault current flows through the cross connection from busbar to busbar with only small back-feed from the line. Under these conditions, the restraint of the differential protection is low and only a slight CT mismatch can cause false operation.

The criterion for preventing relay false operation in this case is high CT dimensioning in relation to the relay pick-up setting. One published rule of thumb recommends that the minimum operate current of the relay should be greater than 1 % of the maximum through-fault current, assuming a CT mismatch of not more than 1 %. [9-15] This high CT symmetry however is difficult to achieve in the short-circuit range with practically designed CTs.

The modern line differential relays (7SD86/87) offer a secure solution for this problem by providing dual current inputs and numerical summation of the currents. In this way each CT current contributes individually to the restraint as in the case of a multi-winding transformer differential relay. (Figure 9.21)

The dual current input makes it also possible to realise a fast and selective "stub protection" (differential protection for the short connections between the circuit breakers and the line isolator). It is automatically released when the line isolator is open.

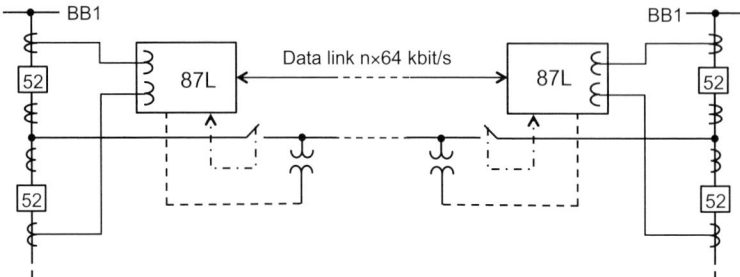

Figure 9.21 Line differential protection, CT connection at 1+1/2 CB substations

The integrated add-on functions further include co-ordinated auto-reclosure, CB failure protection and synchro-check for both circuit breakers.

In this way only one relay provides all protection functions needed for a double breaker line terminal.

9.4 Phase comparison protection with digital communication

The phase comparison protection (PCP) uses the phase angle between the currents for the comparison; the current magnitude is only applied as a release condition. In this manner less information has to be transferred to the opposite end. Using conventional technology, a narrow band analog channel (2.5 or 4 kHz) was sufficient to transfer the modulated voice frequency signal. With digital data transfer, a channel of 64 Kbit/s data rate is equivalent.

The earlier numerical phase comparison relay 7SD51 even only required a simple asynchronous data transfer of 19.2 Kbit/s. [9-6]

The modern relay version 7SD80 uses the faster serial data transmission with the standard rates 512 Kbits/s via dedicated optic fibers and 128 Kbits/s via pilot wire.

Measuring technique

For currents greater than nominal current, the protection functions the same as a conventional phase comparison protection (Figure 9.22).

The sinusoidal current half waves at the line ends are converted to co-phasial square wave signals and checked at each line end with regard to their coincidence. Under ideal conditions, the coincidence angle (φ_C) during external faults, i.e. with through flowing current, is 0° (0 ms at $f_N = 50$ Hz)[1], while during internal faults, having equal phase in-feeds, it is 180° (10 ms).

---

[1] According to the convention that currents flowing into the protection object are counted positive (see section 3.2.2).

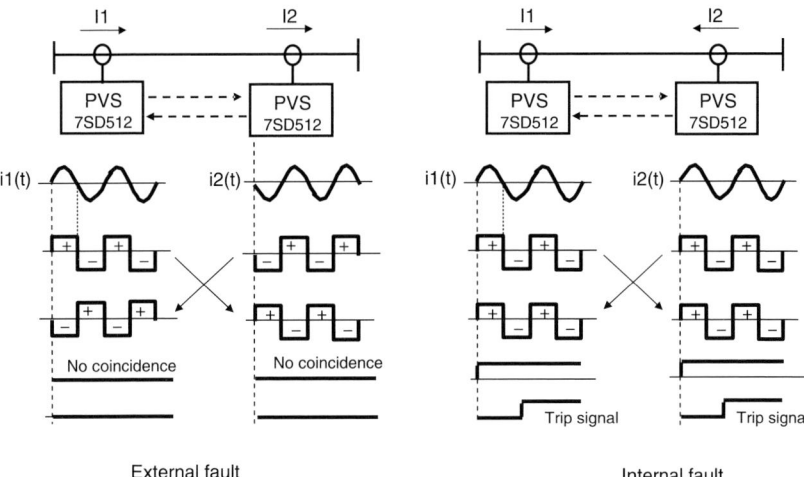

Figure 9.22 Principle of the phase comparison protection 7SD51

Due to the capacitive charging current on the feeder, there however is also a phase angle shift ($\varphi_C = 180° - \varphi$) on the through flowing current (see Figure 9.23 below). The phase angle shift depends on the ratio of the line charging current I_C to the through flowing current I_L: $\varphi_C = 2 \cdot \arcsin(I_C/2I_L)$.

In the case of load or small short-circuit currents on longer cables this angle deviation may become very large. The measurement must therefore only be released when the through flowing current is sufficiently large in comparison with the charging current. (This does not concern the supplementary dynamic PCP method described below.)

In the relay 7SD80, the current waves of the three phases are sampled every millisecond (at $f_N = 50$ Hz) and processed by Fourier filters. The direction of the extracted fundamental frequency component is then used for phase comparison.

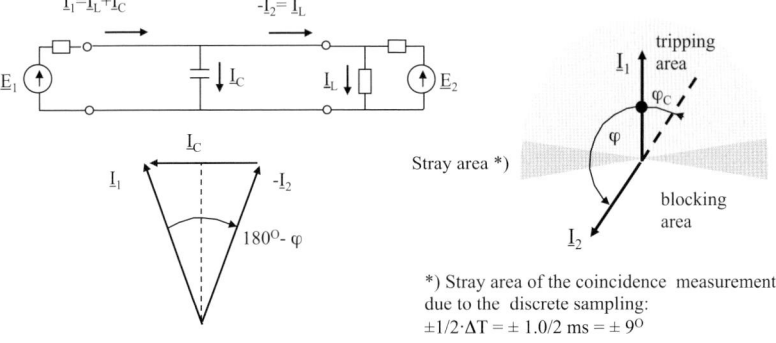

Figure 9.23 Phase comparison protection: Pick-up range and charging current influence

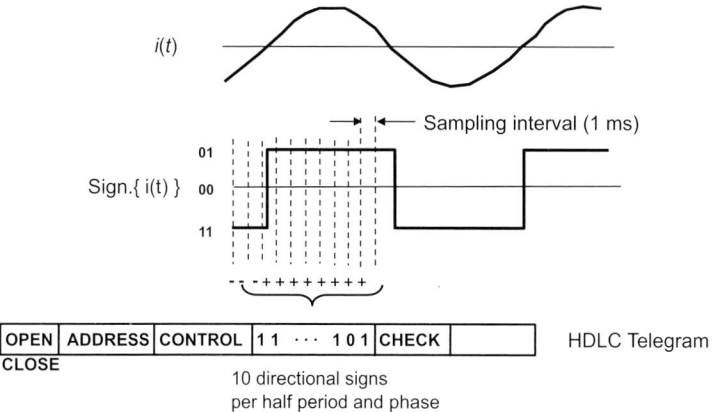

Figure 9.24 Generation of direction signal in the relay 7SD80 (principle)

The direction signals of the three phases are locally buffered and additionally transferred to the remote end, together with further protection and control data. (Figure 9.24)

The polarities of the local and remote directional signals are then compared per phase at both ends. By continuous measurement of the channel delay time[1] and corresponding delay of the local directional signals, a synchronous comparison is ensured.

From the number of matching directional signals (n) and the sampling interval (ΔT), the measured coincidence angle can be obtained: $\varphi_{C\text{-meas.}} = [(n+1/2)\pm1/2]\cdot\Delta T$.

If we assume $\Delta T = 1$ ms and a tripping threshold of $n = 5$, we get: $\varphi_{C\text{-Trip}} =$ 5.5 ±0.5 ms or 99 ±9°. The corresponding limiting angular displacement between the currents at the two line ends is $\varphi_{\text{Trip}} = 180° - \varphi_{C\text{-Trip}}$.

The resulting pick-up characteristic is shown in Figure 9.23.

With small load current and large charging current (longer cable) the phase displacement between the currents at both line ends (φ) shifts from 180° to much smaller angles. Consequently the coincidence angle φ_C increases and may approach the pick-up threshold. Therefore the tripping should only be released if a larger current is flowing and the appearing coincidence angle φ_C is smaller than approx. 45°, in other words when φ is clearly outside the tripping range.

For this purpose, the 7SD80 contains an overcurrent threshold $I_{\text{static}}>$, which must be set suitably high (normally $\geq 1.3 \cdot I_n$).

Directional signals are only released and sent if a settable minimum threshold current is exceeded.

---

[1] The used Ping-Pong method is described in section 4.2.3.

In the case of an internal short-circuit with single-sided strong infeed, no directional signals may therefore be received from the weak infeed end. Under these conditions, the protection trips on overcurrent at the strong infeed side and sends a transfer trip signal to the weak infeed (or only load) end.

In addition to the traditional phase comparison principle, the relay 7SD80 includes a supplementary phase comparison protection based on delta quantities. This means that the protection responds dynamically to changes in current flow. For this purpose, the current values sampled two cycles earlier are subtracted from the presently sampled values. This is done with a suitable digital filter which processes the present sampled values and the previously sampled historic values stored in a buffer. (Figure 9.25)

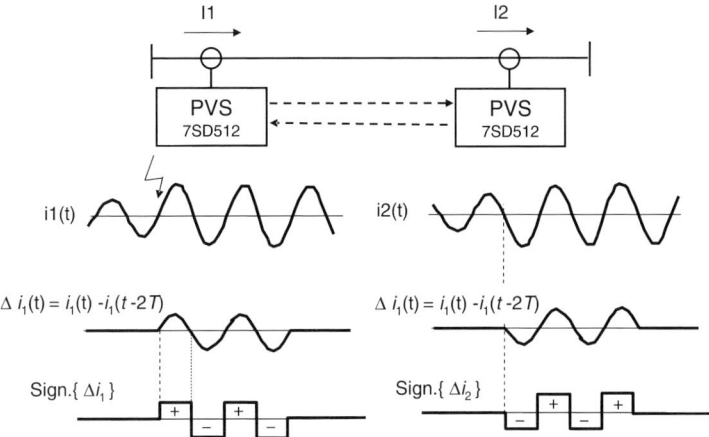

Figure 9.25 Phase comparison protection with delta quantities

The current change $\Delta i(t)$ is only available for the duration of the filter (2 cycles) and disappears thereafter. This delta current is processed along the same phase comparison principle as described above for the static phase comparison protection.

Due to the processing of the current change, the dynamic phase comparison protection is independent of the load and charging current on the feeder. The measurement with delta currents corresponds to a measurement with pure fault current without load influence. High pick-up sensitivity can be obtained thereby. This can be illustrated by means of the superposition principle (Figure 9.26).

By subtracting the load current that flows prior to fault inception from the total measured current, the obtained result is equivalent to the pure short-circuit current supplied by a virtual voltage source at the fault location. The driving voltage U_{FL} thereby corresponds to the voltage prior to fault inception however with inverted polarity.

Total currents after fault inception

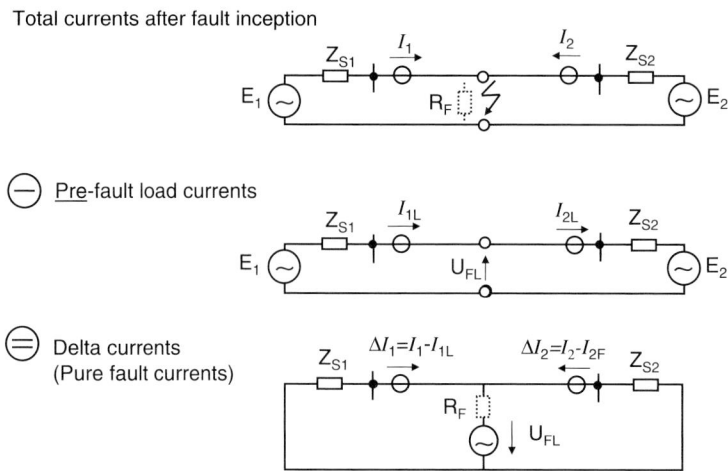

Figure 9.26 Superposition principle applied to the short-circuit calculation

It is recommended to set the dynamic tripping threshold of the relay 7SD80 larger/equal 3 times the charging current I_C of the total feeder to avoid response to transient charging and discharging currents. The pre-setting is $I_{dyn} > = 0.33 \cdot I_n$. For cable feeders up to 10 km and overhead lines up to 100 km in systems up to 110 kV this setting may remain unchanged.

Main features of the current comparison relay 7SD80

This SIPROTEC relay is designed as compact version for application mainly in distribution systems. (Figure 9.27)

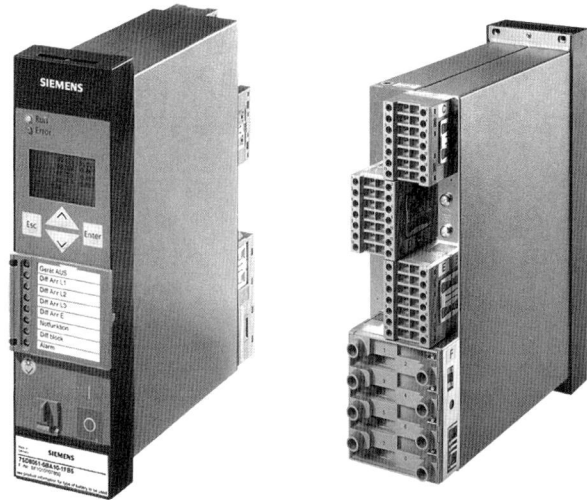

Figure 9.27 Current comparison relay 7SD80

It can be used with dedicated optic fibers (mono-mode, 1300 nm) up to about 40 km, but can also directly communicate through pilot wires up to about 20 km via the integrated PAM[1] modem (40 kBaud).

Even redundant operation of both communication modes in parallel is possible.

Short-circuits are covered by the phase segregated phase comparison protection described above.

Additionally sensitive earth fault protection is included in two variants:

− For *solid and low impedance earthed systems* a biased earth-current differential protection can be selected (same measuring technique as described under section 9.3.2).

− For *isolated or Peterson coil earthed systems* a directional comparison scheme can be optionally chosen. It compares the direction of Var or Wattmetic zero-sequence power measured at both line ends.

Application notes

The relay 7SD80 can be used as complete feeder protection unit. Tap transformers along the feeder are allowed due to the included second harmonic inrush restraint.

In addition to the described comparison protection, 7SD80 contains further functions including O/C-backup protection, auto-reclosure, overload protection, and with VT connection various voltage and frequency monitoring functions.

The 7SD80 can also transfer alarms, and commands for remote tripping.

It further provides 15 s fault recording (8 fault events) and it can be integrated in substation automation systems compatible to current standards, including IEC 61850.

Different CT ratios at both line ends can be numerically adapted in the range of 1 to 4.

The CT requirements are moderate:

Accuracy limit factor in operation $ALF' \geq 30$ and transient dimensioning factor $K_{TD} \geq 1.2$.

The use of the sensitive Wattmetric directional earth-current comparison in compensated networks however requires an additional core-balance (window-type) CT of low ratio (typically 60/1 A) at each line end.

---

[1] PAM: Pulse Amplitude Modulation

9.5 Differential protection of feeders including transformers

On transformer feeders, the transformer and line or cable are connected in series and form a unit. In this manner, one CB can be saved.

On feeders with tee-offs, transformers along the feeder are directly connected. In this case, a substation is saved. In both cases the protection is faced with particular difficulties.

9.5.1 Protection of transformer feeders

The differential protection for this application must include the special features required in transformer differential protection: ratio and vector group adaptation, as well as in-rush blocking and stabilising against over-fluxing. In addition the protected object must be allowed to extend over larger distances, which means that the measured value transferral must have the properties of a feeder differential protection.

With conventional technology for distances up to approx. 1 km the normal transformer differential protection was applied. To reduce the burden applied on the remote CT, the current across the pilot cable was reduced to approx. 100 mA by means of interposing CTs at the switch-gear and at the transformer.

The normal feeder differential protection was however also applied. For this purpose the CTs at one end were connected in delta to provide the vector group adaptation. The in-rush blocking was provided at both ends by means of supplementary relays. [A-15, A-22]

With numerical technology, a relay is now available which includes the properties of a transformer and feeder differential protection in a single unit (7SD84). Figure 9.28 states an application example.

It must also be noted that the zero sequence current elimination must be activated if the winding star point is earthed. The communication may be implemented by means of direct FO connection. For wire bound or directional radio links, additional converters 7XV56 are required (see section 9.3).

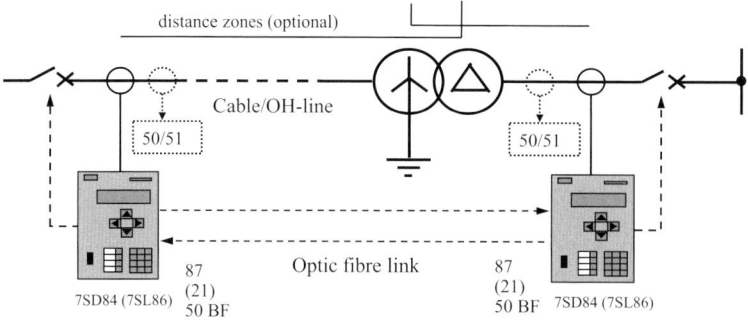

Figure 9.28 Differential protection for a transformer feeder (example)

Optionally, a differential relay with integrated distance zones (7SL86) could be chosen. In this case underreaching zones would be graded into the transformer from both sides. This would provide fast protection even when the signal transmission fails, and also allow a rough discrimination between primary and secondary winding faults. Voltage transformers would however be needed at the side(s) where the 21 function is applied.

In the example overcurrent protection (50/51) is provided at both sides as back-up protection.

9.5.2 Differential protection for feeders with tee-offs (tapped lines)

Direct tees off the feeder without switchgear are common in distribution networks (132 kV and below). In many cases several tees are present. These may be pure consumers (load) as well as back-feeds (distributed generation). The connection may be applied with or without CB. Earthing of the transformer star point on the feeder side is not always the same. In earthed systems the star point of the transformer at the tap may be isolated or effectively earthed. The protection concept must be individually adapted to the conditions of the application at hand. [9-12, 9-13]

For the application of a feeder differential protection the pick-up threshold must be set above the maximum sum total current of the applied tap loads. As in-rush blocking is not generally available with conventional relays, the in-rush current of the transformer in the event of single ended energising of the feeder must be considered. The maximum short-circuit current during a fault on the secondary side of a tap transformer must be considered in any event. If signal communication is available, the differential protection on the feeder may be blocked during critical conditions if the protection on the secondary side of a transformer tap picks up. Alternatively a non-directional distance protection zone may be applied as release criterion for the feeder differential protection. This zone must cover 100% of the feeder (setting 120% Z_L), may however not reach through the tap transformer. Suitable setting can be found when the rating of the tap transformers is not very large and when the feeder is not too long as shown in the following example.

With the numerical protection the feeder with tee-offs can be optimally protected if broad band communication (FO or directional radio) is available. The numerical feeder differential protection 7SD52/86/87 for up to 6 line ends (that is 4 tee-offs are permitted) can be applied (Figure 9.29).

Transformers may also be included in the protected object (left-hand side transformer in Figure 9.29), as all the above mentioned functions of the transformer protection are integrated in the relay.

If a data link is only available between the two ends of the primary feeder then a numerical differential protection for two line ends may also be applied (7SD61/84). The pick-up threshold must then be set above the short-circuit currents of the tee-offs. If a distance protection zone release is applied as described above, the integrated distance protection (option in 7SL86) may be used for this purpose.

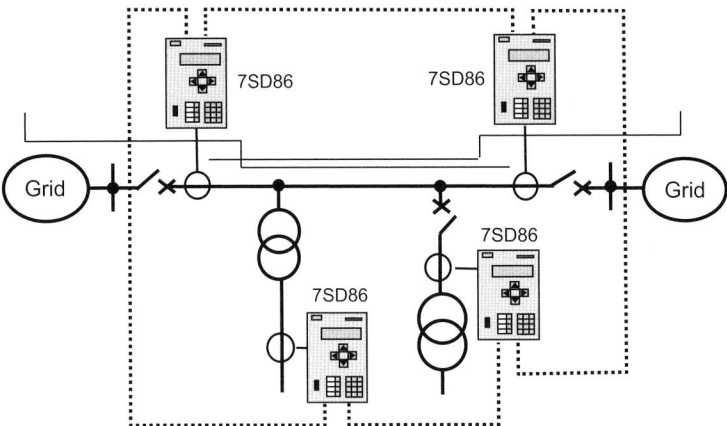

Figure 9.29 Multiple terminal differential protection for a feeder with teedconnections

Example 9-5: Setting of the numerical pilot protection on a line with tee-offs.

Given: HV feeder 110 kV, $l = 27$ km, $X_L = 0.4$ Ω/km

Data of the system in-feed and the tee-offs according to Figure 9.30

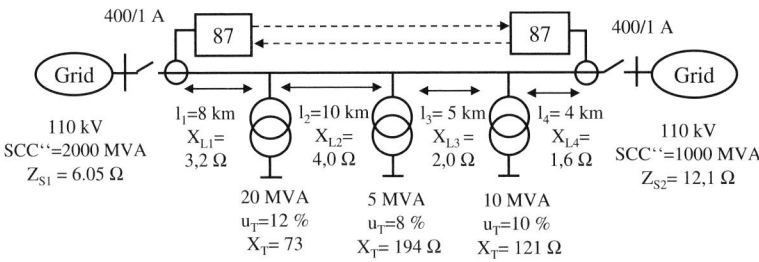

Figure 9.30 Feeder configuration for example 9-5

Wanted: How high must the pick-up threshold of the differential protection be set? Can increased pick-up sensitivity be achieved by applying a distance zone release?

Solution: The short-circuit impedances were calculated and are entered in Figure 9.30. It can be seen that the transformer impedances are an order of magnitude larger than the in-feed and line impedance. These (in-feed and line impedance) may therefore be neglected for an approximate calculation of the transformer short-circuit currents. With this assumption the results obtained are on the safe side.

The largest error current (differential current) for the differential protection results during a short-circuit behind the 20 MVA transformer:

$$I_F = \frac{1.1 \cdot U_n / \sqrt{3}}{X_T} = \frac{1.1 \cdot 110 \cdot 10^3 \, V / \sqrt{3}}{73 \, \Omega} = 957 \text{ A}$$

The pick-up threshold of the differential protection should therefore be set to $I_{\text{Diff}} > = 1.3 \cdot 957 = 1244$ A, providing a security margin of 30%. This setting would then correspond to three times the CT nominal current.

A distance protection zone with 20% overreach would have to be set to $Z_{\text{OR}} = 1.2 \cdot 0.4 \ \Omega/\text{km} \cdot 27 \ \text{km} = 13 \ \Omega$. The smallest transformer short-circuit impedance is 73 Ω. The risk of overreach is therefore safely excluded. Even a setting of approx. 50 Ω is possible to obtain as large as possible reach into the transformers (approx. 70% of the 20 MVA transformer).

For the setting of the distance zone, the in-rush current of the transformer must also be considered, when the feeder is energised from one end. The distance protection then measures the impedance

$$Z_{\text{Inrush}} = \frac{U_n / \sqrt{3}}{0.5 \cdot \sum I_{\text{Rush}}}.$$

For this calculation it was assumed that the numerical protection only evaluates the fundamental and that the fundamental component of the in-rush current is not greater than 50%.

Furthermore it is assumed that for the transformer size of this example, the rush currents do not exceed the value equal to 5 times nominal current so that the following approximation can be made:

$$\sum I_{\text{Rush}} = 5 \cdot \frac{35 \ \text{MVA}}{\sqrt{3} \cdot 110} = 0.918 \ \text{kA} \quad \text{and}$$

$$Z_{\text{Inrush}} = \frac{110 \ \text{kV} / \sqrt{3}}{0.918 \ \text{kA}} = 69 \ \Omega.$$

The measured impedance is therefore far enough outside the intended setting of 50 Ω.

The overreaching zone in this example is therefore perfectly suited for release of the feeder differential protection. The pick-up threshold could be reduced to approx. half CT nominal current; in other words 200 A could be set.

Note: With the conventional protection technology, the direction measurement for close-in faults was not obtainable with absolute security. Voltage memory was relatively expensive and was only applied at the EHV level. The differential protection was therefore superior for providing 100% protection coverage of the feeder when compared to the directional comparison principle.

With the modern distance protection the direction measurement is secure due to the digital voltage memory. A directional comparison protection is therefore comparable with the differential protection in terms of selectivity, especially as the direction comparison can now also be done on a per-phase basis (relay 7SA6). A degree of restriction in terms of selectivity for the distance protection only results in the event of multiple or sequential faults, which are very rare. For the example described above, it would therefore have to be considered whether a distance protection with directional comparison would not be preferable to achieve a simpler solution.

The particular advantage of differential protection is its independence of voltage transformers. The release by a distance zone should in any event not be applied instead of, but in parallel to a current release criterion, so that only the pick-up sensitivity for small current faults is increased.

10 Busbar Differential Protection

Busbar faults are on occasion caused by mechanical or insulator defects, but are frequently caused by incorrect switching operations. Statistically one can count one busbar fault every 10 years (0.63 to 2 faults per feeder in 100 years according to CIGRE survey).

Busbar differential protection has generally been applied in transmission systems because busbar faults must be cleared very fast to maintain system stability. In distribution systems, differential protection has only been installed in important substations and with metal clad, gas-insulated switchgear. In simpler distribution stations protection has been provided in a simpler way, for example by a reverse interlocking scheme, or even dispensed with for cost reasons.

The design of the busbar protection depends essentially on the bus configuration and the arrangement of circuit breakers, isolators and current transformers.

Single buses can be easily protected by muli-terminal differential protection.

Multiple bus configurations with bus couplers and ties however require the replication of the switching state of the isolators and circuit breakers and the implementation of several protection zones to enable selective tripping of only the faulted bus section.

The high impedance principle was applied in Anglo-Saxon countries for protection of the single busbar arrangements (longitudinally separated single bus and 1 ½ CB stations), which are common in those countries. [10-1] The biased (percentage) current differential protection (low impedance measuring principle) for multiple busbar substations is commonly applied in Europe. Due to the complexity in this arrangement it was only implemented relatively late. Siemens was a pioneer in this field: the electro-mechanical version RN23 was introduced in Germany in 1960 [10-2], in 1972 the static analog device generation followed in the form of 7SS1 [10-3, 10-4] and ultimately in 1989 the numerical version 7SS5 [10-5, 10-6] was first applied.

Other suppliers developed a "moderately high-impedance" version [10-7] as well as the phase comparison principle. [10-8]

In digital technology also hybrid solutions with a combination of differential and directional comparison principles are used. [10-10 and 10-16]

The publications [10-11 and 10-12] provide a summary of the basic principles of the busbar protection.

The state of the art in digital busbar protection can be found in the technical papers [10-13 to 10-16].

Very stringent demands regarding security, availability and selectivity are imposed on the busbar protection:

Security

The protection may not under any circumstances trip in the event of an external fault as the loss of a busbar in general causes shut-down of a large section of the system. Tripping is therefore always dependent on a number of criteria. With analog technology, pick-up of the feeder protection was commonly used as an additional criterion. With high impedance protection, and now also in the numerical protection 7SS5, a check-zone which covers the entire busbar arrangement is applied. Furthermore, with the numerical protection 7SS5, the differential criterion is duplicated (with even and odd sampled values) with computation on separate processor systems.

Other manufacturers combine for example differential and phase comparison principles. [10-16]

Availability

The failure to trip or the delayed tripping of a busbar fault creates the risk of system instability and may under critical circumstances cause a system collapse due to the voltage collapse on the busbar. Furthermore, severe damage could be caused at the fault location due to the very high short-circuit current level (especially in GIS switchgear).

Some utilities therefore duplicate the busbar protection at the EHV level.

Selectivity

In multiple busbar substations, only the busbar section which is affected by the fault must be cleared, so that the power supply via the other busbars is ensured. For this purpose, a separate measuring unit for each bus section is required, along with a complex image of the isolator positions for the allocation of the feeder currents and distribution of the trip commands. The numerical protection provides a number of improvements in this regard due to the contact-free numerical processing and complete monitoring of the protection system including the measuring circuits and the isolator positions.

Tolerance to CT saturation

This is probably the most demanding design criterion.

The busbar protection must not false operate even in the case of extreme CT saturation. Modern relays only require two to 3 milliseconds saturation-free time to stabilise the protection during external faults and to ensure fast tripping in case of internal faults.

Fast algorithms using short data windows and sophisticated saturation detectors were developed for this purpose. [10-10, 10-11, 10-14]

Breaker-failure protection

In recent years the need for a local back-up protection (breaker-failure protection) has increased as the availability and tripping times of the remote back-up protection (back-up zones of distance relays) is no longer sufficient. This is of particular relevance close to large power stations.

The numerical busbar protection now provides the convenient opportunity to include a comprehensive breaker-failure protection as supplementary function.

10.1 Low impedance busbar differential protection

The biased differential protection may also be applied to protect busbars. The principle is shown in Figure 10.1 based on electro-mechanical technology.

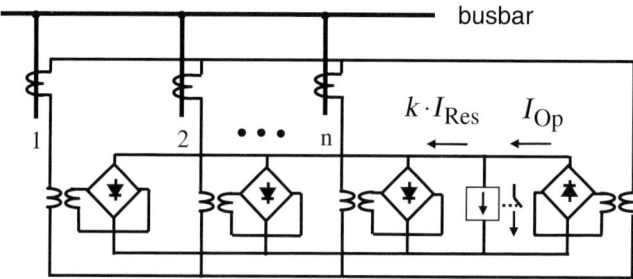

Figure 10.1 Principle of busbar protection

The operating current I_{Op} corresponds to the magnitude of the geometric (phasor) sum of the feeder currents:

$$I_{Op} = |\underline{I}_1 + \underline{I}_2 + \underline{I}_3 + \ldots + \underline{I}_n| \tag{10-1}$$

The restraint current I_{Res} corresponds to the arithmetic sum (sum of absolute values) which is here generated by rectification and addition:

$$I_{Res} = |\underline{I}_1| + |\underline{I}_2| + |\underline{I}_3| + \ldots + |\underline{I}_n| \tag{10-2}$$

The operating current minus a settable component of the restraint current $k \cdot I_{Res}$ flows through the polarity sensitive measuring relay (moving coil relay). For a trip to be issued, the current difference must exceed the pick-up threshold of the moving coil relay I_{DIFF}. Therefore the tripping criterion is as follows

$$I_{Op} > I_B + k \cdot I_{Res} \tag{10-3}$$

The pick-up characteristic shown in Figure 10.2 results.

The rectified pulsating current is smoothed by the inertia of the moving coil relay. The operating time amounts to approximately 2 cycles (approx. 30–40 ms at 50 Hz).

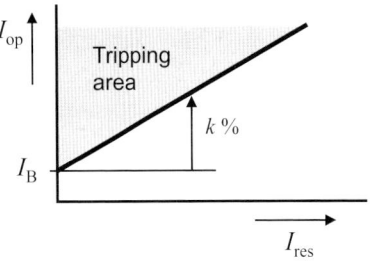

Figure 10.2
Conventional differential protection
tripping characteristic

Isolator replica

On multiple busbar arrangements, each bus section has its own differential relay allocated to it. The currents are routed to the relay via a replica of the isolator switching status. The principle is shown in Figure 10.3 for a double busbar arrangement. The restraining current circuits are not shown for the sake simplicity.

The currents are reduced to a level of 100 mA via interposing CTs (three-phase measurement) or summation CTs (single-phase measurement) and then routed via the isolator replica to the relay. In this manner the main CT always has a low impedance burden via the primary winding of the interposing CT, and the contacts of the switching relays only have to be rated for small current levels.

Figure 10.3 shows the state of the isolator replica with auxiliary supply connected.

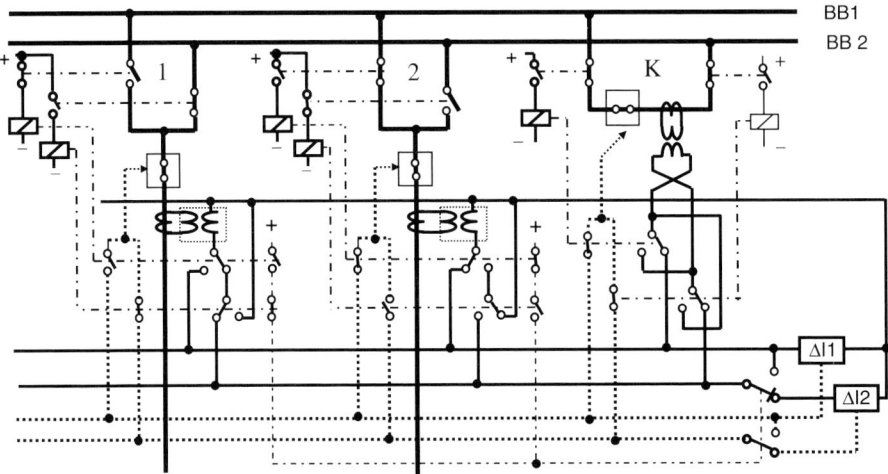

Figure 10.3 Isolator replica of the busbar protection (simplified)

The feeder 1 is connected to BB2 and the feeder 2 to BB1. The coupler is considered to be closed, as both coupler isolators are in the closed position. The circuit breakers in the feeders are closed; their switching state however does not affect the isolator replica. (The special treatment of the coupler is covered in section 10.1.2 below). The currents are routed to the protection systems according to the isolator positions. In this way each busbar section is protected separately and also selectively cleared in the event of a fault.

When changing over a feeder from BB1 to BB2 or vice versa, the bus coupler must be closed (including the CB). For example, for the switching over of bay 1 from BB2 to BB1, the second isolator in bay 1 may be closed for the connection to BB1. The switching over is complete when in the next step the isolator to BB2 is opened in the same bay. Between these two stages a state where both busbars are coupled via isolators exists and therefore cannot be switched off selectively. This state is known as *isolator coupling*. The protection treats the two busbars in this state as a single unit (common protection zone) and the isolator replica routes all the bay currents to the preferred measuring system (ΔI_1 in Figure 10.3). This is done by means of the change over contact in series with the measuring systems. The tripping signals to the bays are switched over in a similar manner.

The isolator replica must be valid for all operating conditions and should revert to a secure default state in every bay in the event of auxiliary supply failure.

For this purpose, an isolator replica contact is required that opens when the isolator primary contact leaves its quiescent position (open position), i.e. the isolator is already assumed to be closed as soon as the isolator primary contact leaves the quiescent open state. By using a normally closed contact, it is ensured that the isolator replica assumes the isolator to be closed in this bay, should the auxiliary supply fail, i.e. the condition isolator coupling (all isolators of a bay closed) is assumed in the isolator replica in this case. The protection in this condition would however clear the entire busbar system in the case of an internal fault, it remains however stable in the case of an external fault in the system which has a much higher probability. If auxiliary supply failure in the bay must leave the isolator replica unchanged, then bi-stable flip-flop (latching) relays must be applied which memorise their position by mechanical or magnetic latching. Latching relays however are costly and were actually only implemented at the EHV level. (In the numerical protection 7SS5 the latching of the isolator position is done via software, see below).

10.1.1 Partially numerical busbar differential protection 7SS600

This protection combines the conventional relay-based isolator replica with a numerical measuring relay. The proven static protection system 7SS10 is upgraded with the modern functionality provided by numerical technology (self-monitoring, oscillographic recording, measured value indication, etc.). Figure 10.4 shows the structure.

The three CT secondary currents are connected via interposing transformers (summation CTs type 4AM5120, refer to section 3.3.2) to reduce their current level (to 100 mA at nominal current); these flow via the measuring input of the relay as

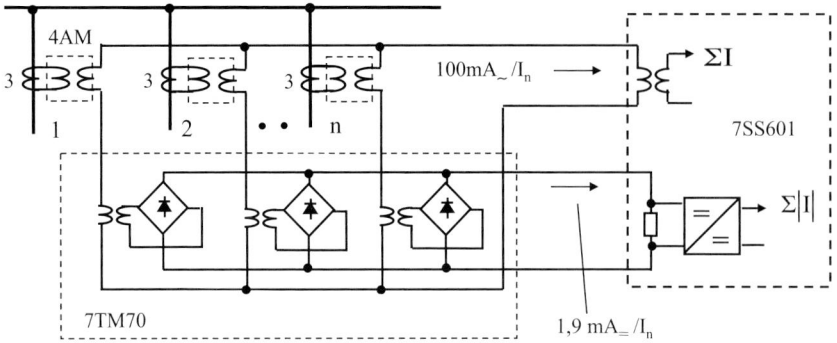

Figure 10.4 Partially numerical busbar protection 7SS600

summated operating current. The restraint current is generated with rectifier modules (7TM70) and fed to the relay as pulsating DC current. For galvanic separation a DC/DC transformer is provided. Further processing of the phasor sum $\sum \underline{I}$ and the absolute value (magnitude) sum $\sum |\underline{I}|$ to provide biased differential protection is then done by the numerical measuring methods (relay 7SS601).

The tripping criterion was slightly modified in comparison with the conventional protection. Pick-up occurs when two criteria are met:

$$I_{Op} > I_{B} \quad \text{and} \quad I_{Op} > k \cdot I_{Res} \tag{10-4}$$

This results in a straight line through the origin for the percentage biased portion of the tripping characteristic (Figure 10.5):

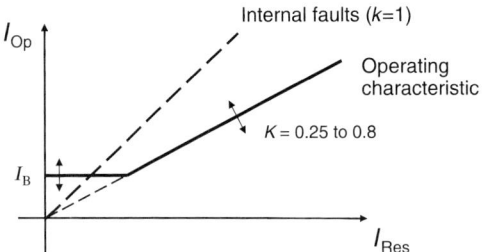

Figure 10.5
Tripping characteristic
of 7SS600

The k-setting depends on the dimensioning of the CTs. The general setting is $k = 0.6$. This requires 4 ms saturation free CT transformation. Furthermore the system time constant may not be greater than 300 ms.

The short tripping time, minimum 15 ms, demands instantaneous sampled value processing of the measured values. The pulsating stabilising current arriving at the relay is smoothed with numerical filtering methods. This is equivalent to the instantaneous charging of a capacitor up to the peak value of the current half wave followed by the time delayed discharge with a time constant of 60 ms.

For the trip condition to be established, the numerically rectified operating current must exceed the smoothed restraint current in at least 3 sequential sampling points: $I_{Op} > k \cdot I_{Res}$.

The advantage of this measuring technique is particularly significant in the case of severe CT saturation. This is shown below (section 10.2).

Application notes

The partially numerical busbar protection 7SS600 consists of plug-in modules mounted in a centralised cubicle similar to the static analog protection 7SS10. This means that the auxiliary units with the restraint modules and the isolator replica relays are located next to the measuring relays. Therefore all the bay currents must be routed to this central location. With a single measuring system application, the summation CTs may however be located at the bay so that only the single phase composite current (100 mA level) has to be routed to the central relay room with a two core cable. In this way, a small CT burden is also achieved.

The low cost protection system 7SS600 in the summation CT version is mostly applied in distribution substations as well as simple high voltage plants. In principle it could also be used with a per-phase measurement. For large plants and at the EHV level the de-centralised fully numerical version 7SS52 with phase segregated measurement and integrated breaker failure protection is preferred (refer to section 10.1.2 below).

The CT dimension requirements are moderate (4 ms saturation free transformation) and the tripping time is below one cycle (minimum 15 ms).

In case of the summation CT version, the following must be observed:

– The pick-up sensitivity for the individual fault types differs.

– For cross-country earth faults with single sided in-feed a very high restraint may occur (see section 3.3.2). A similar situation may arise for single-phase earth faults if only the load side transformers are earthed. This is illustrated in the following example.

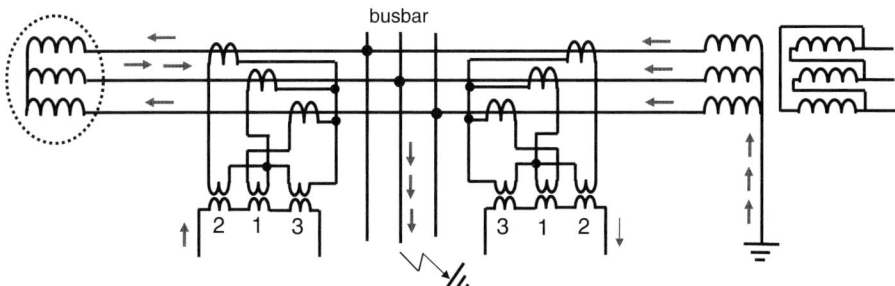

Figure 10.6 Response of protection 7SS600 with special system earthing

Example 10-1: High restraint in the special case with summation CT version

Given: In the shown network (Figure 10.6) a single phase earth fault occurs.

Question: Does the protection 7SS600 summation CT version trip?

Solution: Initially the worst case condition is assumed, which is a fault in phase L2 (smallest composite CT current):

$$I_{M-A} = 2 \cdot I_{L1} + 1 \cdot I_{L3} + 3 \cdot I_E = 2 \cdot (-I_F) + 1 \cdot (-I_F) + 3 \cdot 0 = -3 \cdot I_F$$

$$I_{M-B} = 2 \cdot I_{L1} + 1 \cdot I_{L3} + 3 \cdot I_E = 2 \cdot I_F + 1 \cdot I_F + 3 \cdot 3 I_F = +12 \cdot I_K$$

The following operating and restraining currents are obtained:

$$I_{Op} = |I_{M1} + I_{M2}| = 9 \cdot I_F$$
$$I_{Res} = |I_{M1}| + |I_{M2}| = 15 \cdot I_F$$

The ratio $I_{Op}/I_{Res} = 9/15 = 0.6$

For faults in other phases we get:

Fault L1-E: $I_{Op} = 3 + 12 = 15$, $I_{Res} = 3 + 12 = 15$, $I_{Op}/I_{Res} = 1$

Fault L3-E: $I_{Op} = 0 + 12 = 12$, $I_{Res} = 0 + 12 = 12$, $I_{Op}/I_{Res} = 1$

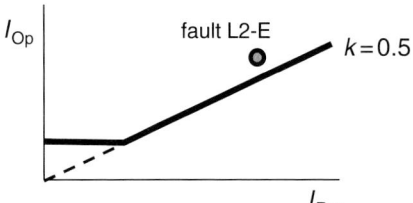

Figure 10.7
Results of example 10-1

The fault L2-E is the worst case, as expected. The bias factor may not be set greater than $k = 0.5$ to achieve secure tripping under these circumstances. The CTs must be adequately dimensioned for this low setting.

Response in systems with earth fault current limiting equipment

The normal summation CT connection (Figure 3.18) provides increased sensitivity for earth faults due to the weighting of the earth current by the factor 3.

This is illustrated by the following example:

Example 10-2: Earth short-circuit in system with earth current limiting equipment

Given: Medium voltage station with neutral earthing via a resistance and composite current busbar protection 7SS600. (Figure 10.8)

Question: Does the protection trip for a busbar fault L2-E?

Solution: The fault L2-E was chosen because the single-phase earth short circuit current in this phase results in the smallest composite current. The ratio of this to the symmetrical three phase current is

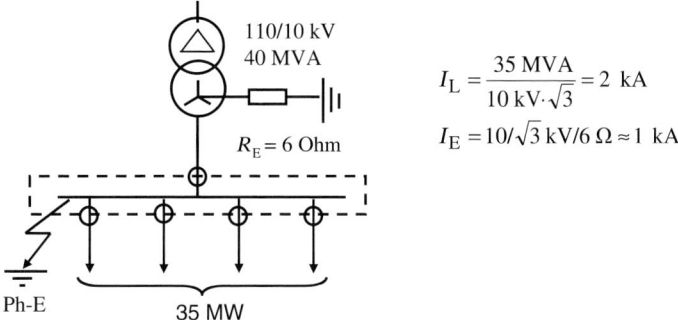

$$I_{L} = \frac{35\ \text{MVA}}{10\ \text{kV}\cdot\sqrt{3}} = 2\ \text{kA}$$

$$I_{E} = 10/\sqrt{3}\ \text{kV}/6\ \Omega \approx 1\ \text{kA}$$

Figure 10.8 System configuration for example 10-2

$I_{\text{M-L2-E}} = \sqrt{3}\cdot I_{\text{M-3-Ph}}$. The composite current on the in-feed side results from the superposition of the earth current on the load current that is flowing. A graphic representation for this situation can be constructed (Figure 10.9).

The composite current that corresponds to a three phase load current of 2 kA is $I_{\text{M-L}}$. The composite current component of the earth current is added thereto. It has the magnitude $I_{\text{M-L2}} = 3\cdot1$ kA and the same direction as the load current in phase L_2. The resulting composite current at the in-feeding end therefore is $I_{\text{M1}} = 1/2\cdot I_{\text{M-L}}$. At the load side the sum of the load current results in the value $I_{\text{M2}} = I_{\text{M-L}}$. The restraining current therefore is $I_{\text{Res}} = \left|I_{\text{M1}}\right| + \left|I_{\text{M2}}\right| = 3/2 \cdot I_{\text{M-L}} = 3\cdot\sqrt{3}$ kA. The operating current corresponds to the earth current and is $I_{\text{Op}} = \left|I_{\text{M-L2}}\right| = 3\cdot1$ kA . The ratio $I_{\text{Op}}/I_{\text{Res}}$ is 0.58.

With a setting of $k = 0.5$ the protection would pick up as shown in Figure 10.9.

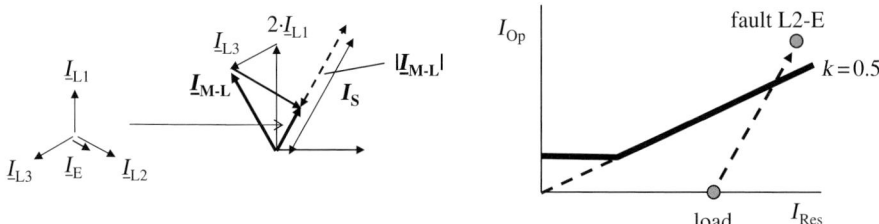

Figure 10.9 Busbar fault in a system with earth current restriction

10.1.2 Fully-numerical busbar protection 7SS52

The first fully-numerical version of busbar protection 7SS50 was introduced in 1988 and upgraded to the decentralied system 7SS52 in the early 1990s. [10-11 and 10-12]

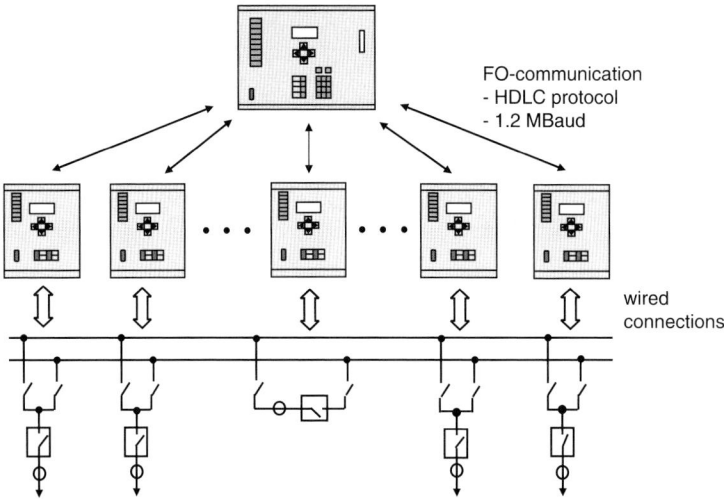

FO-communication
- HDLC protocol
- 1.2 MBaud

wired
connections

Figure 10.10 Structure of the numerical busbar protection 7SS52

This protection system is fully numerical and consists of a central unit with feeder dedicated bay units. The communication is done via FO cables. (Figure 10.10)

The measured values and isolator switching status obtained in the bay units are transferred to the central unit and are processed there according to the differential protection principle. In the opposite direction, the trip commands are serially transmitted to the bay units which issue the trip command to the circuit breaker at the bay level.

This concept has a particular advantage in expansive substations because the bay units can be installed directly in the bay thereby minimising the cabling in the substation. In the extreme with large open-air substations the bay unit may be located in a relay kiosk next to the switchgear.

As the protection uses 20 samples per cycle (sampling rate of 1 or 1.2 kHz at 50 or 60 Hz respectively), and the synchronously sampled measured values are processed in the central unit in real time (comparison every ms respectively 0.833 ms), a very high data rate over the FO channels is required (1.2 MBaud). The data transfer is synchronous, with high security (HDLC protocol, CRC-32). FO bridging lengths of up to 1.5 km are permitted.

The protection has phase segregated measurement and is suitable for the following maximum configuration:

– Triple busbar with transfer bus or quadruple busbar

– 48 bays

– 12 busbar sections

The switching state of isolators and circuit breakers is provided for the protection via auxiliary contacts connected to binary inputs of the bay units. (Figure 10.11)

Making and breaking contacts are used in each case in parallel to allow the monitoring of the corresponding d.c. circuits.

In the isolator end positions, one auxiliary contact must be closed and the other open. Generally a breaking contact is used which opens when the isolator leaves its open position, and the protection assumes the isolator to be closed already when it leaves the open position. The making contact later closes when the isolator reaches the closed position. Specially calibrated contacts are not required.

During the transition of the isolator from the open to the closed state or vice versa both contacts are open for a short time (some seconds in case of open air HV switchgear). The monitoring of the d. c. circuits recognises this transient state and prevents false alarm.

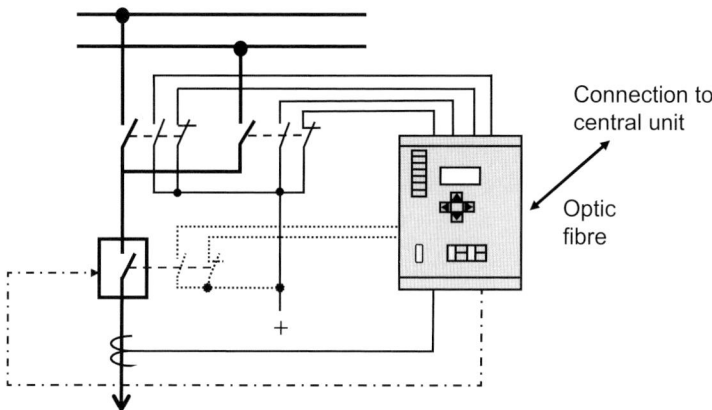

Figure 10.11 Connection of 7SS52 bay unit

The auxiliary contacts of circuit breakers are needed for zone selective tripping of dead end faults in coupler bays and fast clearance of end faults. (See below)

The protection 7SS52 can be adapted with a plant editor (similar to CAD program) on a PC (program DIGSI), to obtain the required configuration and settings. (Figure 10.12) For the common bay types (feeders, bus couplers and bus sectionalisers) symbols and typicals are provided in a library. The parameter set for the protection is generated completely automatically [10-13]

Measuring technique

The differential protection is fully numerical including a numerical isolator replica so that no relay switching is required. The currents are processed numerically with an instantaneous value comparison every ms. In this manner, extremely short tripping times of typically 15 ms including the trip relay time are achieved.

The measuring principle is the same as that of the 7SS600, however with fully numerical measured value processing, adaptation and evaluation.

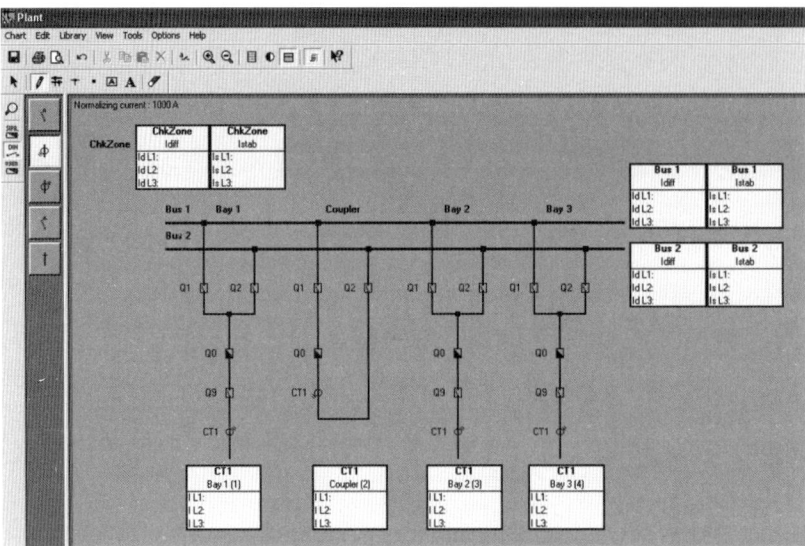

Figure 10.12 Configuration of busbar protection using the DIGSI plant editor

The per-phase measurement ensures that the pick-up sensitivity is the same for all fault types. External faults are recognised with only 2 sampled values so that a saturation-free time of 3 ms is sufficient to obtain a secure decision on whether the fault is external or internal. In the event of an internal fault with very high current level the trip command is released after duplicated measurement with three sample values from each bay.

The pick-up characteristic corresponds to that of the 7SS600 whereby an optional characteristic for sensitive earth fault detection may be set (Figure 10.13). The latter is supervised by an additional criterion (usually U_0>) for security, to ensure that the sensitive characteristic only responds to small earth currents.

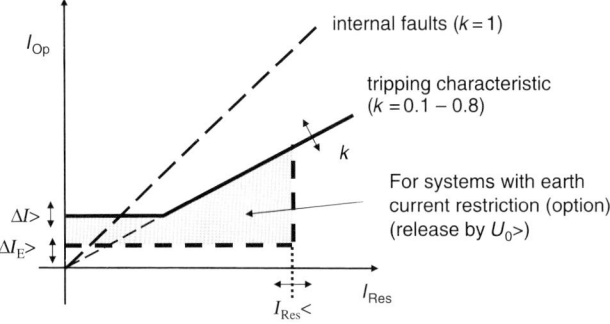

Figure 10.13
Tripping characteristic of
the differential protection

Check-Zone

The application of this zone which covers the entire busbar system increases the security against incorrect tripping in particular for errors in the isolator replica. (Figure 10.14)

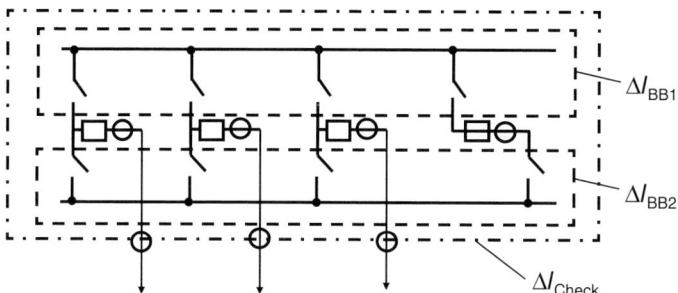

Figure 10.14 Check-zone of busbar protection

In the past, the check-zone was primarily applied with high-impedance protection. In this case it was connected to dedicated CT cores so that a completely separate system existed. The tripping contacts of the zone-selective measuring systems and the check-zone were connected in series.

With the traditional low impedance differential protection, based on conventional technology, the check-zone was only applied in special cases due to the extra expense. The numerical protection 7SS52 on the other hand contains the check-zone as a supplementary criterion, as it can be provided at reasonable cost as an integrated function in the software.

If the busbars are operated individually in a multiple busbar substation, and a fault occurs on one busbar, the load-current flowing through the busbar(s) that is (are) not affected by the fault acts as additional restraint for the check-zone. Under heavy load conditions and small short-circuit currents this may result in over-stabilising and failure to trip.

A similar condition arises when the busbars are coupled with two feeders via a neighbouring station. In this case the short-circuit current can flow from the in-feed through the healthy busbar to the neighbouring station and back to the faulted busbar. The restraint current then contains several summations of the short-circuit current which again results in over-stabilising (in-out-in-effect).

With the numerical protection 7SS52 this problem of over-stabilising was solved by special treatment of the restraint current. The currents flowing towards the check-zone (substation) and flowing away are summated separately based on their direction of flow. Only the smaller summation result is then used for the stabilising. (Figure 10.15)

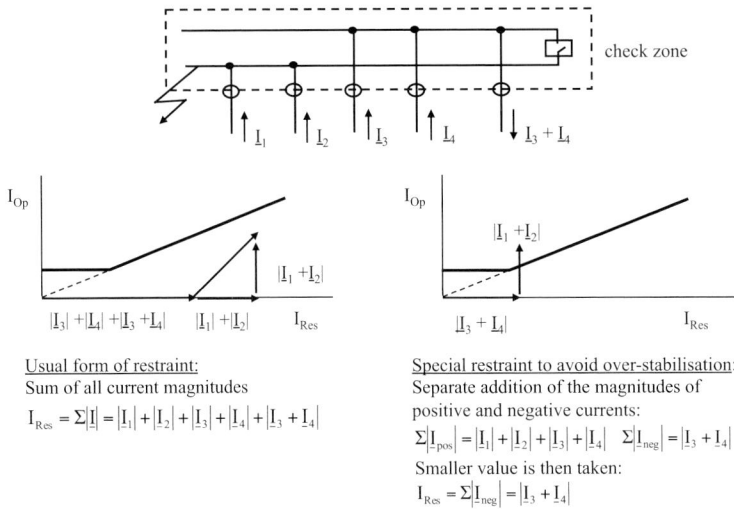

Figure 10.15 Restraint of the check-zone with the numerical protection 7SS52: Special method to prevent over-stabilising

Impact of CT location

The CTs may be located at both sides of the circuit breaker (mostly the case with dead tank breakers) or at one side only (normal with life tank breakers). This has an influence on the protection against faults between CB and CTs (end fault protection) and the protection against CB failure.

With CTs on both sides, the circuit breaker and the end zones lie in the overlapping zone of feeder and busbar protection. Both protections detect and trip the end zone faults.

With one set of CTs only, the CB and the end zones may lie either in the zone of the feeder protection or in that of the busbar protection depending on the CTs being located on the bus side or the feeder side respectively.

If for example the CTs are located on the line side, a fault in the end zone would only be detected by the busbar protection and transferred tripping would be necessary to clear the infeed to the fault from the remote line end.

If the CTs are however located on the bus side, the feeder protection would detect the end fault and forced tripping of the local busbar infeeds would be necessary.

Fault in the "dead zone" of the coupler

Faults at the coupler between circuit-breaker and CT require special measures to obtain selectivity. It must be differentiated between the case where there is a CT only on one side and the case with CTs on both sides of the circuit breaker as well as the condition where the circuit-breaker is closed or open when the fault occurs.

The switching state of the circuit-breaker can be interrogated by the busbar protection via a binary input. This option should be used because the response of the busbar and breaker-failure protection is then automatically adapted to the circuit-breaker switching state. In this way selectivity during faults in the "dead zone" is substantially improved as shown with the following examples.

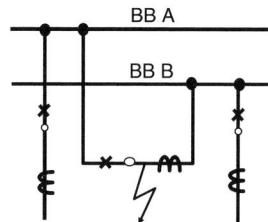

Figure 10.16

CTs on one side, bus-coupler closed:

A) Bus coupler CB auxiliary contacts not connected:

Busbar protection A detects an internal fault and trips busbar A and the bus-coupler CB.

Bus coupler CB opens but the fault still hangs on and protection A remains picked-up. The breaker-failure protection initiates time-delayed (T_{BFP}) reversal of the coupler current in the busbar protection B. As a result, this system will also clear the busbar B.

B) Bus coupler CB auxiliary contacts connected

The busbar protection B will on its own trip immediately after the CB opens because the coupler currents are set to zero, i.e. the time delay introduced by the breaker-failure protection (T_{BFP}) falls away.

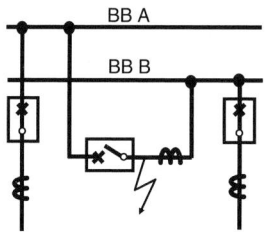

Figure 10.17

CTs on one side, bus-coupler open

A) Bus coupler CB auxiliary contacts not connected:

Busbar protection A issues an unnecessary trip.

Busbar protection B detects an external fault and only trips after the breaker failure delay time T_{BFP}.

B) Bus coupler CB auxiliary contacts connected

Busbar protection A does not trip as the bus coupler current for the busbar protection A is set equal to zero when the coupler CB is open.

Busbar protection B trips immediately as the bus coupler current is also set equal to zero for busbar protection B when the bus coupler is open

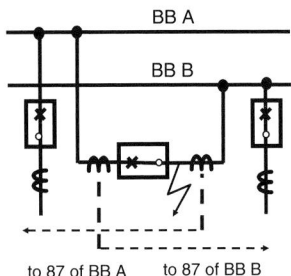

CTs on both sides, bus coupler CB closed

Overlapping protection zones exist

Busbar protection A and B trip immediately.

Both busbars are cleared.

Figure 10.18

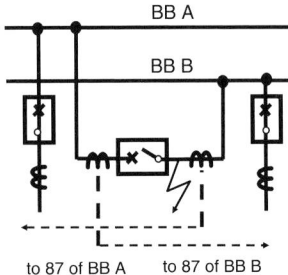

CTs on both sides, bus coupler CB open

Overlapping protection zones exist.

No auxiliary contacts connected:

Busbar protection A trips immediately

Busbar protection B trips via the breaker failure protection function (current reversal) after the time T_{BFP}.

Auxiliary contacts connected:

Busbar protection B trips as the bus coupler current is set equal to zero when the coupler CB is open.

Busbar protection A remains stable.

Figure 10.19

Breaker-failure protection

The basic principle is the same as that in classic breaker-failure protection:

On each circuit breaker a current relay ($I>$) monitors whether current flow resets within a set time (e.g. 150 ms) while a trip command is present. Otherwise the breakers of the feeding bays are tripped with a common bus trip command.

The protection 7SS52 contains a comprehensive breaker-failure protection which allows a bus-zone selective clearance via the isolator replica in the event of feeder faults followed by breaker-failure. In the case of internal busbar faults, followed by breaker-failure, the 7SS52 also generates a command to trip the breaker at the opposite end of the feeder bay. For this purpose, a secure data communication channel must be available.

Two basic variants must be considered:

The breaker-failure protection is provided by the feeder protection

In this case, the feeder protection monitors to detect breaker-failure conditions for example with the breaker-failure protection that is integrated in the numerical feeder protection devices. The busbar protection is used for the selective distribution of trip commands to the circuit breakers of the affected busbar zones via the isolator replica (Figure 10.20). For reasons of security, the initiation is done via two channels for example with the pick-up (fault detection) signal from the feeder protection utilised as second criteria.

The following effect has to be considered when low current setting (e.g. 20% I_n) and short BF tripping time (e.g. 120 ms) are applied:

After interruption of the primary fault current, a transient de-magnetising DC current occurs in the CT secondary circuit. It is significant with linear core CTs and may be very high when the CT has been magnetised single sidedly by an offset fault current. The current detector of the BF protection may then incorrectly see the short circuit current flowing on and cause false tripping. This was a problem with older relays which used r.m.s. or d.c. mean-value based measurement.

Numerical relays now use d.c. suppressing filters and additional current zero-crossing monitors to ensure short reset time (< 20 ms) of the BF current detector.

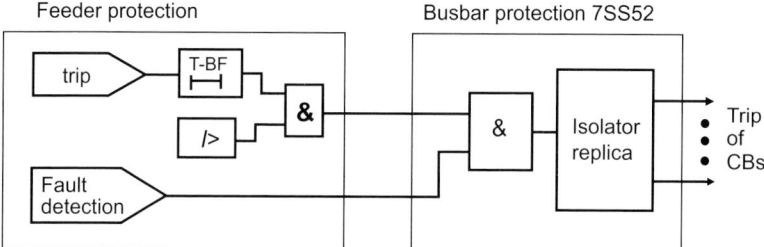

Figure 10.20 External breaker failure protection

The breaker-failure protection is integrated in the busbar protection.

In this case the feeder protection provides its trip command to the busbar protection (7SS52) which contains the entire breaker-failure protection functionality. For security, the initiation is again done via two channels (Figure 10.21). Tripping occurs via de-stabilising the busbar protection in other words, operation is achieved by reversing the current of the faulty feeder (emulation of an internal fault). This has the advantage that the measurement of the busbar protection is used as a further criterion thereby providing additional security against mal-operation.

The drop-off time of the trip commands originating from the feeder protection is in this case not critical. If the circuit breaker opens correctly while the trip command of the feeder protection does not reset, incorrect operation will not occur as the short circuit current disappears so that the measurement without current will not issue a trip command.

To also ensure breaker-failure detection to function correctly when the short-circuit current is small, the differential protection is applied with a separately settable

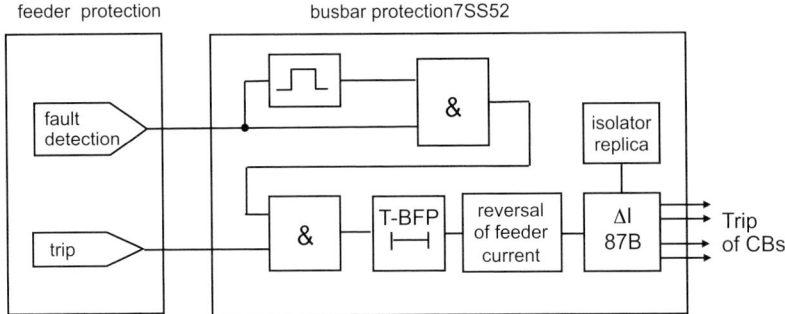

Figure 10.21 Breaker-failure protection in the 7SS52

characteristic following inversion of the current. A smaller pick-up threshold and lower stabilising factor may be chosen for this characteristic.

10.2 Response of the numerical busbar protection in the case of CT saturation and the demands placed on CT dimension

The measurement principle of the fully numerical 7SS52 and the partially numerical 7SS600 is the same. The restraint current is the smoothed peak value of the summated current magnitudes $I_{Res} = |I_1| + |I_2| + |I_3| + \dots + |I_n|$ while the operating current is the unsmoothed magnitude of the phasor summation of the currents $I_{Op} = |I_1 + I_2 + I_3 \pm \dots + I_n|$ (Figure 10.22). For tripping to result, the instantaneous values of the operating current must exceed, for a short minimum duration (with 7SS5 at least 3 sampling points, i.e. 3 to 4 ms), the rectified value of the weighted restraint current: $I_{Op} > k \cdot I_{Res}$.

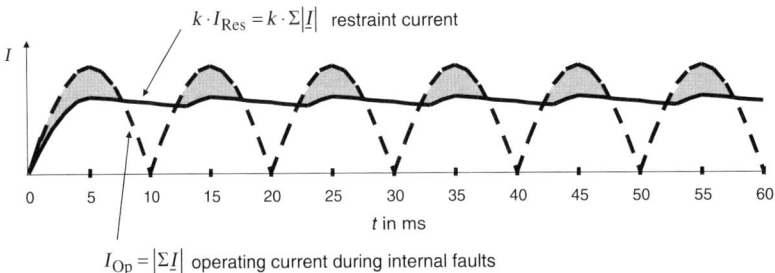

Figure 10.22 Measured values during internal fault (no CT saturation)

The measuring principle was developed with CT saturation in mind, which may occur after a very short time with busbar protection due to the concentration of the short circuit current in-feeds. The principle using additional biasing in the event of

CT saturation (saturation detector) and the measuring logic was explained in section 4.2.4 .

The response in the event of CT saturation and the CT dimensions resulting will be discussed here.

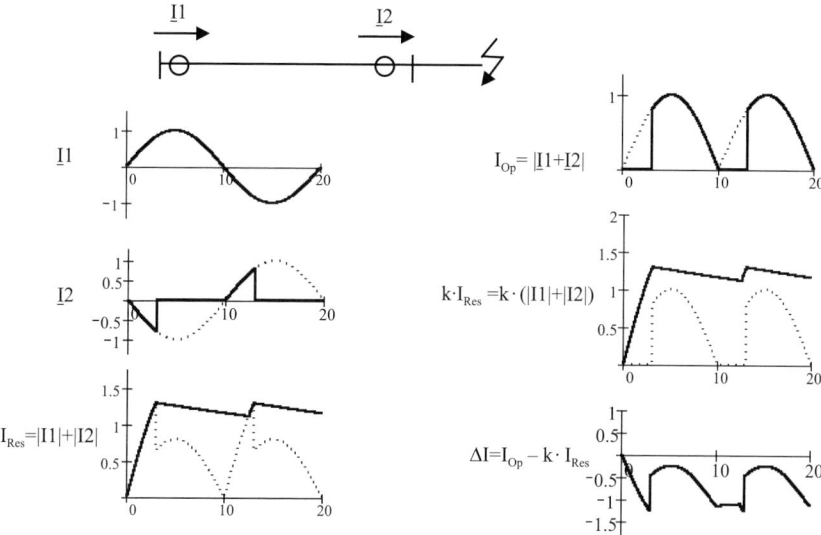

Figure 10.23 Stability during external fault with steady state CT saturation

Initially the external fault with steady state CT saturation will be considered. (Figure 10.23) It can be seen that despite extreme saturation, no tripping results as the biasing current is greater than the tripping current (ΔI negative!).

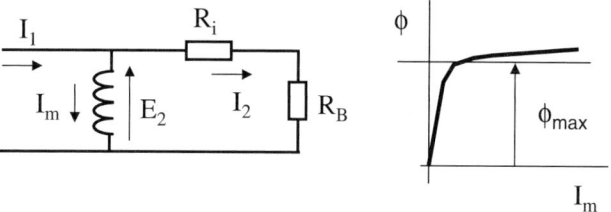

Figure 10.24 Equivalent circuit of a CT

The CT dimension requirements (over-dimensioning factor) with symmetrical short-circuit current (no DC component) depend on the setting of the biasing factor k and can be calculated.

For this purpose the response of the CTs with saturation must be analysed:

The equivalent CT circuit shows Figure 10.24. The flux in the CT is proportional to the area under the induced EMF $e_2(t)$ voltage curve in the time domain:

$$\Phi = \text{prop.} \int e_2(t) \cdot dt = (R_i + R_B) \cdot \int i_2 t \cdot dt \qquad (10\text{-}5)$$

The maximum flux increase, until saturation is reached, therefore corresponds to the maximum EMF that appears with AC voltage according to the operational over-current factor ALF' (Refer to section 5.1).

$$\Phi_{\text{max.}} = \text{prop.} \int_{0}^{180°} \text{ALF}' \cdot e_{2n}(x) \cdot dx = (R_i + R_B) \cdot \int_{0}^{180°} \text{ALF}' \cdot i_{2n}(x) \cdot dx \qquad (10\text{-}6)$$

$$\Phi_{\text{max.}} = \text{prop.}(R_i + R_B) \cdot \text{ALF}' \cdot \hat{I}_{2n} \cdot \int_{0}^{180°} \sin x \cdot dx = \qquad (10\text{-}7)$$

$$= (R_i + R_B) \cdot \text{ALF}' \cdot \hat{I}_{2n} \cdot 2$$

If the short-circuit current is greater than $\text{ALF}' \cdot I_n$, then the available flux change is not sufficient and the CT goes into saturation.

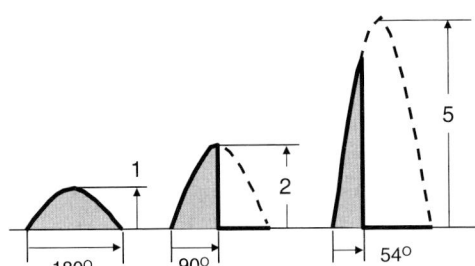

Figure 10.25
Instant of saturation depending on current magnitude

Saturation occurs when the area under the voltage curve in the time domain corresponding to the flux ϕ_{max} is used up (Figure 10.25). With AC short-circuit current that is 5 times greater than the limiting current defined by ALF', saturation will already occur after 54° or 3 ms. The instant x at which saturation occurs can now be determined in relation to the short-circuit current using the following area relationship:

$$\hat{I}_{2F} \cdot \int_{0}^{x} \sin x \cdot dx = \text{ALF}' \cdot \hat{I}_{2n} \cdot \int_{0}^{180°} \sin x \cdot dx$$

or

$$\frac{I_{2F}}{I_{2n}} = \text{ALF}' \cdot \frac{2}{\int_{0}^{x} \sin x \cdot dx}$$

The necessary transient dimensioning factor (K_{TD}), depending on the saturation free time therefore is:

$$K_{TD} = \frac{ALF'}{I_F / I_n} = \frac{1}{2} \cdot \int_0^x \sin x \cdot dx \qquad (10\text{-}8)$$

This equation may now be used to determine the setting of the stabilising factor k, depending on the CT dimension.

The conditions for stability can be obtained from Figure 10.26:

$$I_{Op} < k \cdot I_{Res}, \quad \text{i.e.} \quad I_F < k \cdot 2 \cdot I_F \cdot \sin x \quad \text{or} \quad \sin x > \frac{1}{2 \cdot k} \qquad (10\text{-}9)$$

The equation 10-8 can now be combined with 10-9 to obtain the following relationship:

$$k > \frac{1}{4 \cdot \sqrt{K_{TD} - K_{TD}^2}} \qquad (10\text{-}10)$$

In this case K_{TD} is the transient dimensioning factor.

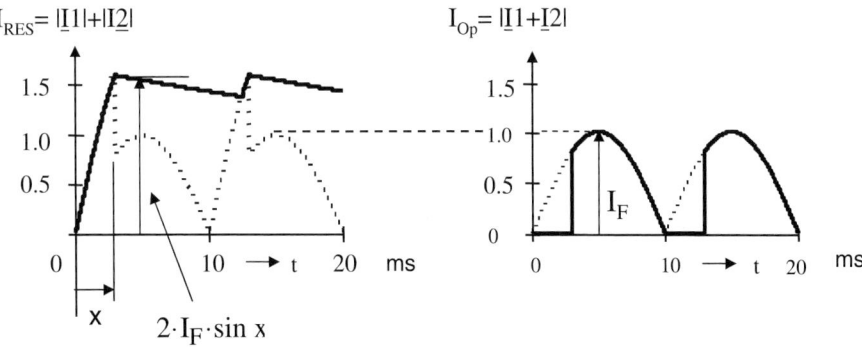

Figure 10.26 Restraint and tripping current (7SS52 and 7SS600)

The reverse is of course also possible whereby with given k the required over-dimensioning factor K_{TD} can be determined. In this way the following equation can be derived from (10-10):

$$K_{TD} > \frac{1}{2} \cdot \left(1 - \sqrt{1 - \left(\frac{1}{2k}\right)^2}\right) \qquad (10\text{-}11)$$

For the lowest practical setting being $k = 0.5$ the value $K_{TD} \geq 0.5$ results.

With the common setting of von 0.65 the result is $K_{TD} > 0.2$.

Note:

This condition (equation 10-11) relates to stability for short-circuit currents without DC component (pure AC short-circuit current). For short-circuit currents including DC component an additional constraint applies which states that a minimum saturation free transformation after fault inception must exist (3 ms for 7SS52 and 4 ms for 7SS600), see below. This corresponds to a transient factor of $K''_{TF} = 0.5$ for 7SS52 and $K''_{TF} = 0.75$ for 7SS600. For the common settings $k \geq 0.5$ the criterion for a minimum saturation free time applies for the dimensioning of the CTs (refer to practical CT dimensioning Table 5.3 in section 5.7).

Internal fault

To reach a trip decision during internal faults, the tripping current must be greater than the stabilising current. With poor CTs only the short saturation free time following fault inception is available for this purpose. The protection must therefore reach a decision very fast to minimise the CT dimensioning requirement. The numerical relays can reach a decision in 3 ms (7SS52) or 4 ms (7SD600) due to the instantaneous value measuring method. Thereby severe saturation during internal faults is permissible. Tripping times of below one cycle (minimum 13 ms) are achieved. (Figure 10.27)

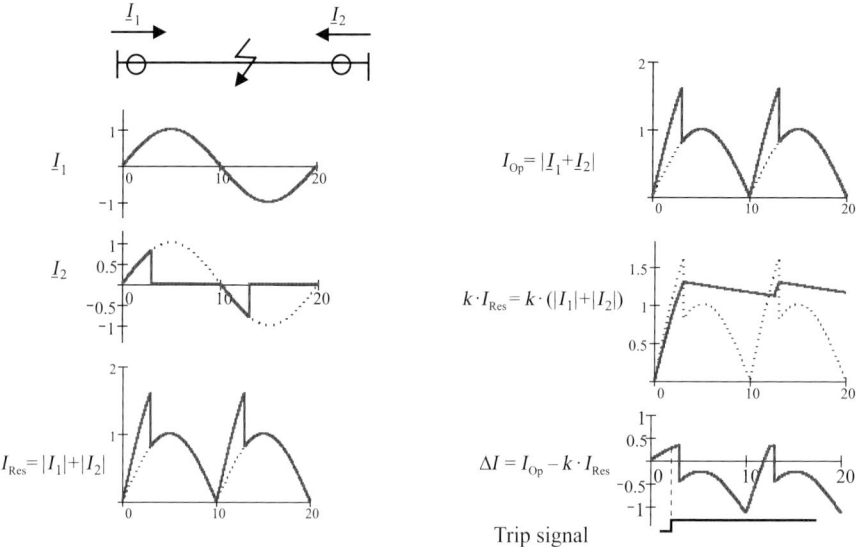

Figure 10.27 Tripping for internal fault with steady state CT saturation

External faults with short-circuit current including DC offset

If the fault inception is close to the voltage zero crossing (this type of fault is usually only caused by lightning strike) short-circuit currents with DC off-set will exist.

As a result of the DC offset the CTs are highly magnetised in one direction and severe saturation may occur, in particular during external faults, when the sum (n – 1) of the feeder currents flow through one feeder to the external fault location. In this case, the integrated saturation detector of the numerical protection is called upon. It detects the external fault before saturation ensues and switches over to a secure 2-out-of-2 measurement. The method of operation was described in section 4.2.4. Due to the DC off-set, the short-circuit current is only affected by saturation in every second half-wave so that the protection remains stable (Figure 10.28).

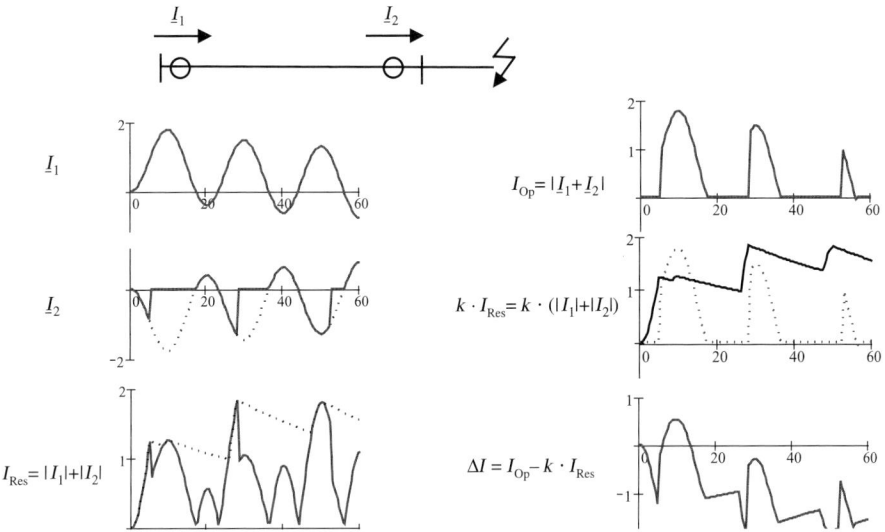

Figure 10.28 Stability during external fault with non-symmetrical CT saturation

In this diagram one sees that the tripping current sharply rises above the stabilising current $k \cdot I_{\text{Res}}$ in the first half wave. Tripping however does not occur because the saturation detector has already switched over to the 2-out-of-2 measurement. In the case at hand, the tripping current thereafter remains below the stabilising current because the system time constant was not very large (40 ms). With longer time constants the tripping current may also remain above the stabilising current for a longer time, however only in every alternate half wave.

In section 5.7 it was stated that the saturation free time of a CT following fault inception depends on the transient factor K''_{TF}. The diagram below (Figure 10.29) indicates the required dimensioning factor K''_{TF} depending on the system time constant (DC time constant of the short-circuit current) T_{N}. It applies to the short saturation free times that the numerical protection allows (4 ms with 7SS600, 3 ms with 7SS5).

The increase of the K''_{TF}-value for system time constants below 5 ms is only of theoretical interest.

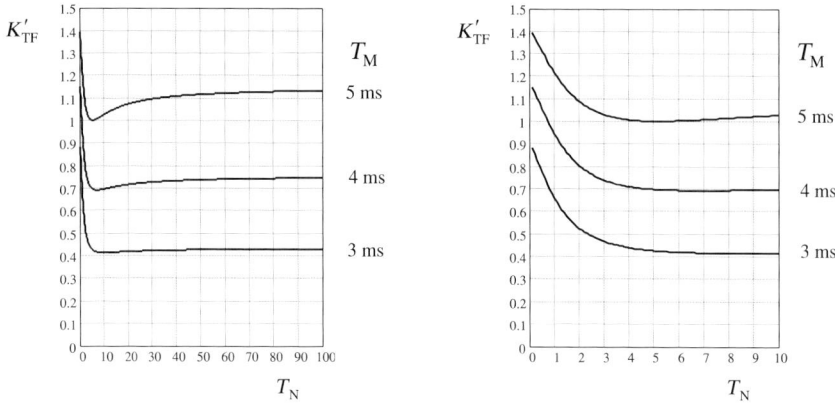

Figure 10.29 CT over-dimensioning factor K''_{TF} for short saturation free time depending on the system time constant

In practice the time constants are above this value so that for general purposes the following transient dimensioning factors $K_{TD} = K''_{TF}$ may be applied independent of the system time constants: $K_{TD} \geq 0.45$ for 7SS5 and $K_{TD} \geq 0.75$ for 7SS600.

To determine these values, the worst case condition for fault inception was considered.

For short saturation free times the fault inception close to the voltage maximum is critical because the non off-set short-circuit current has a very steep initial slope due to the shape of the sin-wave. Although with faults close to the voltage zero crossing the current initially has a slower rise (given by $e^{-t/T_N} - \cos \omega t$), the DC off-

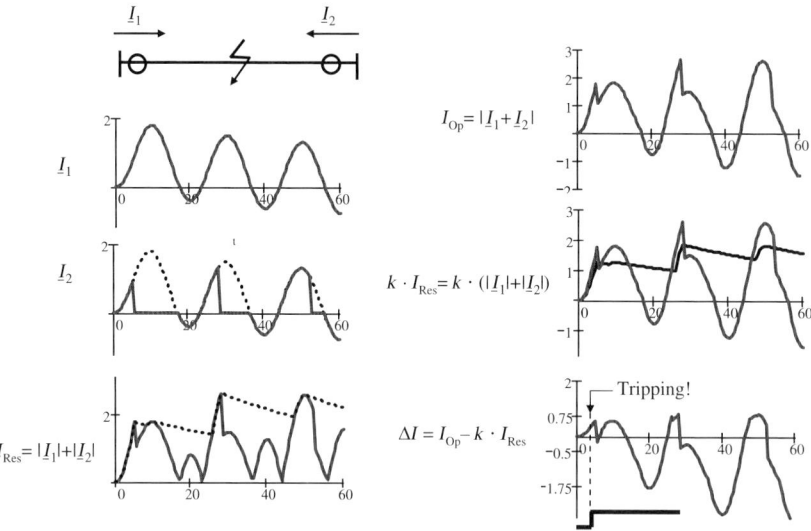

Figure 10.30 Tripping with internal fault and non-symmetrical CT saturation

set with high time constants will have a more severe effect as the fault duration continues.

Internal faults with short-circuit current including DC offset

With regard to CT saturation, the internal fault is not quite as critical as the external fault, as only the infeed of each individual bay contributes to the current flowing via the CT. For CTs having small dimensions, saturation may however also occur under these conditions.

Due to the short measuring times, this however does not present any difficulties to the numerical protection. It reaches a decision in the saturation free time following fault inception as was already the case with pure AC short-circuit currents described above (Figure 10.30).

Example 10-3: Dimensioning of a CT for the numerical busbar protection

Given: Substation with data according to Figure 10.31:

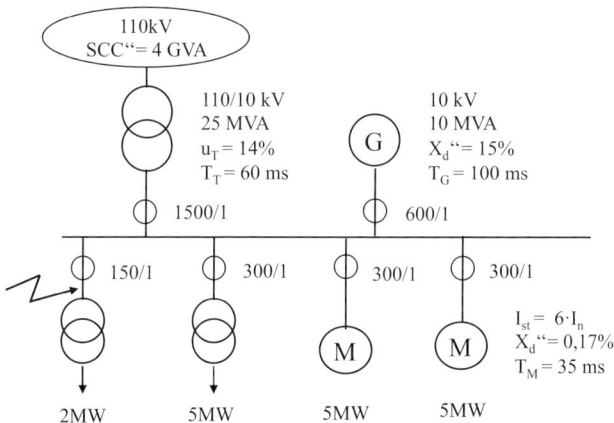

Figure 10.31 Substation data for example 10-3

Task: Dimensioning of the CT for bay 1 (150/1 A)

Solution: Initially the short-circuit currents must be calculated.

In this regard it must be noted that the motors also contribute short-circuit current by back-feed (see section 7.2). Only the high voltage motors will be considered in this regard.

Short-circuit currents:

$$I_{\text{n-T}} = \frac{25 \cdot 10^6}{10 \cdot 10^3 \cdot \sqrt{3}} = 1440 \text{ A}$$

$$I_{\text{n-G}} = \frac{10 \cdot 10^6}{10 \cdot 10^3 \cdot \sqrt{3}} = 577 \text{ A}$$

$$\sum I_{\text{M-n-HV}} = 2 \cdot \frac{5 \cdot 10^6}{10 \cdot 10^3 \cdot \sqrt{3}} = 577 \text{ A}$$

$$I_{\text{F-T}} = \frac{1.1 \cdot 1440}{0.14} = 11.3 \text{ kA}$$

$$I_{\text{F-G}} = \frac{1.1 \cdot 577}{0.15} = 4.2 \text{ kA}$$

$$\sum I_{\text{F-M}} = \frac{1.1 \cdot 577}{0.17} = 3.7 \text{ kA}$$

The total fault current for a fault at the feeder transformer terminals is:

$$\sum I_{\text{F}} = I_{\text{F-T}} + I_{\text{F-G}} + \sum I_{\text{F-M}} = 11.3 + 4.2 + 3.7 = 19.2 \text{ kA}$$

A CT type 5P?, 30 VA, internal burden $P_i = 15\%$ (4.5 VA) is intended:

The connected burden is approx. 1 VA.

Dimensioning for short-circuit current with DC off-set:

The 7SS52 requires a dimensioning for 3 ms saturation free transformation. The corresponding transient dimensioning factor obtained from the diagram in Figure 10.29 is $K_{\text{TD}} = K''_{\text{TF}} = 0.45$.

The result for the accuracy limit factor of the CT obtained thereby is:

$$\text{ALF} = \frac{P_{\text{B}} + P_i}{P_n + P_i} \cdot \text{ALF}' = \frac{P_{\text{B}} + P_i}{P_n + P_i} \cdot K_{\text{TD}} \cdot \frac{I_{\text{F}}}{I_n} =$$

$$= \frac{1 + 4.5}{30 + 4.5} \cdot 0.45 \cdot \frac{19{,}200}{150} = 9.2$$

The CT is then specified as follows: 150/1 A, 5P10, 30VA, $P_i \le 15\%$

Check of stability for pure AC short circuit current:

The k setting should be:

$$k > \frac{1}{4 \cdot \sqrt{K_{\text{TD}} - K_{\text{TD}}^2}} = \frac{1}{4 \cdot \sqrt{0.45 - 0.45^2}} = 0.5$$

A fairly sensitive setting may therefore be chosen. With some safety margin $k = 0.6$ is selected.

10.3 High-impedance busbar protection

The high-impedance busbar protection is suitable for single busbars or single busbars with bus sectionalisers as well as the two busbars in a station with 1 ½ circuit breaker configuration. In other words, it is applicable to all cases where it is not necessary to switch the measured currents according to the isolator switching status. [3-5 and 3-6]

The switching of secondary CT currents via auxiliary contacts of the isolators, as was done in the past on double busbars, presents problems and is prone to errors.

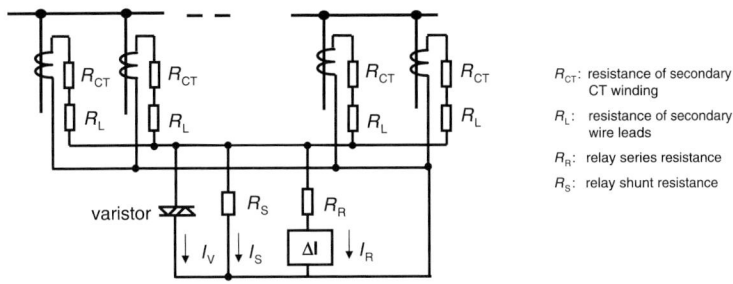

Figure 10.32 HI-busbar protection schematic diagram

Even in countries that traditionally applied high-impedance protection, the low impedance numerical protection is nowadays applied for these applications.

The HI-measuring principle itself cannot be implemented numerically. Mechanical and static analog relays (7VH60) are still applied.

The basic principle described in section 3.4 applies. The special condition for busbar protection is the large number of CTs that are connected in parallel. They must all have the same ratio and should be Class PX according to IEC 60044-1. The schematic connection is shown in Figure 10.32.

Furthermore it is common practice to set the pick-up threshold of the protection so that it does not pick up due to interruptions of the secondary connections when load current is flowing. (In EHV substations the parallel connection of the CTs is partially done by ring connections to prevent a simple interruption from affecting the protection.)

The secondary connections (bus wire) is monitored by a more sensitive relay that is connected in parallel to the tripping relay (setting of 5-10% of the tripping relay threshold). With the static relay 7VH60 the monitoring is integrated.

The following values from the CEGB Standard 993513 (1975) may be used for reference purposes.

"Busbar protection for 400/275/132 kV", applies:

System voltage	400 kV	132 kV
Short-circuit rating	35 GVA	5 GVA
CT	2000/1A	1000/500/1 A
Maximum secondary resistance of the CT R_{CT}	5 Ohm	2.4 Ohm
Maximum CT connection resistance R_L	5 Ohm	2.5 Ohm
Maximum magnetising current of the CT at relay pick-up threshold or at $U_{KN}/2$: I_{mR}	40 mA	60 mA
Minimum knee-point voltage	$60 \cdot (R_{CT} + R_L)$	$95 \cdot (R_{CT} + R_L)$
Minimum knee-point voltage for the CT with the largest secondary connection resistance	600 V	465 V

Example 10-4: Setting a HI busbar protection

Given: System:

$n = 8$ feeders

CT:

$r_{CT} = 600/1$ A, $U_{KN} = 500$ V, $R_{CT} = 4$ Ohm

$I_{mR} = 30$ mA (at the relay pick-up threshold)

CT connection resistance:

$R_L = 3$ Ohm (max.)

Protection:

$I_R = 20$ mA (fixed value), $R_R = 10$ kOhm, $R_S = 250$ Ohm

Varistor:

$I_v = 50$ mA (at the relay pick-up voltage)

Task: Check the pick-up sensitivity and the stability

Solution: With the equations given in section 3.4 the following is obtained:

Primary pick-up current:

$$I_{\text{F-min.}} = r_{CT} \cdot (I_R + I_S + n \cdot I_{mR})$$

$$I_{\text{F-min.}} = \frac{600}{1} \cdot (0.02 + 0.80 + 0.05 + 8 \cdot 0.03) = 666 \text{ A, i.e. } 111\% \, I_{\text{n-CT}}$$

Stability during external faults:

$$I_{\text{F-through-max.}} < r_{CT} \cdot \frac{R_R}{R_L + R_{CT}} \cdot I_R$$

$$I_{\text{F-through-max.}} < \frac{600}{1} \cdot \frac{10.000}{3 + 4} \cdot 0.02 = 17 \text{ kA} = 28 \cdot I_{\text{n-CT}}$$

11 Relay Design

The first fully numerical protection relays with integrated communication interfaces were supplied by Siemens in 1985. They were already designed to be integrated in the substation control system (LSA). As a result of the continuing developments in the micro-processor and communication technology, the performance of the devices continuously increased thereafter.

In the following a short summary of the design and functionality of the numerical devices is provided. Detailed information is given in the system and relay manuals provided by the manufacturers. [11-1, 11-2]

State of technology

Modern relays are intelligent electronic devices (IEDs).

They are of compact design and in addition to the principle function (in this case differential protection) contain a number of supplementary protection functions (e.g. over-current and over-load protection) as well as additional functions for measuring and control.

Keypad and display are located on the device front. Combined protection and control devices are available with a graphic display bay mimic diagram and control keys for local control. These devices are also applied as bay units for the substation control system.

All devices are largely maintenance-free due to the integrated self-monitoring.

The hardware is scalable. The device dimension can vary depending on the scope of functions and the number of interfaces. At the medium and high voltage level, devices with a width of 1/6 to 1/2 of 19" are usually sufficient. The full 19" versions are intended for the EHV level where a large number of binary inputs and relay contact outputs are required.

The connection is basically the same as with conventional relays: the CTs and the voltage transformers (if applied) are isolated via input transformers. Status inputs are coupled via binary inputs (opto-couplers), providing a potential barrier. Output relays are provided for output signals and trip commands.

The serial interfaces for operating and servicing the device, as well as connection to substation control system, are new to the numerical protection. In devices with numerical communication between the line ends protection data interfaces are also provided for the transmission of protection data (measured values, commands).

The devices can be operated locally via the integrated keypad and display. Usually a PC (Laptop) is however used for this purpose applying the user-friendly operating program DIGSI. Via a modem connection, setting and retrieval of load and fault data is also possible from remote. For this purpose the devices in the substation are connected via a central star-coupler to the modem.

SIPROTEC 3 devices

The devices for feeder differential protection with pilot wire connection (7SD502 and 7SD503) described in chapter 9 and the current comparison protection (7SD511/12) have this version. Figure 11.1a shows the front view of these devices with 1/3rd of 19" width. The PC used for dialog with the device is connected to the socket at the right hand bottom side. The device illustrated is intended for flush mounting with terminals on the rear side. For the surface mounting version, the connection terminals are located at the top and bottom side of the device. The device generation SIPROTEC 3 has a serial interface (information interface according to IEC 60870-5-103). This interface is usually provided for optic fiber connection.

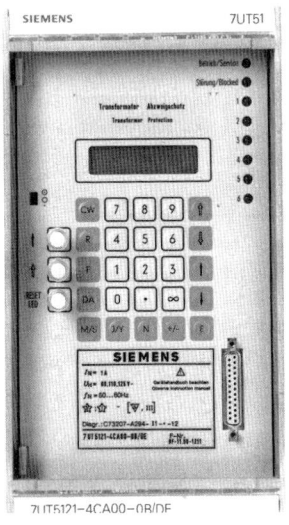

Figure 11.1

a) SIPROTEC 3   b) SIPROTEC 600   SIPROTEC device ranges

Figure 11.1b shows a device in the SIPROTEC 600 range. These compact devices have a width of only 1/6th of 19" and are intended for simpler protection tasks. Indications and setting elements are reduced to a minimum. The number of in and outputs are reduced due to the small number of terminals and only one serial interface is provided. This device version is used in large numbers as overcurrent protection 7SJ6, especially in distribution networks.

The feeder differential protection 7SD600 with external summation CT and the central unit of the partially numerical busbar protection 7SS600 are also devices in the SIPROTEC 600 range.

The analog static high-impedance relay 7VH600 is provided in the same device housing.

SIPROTEC 4 devices

This upgraded device range was introduced in 2000. (Figure 11.2)

The numerical processing unit is standardised and concentrated onto one module. The input/output periphery (measuring inputs, binary inputs, relay outputs) is scaleable and may therefore be adapted to the particular protection application.

The size of the devices varies from 1/3rd to a full 19" width. The connection is with well-proven screw terminals.

For communication, several plug-in modules are available. They allow simple adaptation to existing and future communication standards. In total 4 interfaces can be integrated. (Figure 11.3)

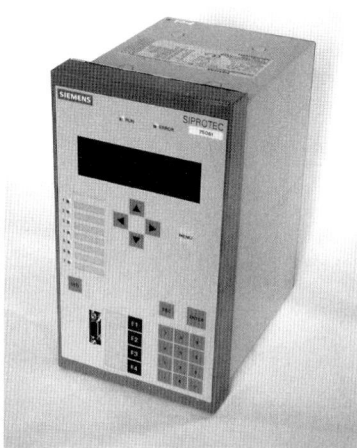
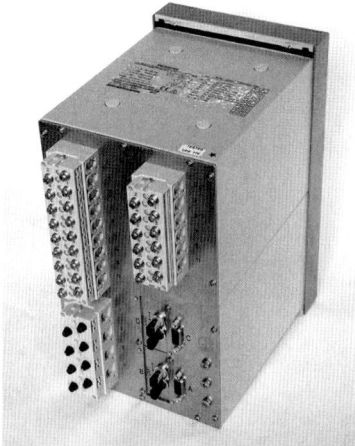

Figure 11.2 Numerical protection, device range SIPROTEC 4

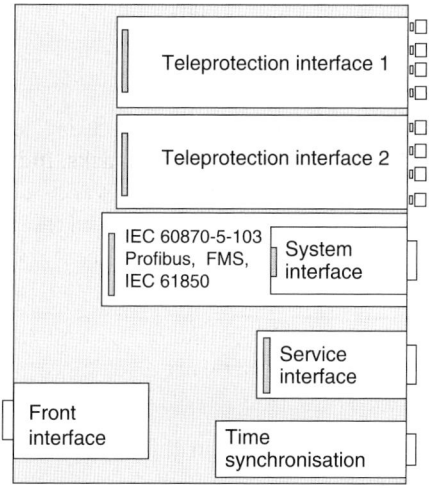

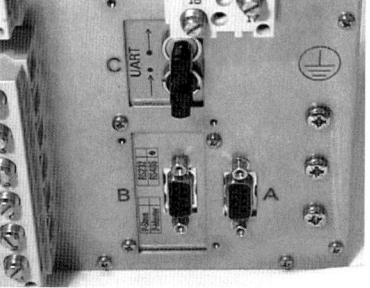

Figure 11.3
SIPROTEC 4 relay, serial interfaces

The protection device can be connected to the substation control system via a system interface and in parallel via the service interface and a star-coupler with a separate remote communication to a PC. The two protection data interfaces allow for protection data communication with two opposite line end stations (for example to protect a three-terminal line). With a further interface, a time synchronisation device (DCF77 of the PTB in Germany or IRIG B via GPS satellite system) can be connected for exact time synchronisation.

Integration of the devices in a substation automation system is shown in Figure 11.4.

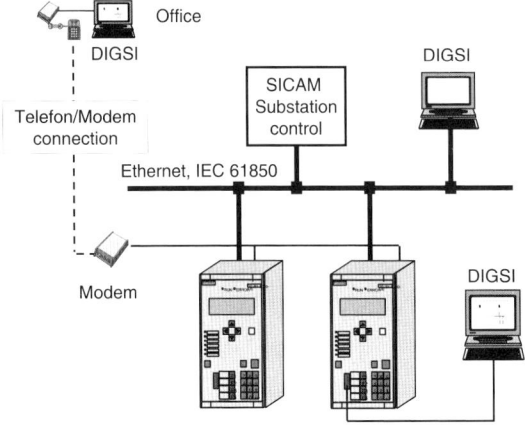

Figure 11.4
Substation automation system
with SIPROTEC 4 devices

The combined protection and control devices have a graphic LCD display on the device front. It allows for local indication of plant and device information. A bay mimic control indication or measured value indication as well as various annunciation lists can be selected in the display (Figure 11.5).

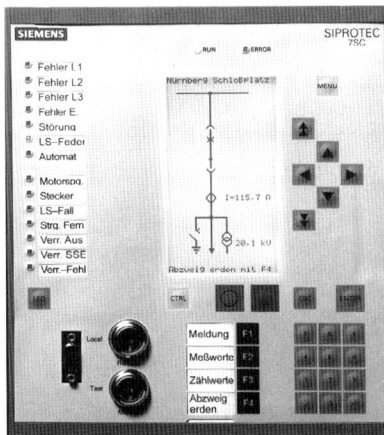

Figure 11.5
SIPROTEC 4, combined protection
and control device

The control diagram can be configured to suit the user's requirements with the software DIGSI 4. Control of the feeder is done via the cursor and the open/close push-buttons.

The SIPROTEC 4 range also provides the option to apply additional user-defined logic. Thereby the binary input signals and internal signals (threshold detectors, protection indications) can be linked and routed to relay outputs and indications. In this way user defined applications without external relay connections can be applied. The logic is implemented with the graphic CFC editor.[1]

SIPROTEC 5 devices

In 2011 the innovated relay generation SIPROTEC 5 was introduced. (Figure 11.6)

Figure 11.6 SIPROTEC 5: relay design (7xx8 series)

This new generation takes the recent progress in micro-processing and communications into account and provides further upgraded relay processing power and storage capacity:

– Modular design: Freely configurable and extendable devices
  (One basic module and up to 9 extension modules)

– Up to 24 current and voltage inputs (1 or 5 A, electronically selectable)

– Up to 40 current inputs when used as busbar protection

– Settable sampling rate 1 to 8 kHz

– Large fault recording capacity: 256 records of up to 24 channels
  (each 20 s, 8 kHz sampled)

– Event record buffer for 2000 alarms, 1 ms accuracy

– Two USB interfaces for PC operation or down/upload of data

---

[1] CFC means continuous function chart

- Add-on functions selectable from library

- Various OF communication interfaces, including IEEE C37.94 compatible interface for direct OF connection to multiplexers of data networks

Using the devices

Development on the software DIGSI over last 20 years (latest release DIGSI 5 in 2011) has turned it into a powerful all-in-one tool for configuring, setting, testing and communicating with the device. The administration as well as the data archiving is also done with this tool.

The operating surface of DIGSI offers a matrix for the routing and allocation of internal functions to the in and output interfaces in addition to the setting menu (Figure 11.7). In this way the internal signals can be allocated to outputs or other destinations via mouse-click. This allows applying clear and concise protection settings.

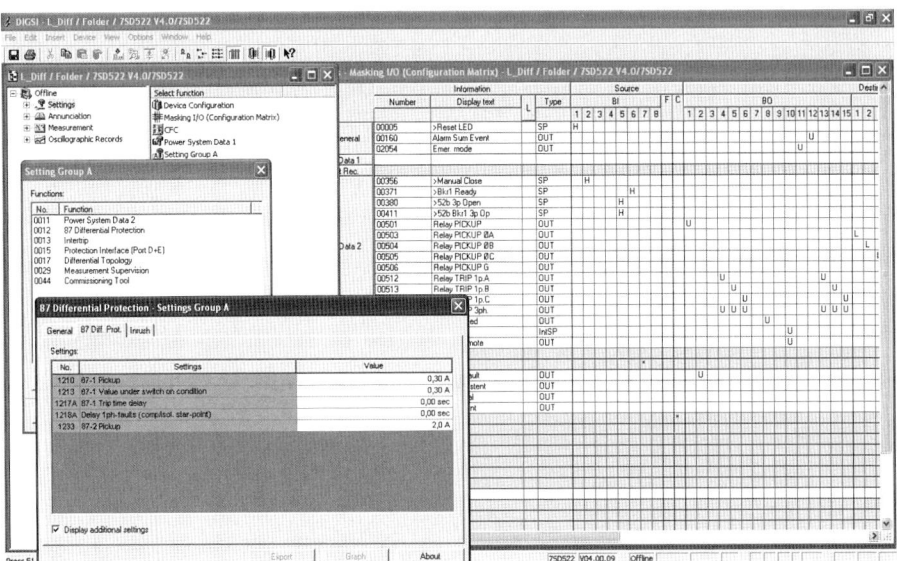

Figure 11.7 Device communication with DIGSI

A further protection tool, SIGRA, is provided to visualise and analyse disturbance records. The measured values can be displayed as records in the time domain or as phasor diagrams Figure 11.8).

Binary traces may be allocated below the measured values so that the alarm and tripping signals can be clearly associated with the course of currents and voltages. In this way the protection response can be analysed in detail.

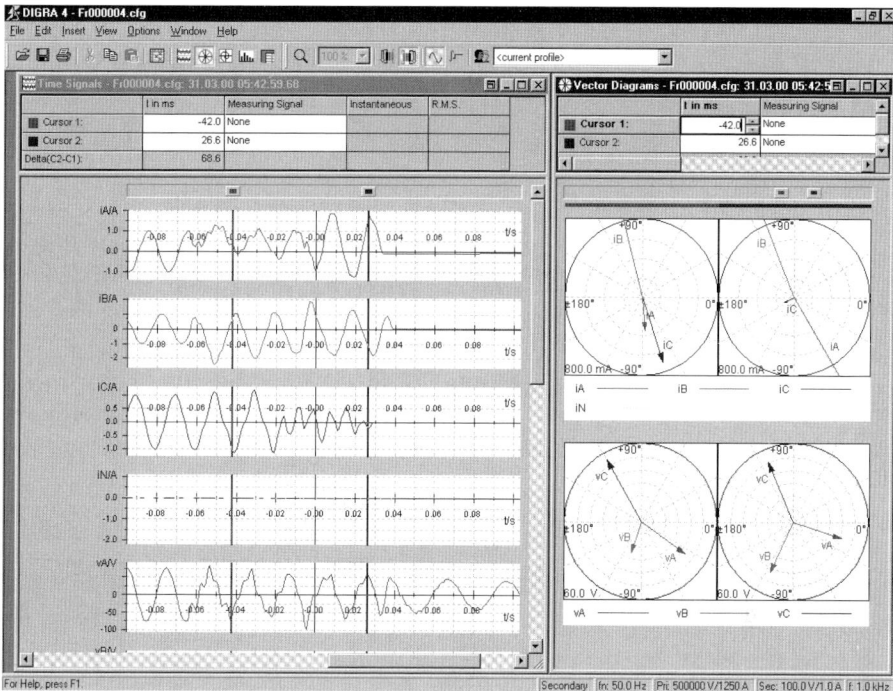

Figure 11.8 Fault record analysis with SIGRA

In applications using the time synchronisation down to the micro-second via GPS, the disturbance records of different substations can be represented in a synchronised manner in one diagram.

Harmonic analysis of the currents and voltages is also possible, as well as the computation of symmetrical components.

12 Commissioning and Maintenance

With numerical protection devices commissioning and maintenance has become far less complicated as a result of the information provided by the devices as well as the integrated self-monitoring.

The information provided here is restricted to general notes regarding the procedures. The special features and new alternatives provided by numerical technology will be illustrated. Specific instructions for the individual devices are provided in the device manuals.

12.1 Commissioning

Pre-testing of the instrument transformers and their connections must be carried out in the same manner as with conventional protection. The measuring functions of the protection devices may already be utilised for this purpose. The binary device outputs can be activated individually by means of the software DIGSI. This largely simplifies the pre-testing of the signalling and tripping circuits, as the internal protection functions do not have to be activated for this purpose. The testing of serial interfaces which are new to the numerical devices can also be carried out in this manner.

Settings are usually applied with the setting program in the office of the protection department, off-line (without protection device) and saved onto a mass storage. In the substation the settings must then only be transferred by PC (Laptop) from the mass storage to the protection device.

To test the protection function with injected signals (current and voltage) PC controlled electronic test equipment is available nowadays which provides almost fully automated test sequences. A three-phase test equipment is recommended as the modern devices monitor symmetry of the three-phase system which may pick up when single-phase tests are carried out.

Primary injection testing is only seldom applied due to cost constraints.

With the feeder differential protection, testing is somewhat more complicated as the currents must be injected at geographically separated locations. In the past, single ended injection was therefore applied for pre-testing by phase synchronous connection of the secondary injection equipment to voltage transformers of an unused feeder in both substations. The test sequence was then simultaneously initiated at both ends when the feeder was energised.

With the electronic test equipment this difficulty no longer exists as the test equipment at both line ends can be synchronised via GPS signals. [12-1]

Commissioning with load currents

The final test, whenever possible, is done with load current or a deliberately created short circuit current.

The integrated overcurrent protection in the differential protection and the separate back-up protection, if available, are set to trip without delay for this test, so that the feeder is immediately cleared if a short-circuit is present.

For the commissioning of generating units, a so-called short-circuit cycle is carried out. For this purpose, the generator is started with a deliberate short-circuit while the system CB is open. The excitation of the generator is then increased; the generator current increases but may not exceed nominal current. In this way stability and tripping of the differential protection can be checked as close to reality as possible. A similar test with short circuit cycle could also be done on a transformer feeder and busbar protection, if a system connection to an available generator can be established. Generally, testing can however only be done with load current. To get a definite indication of the current values and therefore the connection and polarity of the CT circuits, a test current of at least 10% of the nominal device current should be obtained by means of appropriate system switching.

To measure the feeder currents as well as the operating/restraint currents, a large number of measuring instruments had to be connected with conventional protection (12 for a transformer differential protection). With the numerical protection

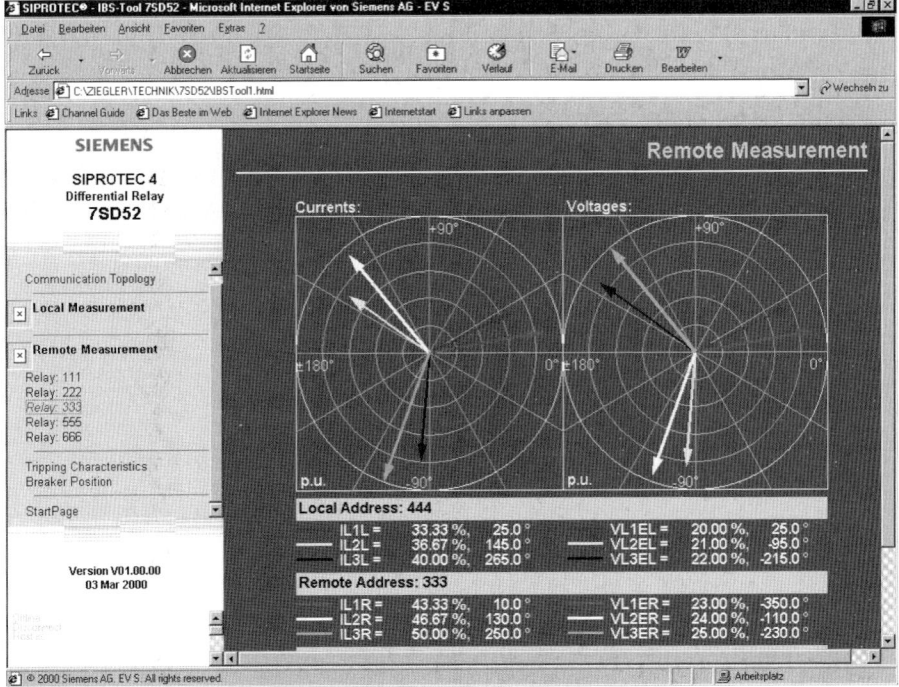

Figure 12.1 Feeder differential protection 7SD52: Representation of the current phasors on a PC with the web monitor (web browser)

the measured values are indicated by the device itself and provided in a summary on the PC monitor via the applied software. Wiring errors (e.g. swapped phase connections) are thereby very quickly identified. When load current is flowing through the system, the operating current (tripping current) should apart from charging currents, be negligibly small and the restraint current should correspond to the sum total of all feeder currents. By reversing the polarity of one current measuring input by means of the corresponding setting parameter, an internal fault can be simulated. Restraint and operating current should in this case have approximately the same magnitude.

An oscillographic record can also be initiated via DIGSI and can then be viewed using SIGRA to calculate the phasors of the current for graphic representation. In this manner, an error in the current comparison can immediately be detected.

SIPROTEC 4 devices provide for a web monitor (web browser). Thereby the phasor diagrams can be called up and visualised online using a common internet browser tool. (Figure 12.1)

The new SIPROTEC 5 line relays 7SD8 and 7SL8 now allow also DIGSI to communicate not only with the local relay but also with the relay at the remote line end(s) through the communication link of the differential protection.

Following the function tests, the final settings should be applied and tripping of the CB must be tested by simulation of an internal fault.

The final settings of the protection for documentation and archiving are extracted locally or from remote via PC.

12.2 Maintenance

The self-monitoring contained in the numerical devices covers 80-90% of the protection equipment. CT circuits are included as long as load current is flowing and the signal communication is also continuously monitored to detect errors. The numerical protection therefore only has to be maintained with fairly long maintenance cycles.

Originally the German Utility Board recommended 4 year intervals for the maintenance cycle on the complete protection equipment [12-2]. At present intervals of between 5 and 6 years are however common and the tendency is towards even larger time intervals. In the periods between the tests plausibility checks with the indicated load values and the stored fault record data are however recommended.

Literature

Essays

[1-1] Kumar, A.; Mainka, M.; Ziegler G.: 20 Years of Digital Protection; Siemens EV-Report 4/94, September 1994, pp. 10-13

[2-1] International Electrotechnical Vocabulary (IEV), Chapter 448: "Power System Protection"; identical with IEC 60050-448

[2-2] International Electrotechnical Vocabulary (IEV), Chapter 447: "Measuring Relays", identical with IEC 60050-447

[3-1] McColl: Automatic Protective Devices for Alternating Current Systems; J. Instn. Engr. 1920, S. 525

[3-2] Neugebauer, H.: Grundlagen des Differentialschutzes; Siemens-Zeitschrift, 26. Jg., Juli 1952, H. 5, Seite 219-224

[3-3] Mathews, P. and Nellist, B.D.: Generalized Circle Diagrams and Their Application to Protective Gear; IEEE Transactions, Febr. 1964, S. 165-173

[3-4] Adamson, C. and Talkhan, E.: Selection of Relaying Quantities for Differential Feeder Protection; Proceedings IEE, 107, Part A, No. 31, February, 1960, S. 37-47

[3-5] Rushton, J.: The Fundamental Characteristics of Pilot-Wire Differential Protection Systems; IEE Paper No. 3645 S, Oct. 1961, p. 409-420

[3-6] Mathews, P.: Protective Current Transformers and Circuits; Chapman & Hall LTD, Lond0n 1955

[3-7] Newcombe, R. W. : The Development of Busbar Protection, The English Electric Journal 14, No. 6, 1956, S. 31-38

[3-8] Neugebauer, H.: Differentialschutz mit Kommandozeiten unter einer Periode; ETZ, Jg. 7, 1955, Heft B4, S. 108 bis 110

[4-1] The 'Merz-Price' system of automatic protection for high-tension circuits; Electrical Review , August 28, 1908

[4-2] Fendt, A.: Statischer Transformatordifferentialschutz; Siemens-Zeitschrift, Jg. 50, Heft 4, 1976, S. 273-276

[4-3] Phadke, A. G. and Thorp, J. S.: Computer Relaying for Power Systems; John Wiley & Sons INC., New York, 1988, ISBN 0 471 92063 0

[4-4] IEEE Tutorial Course: Computer Relaying; IEEE Course Text 79 EH0148-7-PWR, 1979

[4-5] IEEE Tutorial Course: Advancements in Microprocessor based Protection and Communication; IEEE Course Text 79TP120-0, 1997

[4-6] Kwong, W.S., Clayton, M.J. and Newbould, A.: A micro-processor based current differential relay for use with digital communication systems; IEE Conference on Developments in Power system Protection (DPSP), Conference Publication No. 249, London, 1985

[4-7]   Mills, D.L.: Internet time synchronisation: The Network Protocol; IEEE Transactions on Communications, Vol. 39, No. 10, Oct. 199, pp. 1482-1492

[4-8]   IRIG Standard 200-04: IRIG serial time code formats, 2004
        TIMING COMMITTEE , TELECOMMUNICATIONS AND TIMING GROUP of the
        RANGE COMMANDERS COUNCIL, U.S. Army White Sands Missile Range, New
        Mexico, The IRIG web site: www.irigb.com

[5-1]   IEC 60044-1 (1996-12): Instrument Transformers, Part 1: Current Transformers

[5-2]   BS (British Standard) 3938 (1982): Current Transformers

[5-3]   IEEE /ANSI C57.13 (1993): Standard Requirements for Instrument Transformers

[5-4]   Fischer, A. und Rosenberger, G.: Verhalten von linearisierten Stromwandlern
        bei verlagerten Kurschlussströmen; Elektrizitätswirtschaft Jg. 67 (1968), H. 17,
        S. 310-315

[5-4]   Zahorka,R.. Das Verhalten von Stromwandlern bei Einschwingvorgängen mit
        Gleichstromgliedern unter Berücksichtigung der Sättigung; AEG-Mitteilungen
        57, (1967) 1, S. 19-27

[5-5]   Rosenberger, G.: Stromwandler für neuzeitlichen Netzschutz; Siemens-
        Zeitschrift 40 (1966), H. 1, S. 18-23

[5-6]   Hodgkiss : The behaviour of current transformers subjected to transient asym-
        metric currents and the effects on associated protective relays;
        CIGRE Report No. 329, 1960

[5-7]   Korponay, N. E. : The Transient Behaviour and Use of Current Transformers,
        Brown Boveri Review, Vol. 62, June 1975, S. 255-261

[5-8]   IEEE PSRC Report 76 CH 1130-4 PWR: Transient Response of Current Trans-
        formers; IEEE Special Publication, Jan. 1976

[5-9]   IEC 60044-6 (1992-03): Instrument Transformers, Part 6: Requirements for
        protective current transformers for transient performance

[5-10] Iwanusiw, O.W.: Remanent Flux in Current Transformers; Ontario research
        Quarterly , third quarter, 1970, S. 18 -21

[5-11] Bruce, R.G. and Wright, A.: Remanent flux in current transformer cores; Proc.
        IEE, Vol. 113, No. 5, May 1966, S. 915-920

[5-12] Conner, E.E., Greg, R.G. and Wentz, E.C.: Control of Residual Flux in Current
        Transformers; IEEE PSRC Report T 73 037-9

[5-13] Bay, H.H., Halama, W. und Noeller, J.H.: Verhalten von Stromwandlern und Dis-
        tanzrelais bei Kurzschlussströmen mit Gleichstromglied; ETZ-A, Bd. 88 (1967),
        H. 5, S. 113-120

[5-14] Johannesson, T., Ingemansson, D. and Messing L.: The Probability of a Large
        Fault Current DC component; CIGRE Study Committee 34 Colloquium, Sibiu,
        2001, Report 306

[5-15] Siemens Power Engineering Guide – Transmission and Distribution, Part 6;
        Siemens AG, PTD PA, Postfach 4806, D-90026 Nürnberg

[5-16] IEEE WG Report: Relay performance considerations with low ratio CTs and high
        fault currents; IEEE Transactions on Power Delivery, Vol. 8, No. 3, July 1993,
        pp. 884-897

[5-17] IEEE Std C37.110-1996: IEEE Guide for the Application of Current Transformers
        Used for Protective Relaying Purposes.

[5-18] Zocholl, S.E.; Smaha, D.W.: Current transformer concepts; Proceedings of the
        46th Annual Georgia Tech Protective Relay Conference, April 29 – May 1, 1992,
        Atlanta, Georgia (US)

[5-19] Robers, J., Zocholl, S.E.; Benmouyal, G.: Selecting CTs to Optimize Relay Performance; Proceedings of the 24th Annual Western Protective Relay Conference, Spokane, WA (US), 1998

[5-20] Pfundner, R.A.:Accuracy of current transformers adjacent to high current buses; AIEE Transactions, vol. 70, part II, 1951, pp. 1656-1662

[5-21] Gajic, Z., Holst, S., Bonmann, D., Baars, D.A.W.: Influence of stray flux on protection systems; IEE Conference on Developments in Power System Protection, Glasgow, 2008, Conference Publication, pp. 426-431

[6-1]   DIN/VDE 0228, (12/87): Maßnahmen bei Beeinflussung von Fernmeldekabeln durch Starkstromanlagen, Teil 1: Allgemeine Grundlagen; Beuth Verlag GmbH, Berlin (Translation as BSI Version (1987): Proceedings in the case of interference on telecommunication installations by power installations, General)

[6-2]   DIN/VDE 0228, (12/87): Maßnahmen bei Beeinflussung von Fernmeldekabeln durch Starkstromanlagen, Teil 2 Beeinflussung durch Drehstromanlagen; Beuth Verlag GmbH, Berlin

[6-3]   Geise, F. und Vogel, W. : Hilfskabel für den Netzschutz in Hochspannungsnetzen; Siemens Zeitschrift 1961, H. 9. S. 661-665

[6-4]   Kuhnert, E. und Latzel, G. : Schutz von Fernmeldekabeln durch Gasentladungsableiter gegen Beeinflussungsspannungen; ETZ-A, Bd. 85 (1964), H. 20, S. 666-670

[6-5]   DIN VDE 0816, Teil1 (2/1988): Außenkabel für Fernmelde- und Informationsverarbeitungsanlagen: Kabel mit Isolierhülle und Mantel aus Polyethylen in Bündelverseilung; Beuth Verlag GmbH, Berlin

[6-6]   AIEE Committee Report: Protection of Pilot-Wire Relay Circuits, AIEE Transactions, June 1959, S. 205 -215

[6-7]   IEEE Standard 468 (1980): A Guide for the Protection of Wire Line Communications Facilities Serving Electric Power Stations

[6-8]   Hallmark, C.: How to protect communications lines from high substation voltages; Electrical World, October 1998, S. 12-18

[6-9]   Heinhold, L. und Stubbe, R. : Kabel und Leitungen für Starkstrom; Publicis MCD Verlag, München, 1999

[6-10] Mahlke, G. und Goessing, P. : Fiber Optic Cables, Fundamentals, Cable Design, System Planning; Publicis MCD Verlag, 2001

[6-11] Conrads, D.: Daten-Kommunikation, Verlag Vieweg, Braunschweig, 1993 ISBN 3-528-14589-7

[6-12] Wenzel, P.: Daten-Fernübertragung, Verlag Vieweg, Braunschweig, 1986 ISBN 3-528-04369-5

[6-13] Report of CIGRE Working Group 34/35.11: Protection Using Telecommunications; CIGRE Brochure No. 192, 2001

[6-14] Hoffelmann, J. ,Lienart, P., Wellens, F. und Delgado, M.: Protection of a two-circuit three terminal line by means of equipment employing digital microwave links as communication carrier; CIGRE Conference 1990, Report 34-203

[6-15] Enarson, T., Wennerlund, P., Cederblad, L., Lindahl, S. und Holst, S.: Experiences of current differential protections for multi-terminal power lines using multiplexed data Transmission channels; CIGRE Conference 1994, Report 34-203

[6-16] Report of Working Group H9 of IEEE PSRC: Digital Communications for Relay Protection (Jan. 2011)

[6-17] IEC 60870-5-1 to 5: Telecontrol Equipment and Systems – Part 5: Transmission Protocols, Section 1 to 5

[6-18] HDLC (High Level Data Link Control): ISO 3309.2 und 4335, DIN 66221

[6-19] IEC 60834-2, 1993:Teleprotection Equipment of Power Systems –Performance and Testing – Part 2: Analog Comparison Schemes

[6-20] ISO/IEC 7498-1: Information Technology – Open System Interconnection – Basic Reference Model: The Basic Model

[6-21] Report of CIGRE Working Group 34.05: Application of Wide-Band Communication Circuits to Protection, CIGRE Brochure No. 84

[6-22] Koreman,C.G.A., Morren, E. und Nuijs, G.W.: Recommendations for SDH Networks due to Protection Signalling; CIGRE Symposium Helsinki, 1995, Report 400-02

[6-23] Wilson, R.E. and Kusters, J.A.: International Time Keeping for Power System Users; IEE Conference on Developments in Power System Protection, Nottingham, 1997, Conference Publication N. 434, S. 351-354

[6-24] Southern, E.P., Crossley, P.A., Potts, S., Weller, G.C.: GPS Synchronised Current Differential Protection; IEE Conference on Developments in Power System Protection, Nottingham, 1997, Conference Publication N. 434, S. 342-345

[6-25] Li, H.Y. et al: New Type of Differential Feeder Protection Relay Using the Global Positioning System for Data Synchronisation; IEEE transactions on Power delivery, Vol. 12 No. 3, July 1997

[6-26] Serizawa, Y. et al: Wide-Band Digital Communication Requirements for Differential Teleprotection; CIGRE Symposium in Helsinki, 1995, Report 600-03

[6-27] IEEE Power System Relaying Committee Report: A Survey of Optical Channels for Protective Relaying: Practices and Experience, IEEE Trans. On Power delivery, Vol. 10, N0. 2, April 1995, S. 647 – 658

[6-28] EN 870-5-103 (IEC 870-5-13): Telecontrol Equipment and Systems, Part 5-103: Transmission Protocols – Companion Standard for Informative Interface of Protection Equipment

[6-29] Ward, S. et al: Pilot Protection Communication Channel Requirements RFL Electronics Inc., www.rflelect.com/technical_papers.html

[6-30] Ward, S.: Communication Channel Requirements for Pilot Protection, PAC World (www.pacw.org), July 2007, pp. 46 – 51

[6-31] IEEE C37.94 Standard for N Times 64 Kilobit Per Second Optical Fiber Interfaces Between Teleprotection and Multiplexer Equipment , 2002

[7-1] VDEW –Ringbuch " Schutztechnik", Teil 5: Richtlinien für den Blockschutz, VWEW Verlag, Frankfurt, 1994

[7-2] IEEE Guide for AC Generator Protection; ANSI/IEEE C37.102-1987

[7-3] IEEE Guide for Generator Ground Fault Protection, ANSI/IEEE C37.101-1985

[7-4] Sills, H.R. and McKeever, J.L.: Characteristics of Split-Phase Currents as a Source of Generator Protection; AIEE Transactions, Oct. 1953, S. 1005-1016

[7-5] Buttrey, M., Hay, D. und Weatherall, P.M.: Generator Interturn Fault Protection; IEE Conference on Developments in Power System Protection, London,1975, Conference Publication No. 125, S. 42-47

[8-1] Bödefeld, T. und Sequenz, H.: Elektrische Maschinen, Eine Einführung in die Grundlagen; Springer Verlag, Wien, 1962

[8-2] Janus, R.: Transformatoren: VDE Verlag GmbH, Frankfurt

[8-3] Schmidt, W.: Über den Einschaltstrom bei Drehstromtransformatoren; ETZ-A, Bd. 82, 1961, H. 15, S. 471-474

[8-4] Schmidt, W.: Vergleich der Größtwerte des Kurzschluss- und Einschaltstromes von Einphasentransformatoren; ETZ-A, Bd. 79, h.21, 1958, S. 801-806

[8-5] Einschaltstromstöße von Verteiltransformatoren; BBC Mitteilungen, Bd. 52, Nr. 11/12, 1965

[8-6] Specht, T.R.: Transformer Magnetizing Inrush Current; AIEE transactions, 1951, Vol. 70, S. 323-327

[8-7] Sonneman, W.K., Wagner, C.L. and Rockefeller, G,D.: Magnetizing Inrush in Transformar Banks; AIEE Transactions 1958, S. 884-892

[8-8] Nelson, P.Q. and Benko, I.S.: Determination of Transient Inrush Currents in Power Transformers Due to Out-of-Phase Switching Occurrences, IEEE Paper 70 TP 710-PWR, 1970

[8-9] Carlson, A.: Übertragbare Sättigung bei Leitungstransformatoren; ABB Technik 8/9, 1990, S. 29-34

[8-10] Hayward, C.D.: Prolonged Inrush Currents With Parallel Transformers Affect Differential Relaying; AIEE Transactions, Vol. 60, 1941, S. 1096-1101

[8-11] Guzman, A., Zochol, S. and Benmouyal, G.: Performance Analysis of Traditional and Improved Transformer Differential relays; Schweitzer Engineering Laboratories, Pullman, Washington, US, (www.selinc.com/sel-lit)

[8-12] Wisznewski, A., Winkler, W. und Sowa, P. und Marczonek, .: Selection of Settings of Transformer Differential Relays; IEE Conference on Developments in Power-System Protection, 1985, Conference Publication No. 249, S. 204-208

[8-13] Anderson, P. M.: Power System Protection, IEEE Power Engineering Series (Book), McGraw-Hill, New York, Section 17.2

[8-14] Koehler, R., Feser, K., Schiel, L., Schuster, N., Hofsmoen, W.F.: Restricted Earth-Fault Protection for Power Transformers; Conference paper, Siemens E D EA PRO, Postfach 4806, D-90592, Nürnberg

[8-15] Schiel, L. und Schuster, N.: Umfassendes Konzept für den Transformatorschutz, ETZ, B. 115 (1994), Heft 9, S. 496-502

[8-16] Schuster, N. und Schiel, L.: Multifunktionsschutz für Zweiwicklungstransformatoren, ew Jg. 100 (2001), Heft11, s. 40-44

[8-17] ANSI/IEEE C57 135: IEEE guide for the application, specification and testing of phase shifting transformers

[8-18] Verboomen, J. et al: Phase Shifting Transformers: Principles and Applications; International Conference on Future Power Systems, Amsterdam, 2005

[8-19] Hurlet, P. et al:French Experience in Phase Shifting Transformers; CIGRE Conference 2006, Report A2-204

[8-20] Ibrahim, M. A. and Statcom, F. P.: Phase Angle Regulating Transformer Protection; IEEE Transactions, vol. 9, No. 1, January 1994, pp. 394-404 IEEE Special Publication of PSRC Working Group K1: Protection of phase angle regulating transformers (PAR); Oct. 21, 1999

[8-21] Wang, J., Gajic, Z. und Holst, S.: The multifunctional Numerical Transformer Protection and Control System With Adaptive and Flexible Features; IEE Conference on Developments in Power System Protection, Amsterdam, 2001, Conference Publication, S. 165-168

[8-22] Rimez, J. et al: Grid implementation of a 400 MVA 220/150 kV -15°/+3° phase shifting transformer for power flow control in the Belgian network: specification and operation conditions; CIGRE Conference, Paris 2006, Report A2-202

[8-23] Koreman, C,G.A. et al: Protection and Control of Two Phase Shifting Transformers in the Netherlands; 8th International IEE Conference on Developments in Power System Protection, 2004, pp. 156-159

[8-24] Sevov, L. and Wester, C.: Phase Angle Regulating Transformer Protection Using Digital Relays; 8th International IEE Conference on Developments in Power System Protection, 2004, pp. 356-379

[8-25]   138 KV Phase Shifting Transformer Protection: EMTP Modeling and Model Power System Testing; 8th International IEE Conference on Developments in Power System Protection, 2004, pp. 343-347

[8-26]   Gajic, Z. and Holst, S.: Use of 87 Relay Principles for Overall Differential Protection of Phase Angle Regulating Transformers; CIGRE Study Committee B5 Colloquium, Jeju Island, Korea, 2009, Report 206

[9-1]   Ferschl, L.: Der Vergleichsschutz in Hochspannungsnetzen und –anlagen; Elektrotechnik und Maschinenbau (EuM), 81, 1964, S. 557 -564 und S. 596-599

[9-2]   Calero, F. and Elmore, W.A.: Current Differential and Phase Comparison Relaying Schemes; Proceedings of the 19th Annual Western Protective Relay Conference, Spokane, WA, US

[9-3]   Hagen, D., Holbach, J., Schuster, N.: Numerischer Leitungsdifferentialschutz mit Hilfsadern, etz 116(1995), H. 3, 10-14

[9-4]   Schuster, N.: Schutz von Mehrendenleitungen im Hochspannungsnetz; Elektrizitätswirtschaft, J. 99 (2000), Heft 7, S. 12-17

[9-5]   Hartmann, W. und Schuster, N.: Optimizing Security and Sensitivity for Line Differential Protection Applications; Georgia Tech Protective Relaying Conference 2001

[9-6]   Koch, G. und Schmidt, E.: Ein numerischer Stromvergleichsschutz mit digitaler Messwertübertragung, Elektrie 45 (1991), S.272-276

[9-7]   Kwong, WS., Clayton, M.J. and Newbould, A.: A Microprocessor Based Current Differential Relay For Use With Digital Communication Systems; IEE Conference on Developments in Power System Protection, London, 1985, Conference Publication No. 249, S. 65 – 69

[9-8]   I.J. Hall, S. Potts: Unit protection utilising synchronous digital hierarchy (SDH) communication systems – Why a new design approach is necessary, IEE DPSP Conference 2001, IEE Conference publication No. 479, pp. 106 – 109

[9-9]   Ito, H. et al: Development of an Improved Multifunction High Speed Operating Current Differential Relay for Transmission Line Protection; IEE Conference on Developments in Power System Protection, Amsterdam,2001, Conference Publication, S. 511-514

[9-10]   Adamiak, M.G. und Alexander, G.E.: A New Approach to Current Differential Protection for Transmission Lines; 24th Annual Western Protective Relay Conference, Spokane, Washington, 1996

[9-11]   Forsman, S.: High Speed Phase Segregated Line Differential Protection; SIPSEP Conference, Monterey, Mexico1996

[9-12]   Application Guide on Protection of Complex Transmission Network Configurations; CIGRE Working Group Report, Ref. No. 62, 1992

[9-13]   Protection of Multi-terminal and Tapped Lines; AIEE Committee Report, AIEE Proceedings, April 1961, S. 55-66

[9-14]   Ziegler, G.: Numerical Distance Protection, Principles and Applications; Publicis Corporate Publishing, Erlangen, 4th Edition, 2011, ISBN 978-3-89578-381-4, www.publicis-erlangen.de/books

[9-15]   Ballentine, E.L and Hale, E.A.: Transformer differential relays (When used on Breaker-And-One-Half or Ring Busses); 24th Annual Western Protective Relay Conference, Spokane, Washington, USA, Oct. 21-23, 1997

[9-16]   Kiessling, G. and Schwabe, S.: Software solution for fault record analysis in power transmission and distribution; IEE Conference on Developments in Power System Protection, Amsterdam, 2004, Session 6, paper 48

[9-17] Philippot, L.: Parameter estimation and error estimation for line fault location and distance protection in power transmission systems; Ph.D. Thesis, Université Libre de Bruxelles, February 1996

[10-1] Seeley, H.T. and Roeschlaub, F. : Instantaneous Bus Differerential Protection Using Bushing Current Transformers, AIEE Transactions, 1948, Vol. 67, S. 1709-1717

[10-2] Abel, G.: Sammelschienenschutz für 110-kV-Anlagen; Siemens-Zeitschrift, Jg. 34, Mai 1960, Heft 5, S. 310-315

[10-3] Zurowski, E.: Dreifach gesicherter Sammelschienenschutz; Siemens-Zeitschrift, Jg. 46, April 1972, Heft 4, S. 257-259

[10-4] Zurowski, E.: Triple Safe Busbar Protection; Siemens Review, 4/72

[10-5] Meisberger, F. und Schenk, U.: Digitaler Stationsschutz; EET, 1991, H. 2

[10-6] Meisberger, F. und Schenk, U.: Station Protection Has Now Gone Digital, Siemens EV-Report 2/92, S. 12-13

[10-7] Forford, T.: A Half Cycle Bus Differential Relay; IEE Conference on Developments in Power System Protection, London,1975, Conference Publication No. 125, S. 79-85

[10-8] Haug, H. und Foster, M.: Electronic Bus Zone Protection; CIGRE Conference 1968, Report 31-11

[10-9] Andrichak, J.G and Cardenas, J.: Bus Differential Protection; Western Protective Relay Conference, Spokane, 1995

[10-10] Kasztenny, B., Sevov, L. und Brunello, G.: Digital Low-Impedance Bus Differential Protection – Review of Principles and Approaches; 55th Annual Georgia Tech Protective relaying Conference, Atlanta, 2001

[10-11] Funk, H.W. und Ziegler, G.: Numerical Busbar Protection, Design and service experience; IEE Conference on Developments in Power System Protection, Nottingham,1997, Conference Publication No. 434, S. 131-134

[10-12] Funk, H.-W. und Meisberger, F.: Schneller dezentraler Sammelschienenschutz; Elektrizitätswirtschaft, Jg. 97 (1998), H. 22, S. 35-37

[10-13] Schick, M., Reichenbach, G. und Funk, H.-W.: Flexible graphische Konfiguration des digitalen Sammelschienenschutzes, ew, Jg. 101 (2002), Heft 4, S. 86-89

[10-14] Peck, D.M., Nygaard, B. und Wadelinus, K.: A New Busbar Protection System With Bay-Oriented Structure; IEE Conference on Developments in Power System Protection, York (UK),1993, Conference Publication No. 368, S. 228-231

[10-15] Andow, F., Suga, N., Murakami, Y. und Inamura, K.: Microprocessor-based Busbar Protection Relay; IEE Conference on Developments in Power System Protection, York (UK),1993, Conference Publication No. 368, S. 103-106

[10-16] Ilar, M., Reimann, B. und Brunner, D.: Numerischer Sammelschienenschutz REB 500 – Neue Schutzmöglichkeiten durch dezentrale Installation; ABB Technik (1997), S. 24-32

[11-1] Relay Catalogue SIPROTEC (Digital Protection), Siemens E D EA, current edition; www.siprotec.de

[11-2] SIPROTEC, System Description, Siemens E D EA, current edition; www.siprotec.de

[12-1] Omicron electronics GmbH: Literature: End-to-End Testing (www.omicron.at)

[12-2] Prüfempfehlungen für digitale Schutzeinrichtungen mit Selbstüberwachung, herausgegeben vom VDEW-Arbeitskreis "Relais- und Schutztechnik", April 1995

Protection technology reference books

[A-1]   Herrmann, H.-J.: Digitale Schutztechnik (Grundlagen, Software, Ausführungs-beispiele); VDE-VERLAG, Berlin, 1997, ISBN 3-8007-1850-2

[A-2]   Clemens, H. und Rothe, K.: Schutz in Elektroenergiesystemen, 3. Auflage, Verlag Technik GmbH Berlin, 1991, ISBN 3-341-00828-4

[A-3]   Schossig, W.: Netzschutztechnik; VDE Verlag GmbH, Frankfurt am Main, 2001, ISBN 3-8007-2641-6

[A-4]   Doemeland,W.: VEM-Handbuch Relaisschutztechnik, 4. Auflage, Verlag Technik, Berlin, 1989, ISBN 3-341-00491-2,

[A-5]   Ungrad,H., Winkler, W. und Wisznewski, A.: Schutztechnik in Elektroenergie-systemen; Springer Verlag, Berlin, 1991, ISBN 3-540-53385-0

[A-6]   VWEW-Verlag: VDEW-Ringbuch Schutztechnik, 1988 mit laufenden Ergänzungen

[A-7]   Hubensteiner, H. (Editor) and 10 further authors: Schutztechnik in elektrischen Netzen; VDE-VERLAG, Berlin und Offenbach1989

[A-8]   Hubensteiner, H. (Editor) and 8 further authors: Schutztechnik in elektrischen Netzen 2 (Planung und Betrieb); VDE-VERLAG, Berlin und Offenbach1993

[A-9]   Müller, L. und Boog, E.: Selektivschutz elektrischer Anlagen; VWEW-Verlag, Frankfurt am Main, 1990

[A-10] Blackburn, J. L.: Protective Relaying: Principles and Applications, Marcel Dekker, Inc., New York, Basel, Hong Kong, 1987

[A-11] Blackburn, J. L.: Symmetrical Components for Power Systems Engineering, Marcel Dekker, Inc., New York, Basel, Hong Kong, 1993

[A-12] Mason C.R.: The Art & Science of Protective Relaying, John Wiley & Sons, Inc. New York, London, Sydney, 1956 (6th Edition 1967)

[A-13] Wright, A. and Christopoulos, C.: Electrical Power System Protection; Chapman&Hall, London, 1993, ISBN 0-412-39200-3

[A-14] Horowitz, S.H. and Phadke, A.G.: Power System Relaying, 2nd Edition, 1995, Research Studies Press LTD., England, ISBN 0-86380-185-4

[A-15] Network Protection and Automation Guide, Alstom, 2002, ISBN 2-9518589-0-6

[A-16] Elmore, W.A.: Protective Relaying, Theory and Applications (ABB), Marcel Dekker Inc., New York, Basel, 1994, ISBN 0-8247-9152-5

[A-17] Elmore, W.: Pilot Protective Relaying (ABB), Marcel Dekker, New York, 2000

[A-18] Reimert, D.: Protective Relaying for Power Generation Systems, CRC Press,Taylor and Francis Group, Boca Raton (USA), 2006, ISBN 978-0-8247-0700-2

[A-19] Green, J.H.: The Irwin Handbook of Telecommunications, McGram-Hill, New York, 2000, ISBN 0-07-135554-5

Books out of print:

[A-20] Neugebauer, H.: Selektivschutz; Springerverlag Berlin/Göttingen/Heidelberg, 1958

[A-21] Erich, M.: Relaisbuch (VDEW); Franck'sche Verlagshandlung, Stuttgart, 1959

[A-22] Warrington, A.R.C.: Protective Relays. Their Theory and Practice, Chapman and Hall, Volume 1, London, 1962

[A-23] Clarke, E.: Circuit Analysis of A-C Power Systems, Vols. I and II, General Electric Co., Schenectady, N.Y., 1943 und 1950

Addendum

Device designation according to IEC 60617 and ANSI/IEEE C37-2

Protection functions	Designation to IEC	Designation to ANSI
Distance	$Z<$	21
Over-temperature	$\theta>$	26
Power direction	$\vec{P}>$	32
Negative sequence overcurrent	$I2>$	46
Thermal overload	$\Theta>$	49
Overcurrent, instantaneous	$I>>$	50
Overcurrent, delayed	$I>, t$	51
Earth overcurrent, instantaneous	$I_E>>$	50N
Earth overcurrent, delayed	$I_E>, t$	51N
Over-voltage	$U>$	59
Displacement voltage	$U_0>$	59N
Direction earth-fault protection	$\vec{I_E}>$	67N
Under-frequency	$f<$	81U
Over-frequency	$f>$	81O
Differential protection	$I_d> (\Delta I>)$	87

Index

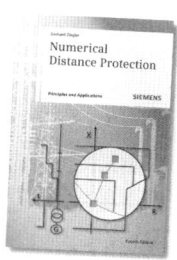

Gerhard Ziegler

Numerical Distance Protection

Principles and Applications

4th updated and enlarged edition, 2011,
419 pages, 275 illustrations, 24 tables, hardcover
ISBN 978-3-89578-381-4, € 59.90

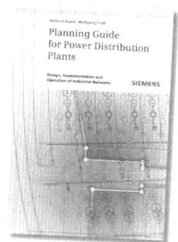

Hartmut Kiank, Wolfgang Fruth

Planning Guide for Power Distribution Plants

Design, Implementation and Operation of Industrial Networks

2011, 427 pages, 178 illustrations,
129 tables, hardcover
ISBN 978-3-89578-371-5, € 69.90

Industry Automation Translation Services (Eds.)

Dictionary of Electrical Engineering, Power Engineering and Automation
Wörterbuch Elektrotechnik, Energie- und Automatisierungstechnik

Part 2 English-German; Teil 2 Englisch-Deutsch

6th extensively revised and substantially
edition, 2009, 994 pages, hardcover
ISBN 978-3-89578-314-2, € 89.90

Wörterbuch Elektrotechnik, Energie- und Automatisierungstechnik
Dictionary of Electrical Engineering, Power Engineering and Automation

Teil 1 Deutsch-Englisch; Part 1 German-English

6., wesentlich überarbeitete und erweiterte
Auflage, 2011, 1.042 Seiten, gebunden
ISBN 978-3-89578-313-5, € 89.90

CD-ROM, Edition 2011

Deutsch-Englisch; Englisch-Deutsch
German-English; English-German

Technical requirements: Windows 7/Vista/XP
ISBN 978-3-89578-315-9, € 189.90

www.publicis-books.de